Organic Nanostructured Thin Film Devices and Coatings for Clean Energy

Handbook of
Nanostructured Thin Films and Coatings

Organic Nanostructured Thin Film Devices and Coatings for Clean Energy

Edited by
Sam Zhang

CRC Press
Taylor & Francis Group
Boca Raton   London   New York

CRC Press is an imprint of the
Taylor & Francis Group, an **informa** business

CRC Press
Taylor & Francis Group
6000 Broken Sound Parkway NW, Suite 300
Boca Raton, FL 33487-2742

International Standard Book Number: 978-1-4200-9393-3 (Hardback)

Visit the Taylor & Francis Web site at
http://www.taylorandfrancis.com

and the CRC Press Web site at
http://www.crcpress.com

Contents

Contents

Preface

The twenty-first century is said to be the century of nanotechnologies. In a way, it is. The development of science and technology has come to a stage where "microscopic" is no longer enough to properly describe or depict a scientific phenomenon or a technological process. With the advance of nanoscience and nanotechnology, the world technological landscape changes not only affect the way scientists do research, technologists carry out development, and engineers manufacture products, but also the way ordinary people go about their daily life, through, for instance, nanomedicine, cell phones, controlled drug delivery, no-pain operations, solar cell–powered gadgets, etc. Thin films and coatings play a very important and indispensable role in all of these. This three-volume book set aims to capture the development in the films and coatings area in relation to nanoscience and nanotechnology so as to provide a timely handbook series for researchers to refer to and for newcomers to learn from, and thus contribute to the advancement of the technology.

The three-volume book set, *Handbook of Nanostructured Thin Films and Coatings*, has 25 chapters where 11 chapters in volume 1 concentrate on the mechanical properties (hardness, toughness, adhesion, etc.) of thin films and coatings, including processing, properties, and performance, as well as a detailed analysis of theories and size effect, etc., as listed here: Chapter 1, The Fundamentals of Hard and Superhard Nanocomposites and Heterostructures; Chapter 2, Determination of Hardness and Modulus of Thin Films; Chapter 3, Fracture Toughness and Interfacial Adhesion Strength of Thin Films: Indentation and Scratch Experiments and Analysis; Chapter 4, Toughness and Toughening of Hard Nanocomposite Coatings; Chapter 5, Processing and Mechanical Properties of Hybrid Sol-Gel- Derived Nanocomposite Coatings; Chapter 6, Using Nanomechanics to Optimize Coatings for Cutting Tools; Chapter 7, Electrolytic Deposition of Nanocomposite Coatings: Processing, Properties, and Applications; Chapter 8, Diamond Coatings: The Industrial Perspective; Chapter 9, Amorphous Carbon Coatings; Chapter 10, Transition Metal Nitride–Based Nanolayered Multilayer Coatings and Nanocomposite Coatings as Novel Superhard Materials; and Chapter 11, Plasma Polymer Films: From Nanoscale Synthesis to Macroscale Functionality.

Volume 2 contains eight chapters focusing on functional properties, i.e., optical, electronic, and electrical properties, and the related devices and applications: Chapter 1, Large-Scale Fabrication of Functional Thin Films with Nanoarchitecture via Chemical Routes; Chapter 2, Fabrication and Characterization of SiC Nanostructured/Nanocomposite Films; Chapter 3, Low-Dimensional Nanocomposite Fabrication and its Applications; Chapter 4, Optical and Optoelectronic Properties of Silicon Nanocrystals Embedded in SiO_2 Matrix; Chapter 5, Electrical Properties of Silicon Nanocrystals Embedded in Amorphous SiO_2 Films; Chapter 6, Properties and Applications of Sol-Gel-Derived Nanostructured Thin Films: Optical Aspects; Chapter 7, Controllably Micro/Nanostructured Films and Devices; and Chapter 8, Thin Film Shape Memory Alloy for Microsystem Applications.

Volume 3 focuses on organic nanostructured thin-film devices and coatings for clean energy with six chapters discussing the processing and properties of organic thin films, devices, and coatings for clean energy applications: Chapter 1, Thin Film Solar Cells Based on the Use of Polycrystalline Thin Film Materials; Chapter 2, Anodized Titania Nanotube Array and its Application in Dye-Sensitized Solar Cells; Chapter 3, Progress and Challenges of Photovoltaic Applications of Silicon Nanocrystalline Materials; Chapter 4, Semiconductive Nanocomposite Films for Clean Environment; Chapter 5, Thin Coating Technologies and Applications in High-Temperature Solid Oxide Fuel Cells; and Chapter 6, Nanoscale Organic Molecular Thin Films for Information Memory Applications.

A striking feature of these books is that both novice and experts have been considered while they were written: the chapters are written in such a way that for newcomers in the relevant field, the handbooks would serve as an introduction and a stepping stone to enter the field with least confusion, while for the experts, the handbooks would provide up-to-date information through the figures, tables, and images that could assist their research. I sincerely hope this aim is achieved.

The chapter authors come from all over the globe: Belgium, China, the Czech Republic, Egypt, Germany, India, Korea, Singapore, Taiwan, the Netherlands, the United Kingdom, and the United States. Being top researchers at the forefront of their relevant research fields, naturally, all the contributors are very busy. As editor, I am very grateful that they all made special efforts to ensure timely response and progress of their respective chapters. I am extremely indebted to many people who accepted my request and acted as reviewers for all the chapters—as the nature of the writing is to cater to both novice and experts, the chapters are inevitably lengthy. To ensure the highest quality of the chapters, more than 50 reviewers (at least two per chapter) painstakingly went through all the chapters and came out with sincere and frank criticism and suggestions that helped make the chapters complete. Though I am not able to list all the names, I would like to take this opportunity to say a big thank you to all of them. Last but not least, I would like to convey my gratitude to many CRC Press staff, especially Allison Shatkin and Jennifer Ahringer at Taylor & Francis Group, for their invaluable assistance rendered to me throughout the entire endeavor that made the smooth publication of the handbook set a reality.

Sam Zhang
Singapore

Editor

Sam Zhang Shanyong, better known as Sam Zhang, received his BEng in materials in 1982 from Northeastern University (Shenyang, China), his MEng in materials in 1984 from the Central Iron and Steel Research Institute (Beijing, China), and his PhD in ceramics in 1991 from the University of Wisconsin-Madison (Madison, Wisconsin). Since 2006, he has been a full professor at the School of Mechanical and Aerospace Engineering, Nanyang Technological University (Singapore).

Professor Zhang serves as editor in chief for *Nanoscience and Nanotechnology Letters* (United States) and as principal editor for the *Journal of Materials Research* (United States), among other editorial commitments for international journals. He has been involved in the fields of processing and characterization of thin films and coatings for the past 20 years, his interests ranging from hard coatings to biological coatings and from electronic thin films to energy films and coatings. He has authored/coauthored more than 200 peer-reviewed international journal articles, 14 book chapters, and guest-edited 9 journal volumes in *Surface and Coatings Technology* and *Thin Solid Films*. Including this handbook, he has authored and/or edited 6 books so far: *CRC Handbook of Nanocomposite Films and Coatings*: Vol. 1, *Nanocomposite Films and Coatings*: *Mechanical Properties*; Vol. 2, *Nanocomposite Films and Coatings*: *Functional Properties*; Vol. 3, *Organic Nanostructured Film Devices and Coatings for Clean Energy*, and *Materials Characterization Techniques* (Sam Zhang, Lin Li, Ashok Kumar, published by CRC Press/Taylor & Francis Group, 2008); *Nanocomposite Films and Coatings—Processing, Properties and Performance* (edited by Sam Zhang and Nasar Ali, Published by Imperial College Press, U.K., 2007), and *CRC Handbook of Biological and Biomedical Coatings* (scheduled for a 2010 publication by CRC Press/Taylor & Francis Group).

Professor Zhang is a fellow at the Institute of Materials, Minerals and Mining (U.K.), an honorary professor at the Institute of Solid State Physics, Chinese Academy of Sciences, and a guest professor at Zhejiang University and at Harbin Institute of Technology. He was featured in the first edition of *Who's Who in Engineering Singapore* (2007), and featured in the 26th and 27th editions of *Who's Who in the World* (2009 and 2010). Since 1998, he has been frequently invited to present plenary keynote lectures at international conferences including in Japan, the United States, France, Spain, Germany, China, Portugal, New Zealand, and Russia. He is also frequently invited by industries and universities to conduct short courses and workshops in Singapore, Malaysia, Portugal, the United States, and China.

Professor Zhang has been actively involved in organizing international conferences: 10 conferences as chairman, 12 conferences as member of the organizing committee, and 6 conferences as member of the scientific committee. The Thin Films conference series (The International Conference on Technological Advances of Thin Films & Surface Coatings), initiated and, since, chaired by Professor Zhang, has grown from 70 members in 2002 at the time of its inauguration to 800 in 2008. It has now become a biannual feature at Singapore.

Professor Zhang served as a consultant to a city government in China and to industrial organizations in China and Singapore. He also served in numerous research evaluation/advisory panels in Singapore, Israel, Estonia, China, Brunei, and Japan. Details of Professor Zhang's research and publications are easily accessible at his personal Web site: http://www.ntu.edu.sg/home/msyzhang.

Contributors

D.C. Ba
Vacuum and Fluid Engineering Research
  Center
School of Mechanical Engineering and
  Automation
Northeastern University
Shenyang, China

I. Forbes
School of Computing, Engineering and
  Information Science (CEIS)
University of Northumbria
Newcastle upon Tyne, United Kingdom

Xiaodong He
School of Aeronautics
Harbin Institute of Technology
Harbin, China

Joe H. Hsieh
Department of Materials Engineering
Ming Chi University of Technology
Taipei, Taiwan

San Ping Jiang
Curtin Centre for Advanced Energy Science
  and Engineering, Department of Chemical
  Engineering
Curtin University of Technology
Perth, Australia

J.C. Li
Vacuum and Fluid Engineering Research
  Center
School of Mechanical Engineering and
  Automation
Northeastern University
Shenyang, China

R.W. Miles
School of Computing, Engineering and
  Information Science (CEIS)
University of Northumbria
Newcastle upon Tyne, United Kingdom

K.T. Ramakrishna Reddy
Department of Physics
Sri Venkataswara University
Tirupati, Andhra Pradesh, India

Jatin K. Rath
Faculty of Science
Debye Institute for Nanomaterials Science,
  Nanophotonics—Physics of Devices
Utrecht University
Utrecht, the Netherlands

Y.L. Song
Institute of Chemistry
Chinese Academy of Science
Beijing, China

Lidong Sun
School of Mechanical and Aerospace
  Engineering
Nanyang Technological University
Singapore, Singapore

Xiao Wei Sun
School of Electrical and Electronic
  Engineering
Nanyang Technological University
Singapore, Singapore

Sam Zhang
School of Mechanical and Aerospace
  Engineering
Nanyang Technological University
Singapore, Singapore

G. Zoppi
School of Computing, Engineering and
  Information Science (CEIS)
University of Northumbria
Newcastle upon Tyne, United Kingdom

1 Thin Film Solar Cells Based on the Use of Polycrystalline Thin Film Materials

R.W. Miles, G. Zoppi, K.T. Ramakrishna Reddy, and I. Forbes

CONTENTS

1.1 INTRODUCTION

As shown in Figure 1.1, most solar cells produced and sold at the present time are based on the use of crystalline or multi-crystalline silicon [1]. Despite the excellent progress in developing these technologies, it has been realized for some time that there is a need to reduce production costs significantly to compete directly with other forms of power generation. This need for solar cells and modules to be produced with much lower manufacturing costs has been the major impetus for developing thin film solar cells [2–6].

Crystalline and multi-crystalline silicon have indirect energy bandgaps, with correspondingly low optical absorption coefficients, such that several hundred microns of silicon are needed to

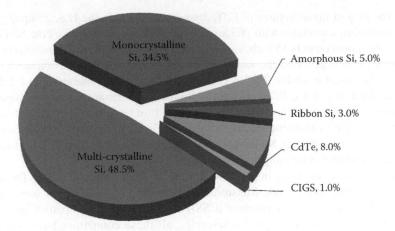

FIGURE 1.1 The market share of PV cell types sold during 2008.

absorb most of the incident light. Carriers generated by the incident light must also diffuse to the junction region to be acted upon by the electric field at the junction to contribute to the power generated. For the "minority carrier diffusion length" to be comparable or greater than the "optical absorption depth," recombination centers in the silicon must be minimized by

1. Purifying the silicon to a high level
2. Producing the silicon with as few crystal defects as possible

Silicon solar cells are costly because of

1. High material usage
2. High processing costs (to purify the material and minimize crystal defects)
3. High "handling costs" in the manufacturing process, as often the processing is not fully automated

Thin film solar cells based on the use of direct-energy-bandgap semiconductors, e.g., hydrogenated amorphous silicon (αSi:H), cadmium telluride (CdTe), and copper indium gallium diselenide (CIGS), minimize these costs because their correspondingly high optical absorption coefficients ($\alpha > 10^4$ cm^{-1}) mean that only a few microns of material are needed to absorb the incident sunlight. This means that a long minority carrier diffusion length is not required, because most of the photocarriers are generated within the depletion region or very near to the edge of the depletion (usually, <1 μm). This means that relatively low-cost methods can be used to make devices. Furthermore, depositing thin films of materials lends itself naturally to continuous production processes that are capable of depositing large areas of material.

The most successful materials used for making photovoltaic (PV) solar cells in the thin film form are αSi:H, CdTe, CIGS, and CuInS$_2$ (copper indium disulfide). Despite the continuing success of deploying αSi:H devices over the past decade, it should be noted that the efficiencies of commercially produced single-junction αSi:H modules are low (<4%) due to the light-induced degradation of efficiency, due to the Stabler–Wronski effect [7]. More complex structures, such as double junction, triple junction, and micromorph tandem devices, have been developed to improve efficiency and stability [7]. However, the very best devices produced only have efficiencies of around 10% and the complexity of manufacturing increases device fabrication costs.

This chapter will concentrate on the much more efficient, stable solar cells made using CdTe, CIGS, and CuInS$_2$. Some novel inorganic absorber-layer materials that are currently of interest are also discussed. These include CuInAlSe$_2$, Cu$_2$ZnSn(Se,S)$_4$, and SnS (tin monosulfide).

Currently, the largest manufacturer of CdTe-based solar cells is the U.S. company, First Solar. First Solar manufactures modules with efficiencies approaching 10% with a cost of 0.74 €/W. The CdTe modules are approximately 15% cheaper for each kW installed than their nearest rivals. Despite needing a larger area to generate the same amount of power, the cost advantage has persuaded many to buy CdTe-based modules [8]. Recent projects include the integration of CdTe modules onto a Logistics Building of a U.S. Army base in Ramstein, Germany (so far the biggest example of building-integrated photovoltaics [BIPV] using a thin film technology), and the fabrication of a large solar power station in Muldentalkreis, Saxony, and Germany. First Solar is currently building four further production facilities in Malaysia (joint capacity of 480 MW) and one further production facility in Frankfurt (Oder), Germany (120 MW facility) [8].

Other companies that have been involved in developing pilot production lines for the manufacture of CdTe-based modules include Matsushita (screen printing), BP Solar (electrodeposition), and Antec GmbH (close-spaced sublimation [CSS]). Although these companies had good success in producing small-area cells with good efficiencies, all these companies have now ceased production. This has partly been because of concerns with the lack of public acceptability of using a toxic metal such as cadmium in a "green product." This is despite detailed studies having shown that the environmental costs are no worse than those with other methods of energy production [4]. It can in fact be argued that combining cadmium (a highly toxic by-product of the extraction of zinc and copper) with tellurium to form CdTe, a nontoxic compound, that can be used to produce "green electricity," is environmentally beneficial. The impressive success of First Solar in selling its products into Germany, one of the most environmentally sensitive countries in the world, suggests that CdTe technology is becoming increasingly acceptable to the general public. Table 1.1 gives the capacity and output of First Solar and a new German company, Calyxo, during the period 2007–2008.

Table 1.2 lists the main producers of chalcopyrite-based (CIGS and $CuInS_2$) solar modules during 2007–2008. It is clearly evident that there are many more manufacturers than for CdTe-based cells and modules. This is because the chalcopyrite solar cells and modules can be made totally Cd free, minimizing environmental concerns. Most of these manufacturers use CIGS as the absorber-layer material. The exceptions are Sulphurcell and Odersun, who use copper indium disulfide. In the laboratory, CIGS-based cells have been produced with efficiencies up to 19.9% using the co-evaporation method [9], just lower in efficiency than the best cells made using multi-crystalline silicon. However, transferring the excellent results obtained in the laboratory into low-cost modules with acceptable efficiencies and yields has proved challenging. The best manufacturing costs are about 2.5 €/Wp, i.e., at present comparable to multi-crystalline silicon [8]. However, it is expected that with economies of scale, the situation will reverse over the next decade. There are currently concerns with respect to the lack of abundance of indium and gallium. This is pushing up the price of the raw materials to very high levels and may limit the large-scale deployment of CIGS-based modules in the longer term. It is however expected that these elements could be replaced by other elements that are more available and cheaper, e.g., Sn and Zn or Al. Work is already ongoing into trying to develop such materials. Figure 1.2 shows an example of BIPV, the 85 kWp Technium

TABLE 1.1

Companies Manufacturing CdTe Modules during 2007–2008

Company	Country	2007 Capacity (MW)	2007 Production (MW)	2008 Capacity (MW)	2008 Production (MW)
Calyxo	Germany	8	1	25	5
First Solar	United States	210	100	210	200

Source: Rentzing, S., *New Energy*, 3, 58, 2008.

TABLE 1.2
Companies Manufacturing Chalcopyrite Modules during 2007–2008

Company	Country	2007 Capacity (MW)	2007 Production (MW)	2008 Capacity (MW)	2008 Production (MW)
Avancis	Germany			20	1
Day Star Technologies	United States			25	1
Global Solar	United States	3	3	75	75
Honda Soltec	Japan	27.5	0	27.5	20
Johanna Solar	Germany			30	3
Odersun	Germany	4.5	1	30	1
Nano Solar	United States	430	0	430	1
Scheuten	The Netherlands	10	3	40	10
Showa Shell Sekiyo	Japan	20	10	20	20
Solibro	Germany			25	1
Sulphurcell	Germany	5	1	5	2
Würth Solar	Germany	15	15	30	30
VHF Technologies	Switzerland	2	0.1	25	5

Source: Rentzing, S., *New Energy*, 3, 58, 2008.

FIGURE 1.2 The 85 kWp Technium OpTIC building in St Asaph, Wales, which is covered with the of CIGS modules manufactured by Shell Solar.

OpTIC (Opto-electronics Technology and Incubation Centre) in St Asaph, Wales, United Kingdom, which is covered with the CIGS modules manufactured by Shell Solar.

The thin film solar cells and modules made using CdTe are usually formed by sequentially depositing layers onto glass substrates in the following sequence: transparent conductive oxide (TCO), buffer layer, absorber layer, and finally the back-contact layer. This structure, in which the TCO layer is deposited first and the back-contact layer last, is known as the "superstrate configuration." The thin film solar cells and modules made using the chalcopyrite compounds are usually formed by sequentially depositing layers onto glass substrates in the following sequence: back contact, absorber layer, buffer layer, TCO, and finally the top grid contact, and are referred to as being in the "substrate configuration."

The following sections will deal with (1) the methods used to produce the CdTe absorber layers, (2) the methods used to produce the chalcopyrite compound absorber layers, (3) novel materials that have the potential to replace the more established materials, (4) the buffer layers, and (5) the TCO layers most commonly used and under development.

1.2 THIN FILM SOLAR CELLS BASED ON THE USE OF CdTe

1.2.1 HISTORICAL DEVELOPMENT OF CdTe SOLAR CELLS

The potential of using CdTe as an absorber-layer material in a solar cell has long been recognized. With a near-optimum, direct energy bandgap of 1.45 eV, CdTe-based solar cells can, in theory, be used to produce PV solar cell devices with efficiencies up to 27% [10]. Historically, CdTe homojunction cells were the first to be investigated. However, the efficiencies were limited to <6% due to surface recombination. This arose because the high optical absorption coefficient of the CdTe meant that the collecting junction had to be located near the surface of the CdTe [11]. The first efficient polycrystalline thin film CdTe-based solar cell was the p-Cu_2Te/n-CdTe heterojunction solar cell. Such devices were also produced with conversion efficiencies up to 6% [12]. However, these solar cells were found to be unstable with the instability found to be associated with the use of the Cu_2Te layer.

In 1969, a p-CdTe/n-CdS heterojunction solar cell was fabricated with an efficiency of 1% [13]. This thin film solar cell device was made by evaporating CdS onto a TCO-coated glass substrate. CdTe was then deposited onto the CdS and then a metallic back contact onto the CdTe. Such a "superstrate configuration device" is shown in Figure 1.3. In 1972, an all thin film CdTe/CdS solar cell was produced by Bonnet and Rabenhorst with an efficiency >5% [14]. The promising efficiency and good stability of these devices stimulated worldwide interest in the development of CdS/CdTe thin film solar cells.

1.2.2 MATERIAL PROPERTIES

CdTe is part of the II^B–VI^A compounds family. With a direct energy bandgap of 1.45 eV and a large optical absorption coefficient ($>10^5$ cm^{-1}) for the visible spectrum, only a few microns of CdTe are needed to absorb >95% of photons with energies >1.45 eV. CdTe is also amphoteric, i.e., it is possible to produce homojunctions with appropriate doping.

The temperature–composition phase diagram of CdTe has been described in detail by Zanio [11]. The melting point of CdTe is 1092°C which is significantly greater than Cd (324°C) and Te (450°C). This results in a large range of deposition temperatures available for the production of CdTe thin films. A detailed examination of the stoichiometric region reveals the presence of a symmetrical region which allows nonstoichiometric doping of the CdTe compound to be easily achieved. This makes the conductivity type of CdTe easily controllable. The crystal structure of CdTe is zinc

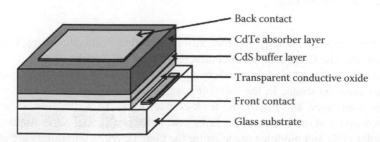

FIGURE 1.3 CdTe/CdS solar cell in superstrate configuration.

blende (Figure 1.4) with a unit cell length of 6.481 Å [15] and a CdTe bond length of 2.806 Å [16].

The most common defect levels in CdTe and their position relative to the conduction or valence band are shown in Table 1.3. Nonstoichiometric doping is achieved by controlling the concentration of native defects in the undoped material. Cadmium-rich growth enhances the formation of tellurium vacancies (V_{Te}), which can become positively charged. These defects act as donors, and the CdTe material grown is therefore n-type. The tellurium-rich film will be p-type, due to the presence of cadmium vacancies (V_{Cd}^- or V_{Cd}^{2-}), which tend to act as acceptor impurities. The CdTe and CdS layers can also be extrinsically doped in order to increase carrier concentration in each layer by adding appropriate donor or acceptor impurities. However, doping is usually achieved during a post-deposition heat treatment of the CdTe in air, following a dip in a solution of $CdCl_2$ in methanol or by the direct incorporation of $CdCl_2$ into the growing CdTe, followed by an anneal in air [11] (see Section 1.2.4.5 for further details).

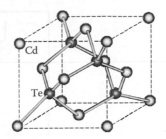

FIGURE 1.4 Unit cell of the CdTe zinc blende crystal structure.

1.2.3 Deposition Techniques for CdTe

1.2.3.1 Thermal Evaporation

CdTe powder, or Cd or Te in the elemental form, can be heated to sublime onto an appropriate substrate, e.g., a glass slide. A schematic diagram of thermal evaporation is shown in Figure 1.5. Deposition is usually carried out in high vacuum, using a source temperature in the range 600°C–800°C and a substrate temperature in the range 100°C–400°C. A deposition rate of 1 μm/min can be obtained using a source temperature of 800°C and a substrate temperature of 100°C. Higher substrate temperatures result in a lower deposition rate. The as-deposited films often exhibit a [111] preferred orientation [24] and also a columnar grain structure [25]. The grain size strongly depends on film thickness and substrate temperature. Typically, for a 2 μm thick film, grain sizes range from 100 nm for a substrate temperature of 100°C to 1 μm when the substrate temperature is 350°C.

1.2.3.2 Sputtering

CdTe films can be successfully deposited by radio-frequency (RF) magnetron sputtering from compound targets. In the case of CdTe, the mass transfer of Cd and Te occurs via the bombardment of the CdTe target by Ar^+ ions, followed by diffusion to the heated substrate (200°C–300°C) and

---

TABLE 1.3
Ionization Energy (eV) of Common Defects and Impurities in CdTe

Type	Acceptors						Donors				
Name	As_{Te}^-	$V_{Cd}Cl_{Te}^-$	V_{Cd}^-	V_{Cd}^-	V_{Cd}^{2-}	$V_{Cd}Te_{Cd}$	Cl_{Te}^+	V_{Te}^+	Cl_{Te}^+	V_{Te}^{2+}	Cd_i^{2+}
Ionization energy (eV)	0.10	0.12	0.14	0.45	0.60–0.74	0.74	0.014	0.04	0.07	0.40	0.64

Sources: Hoschl, P. et al., *Mater. Sci. Eng. B*, 16, 215, 1993; Abulfotuh, F.A. et al., Study of the defects levels, electrooptics and interface properties of polycrystalline CdTe and CdS thin films and their junction, in *Proceedings of the 26th IEEE Photovoltaic Specialists Conference*, Anaheim, CA, 1997, p. 451; Emanuelsson, P. et al., *Phys. Rev. B*, 47, 15578, 1993; Wienecke, M. et al., *Semicond. Sci. Technol.*, 8, 299, 1993; Krsmanovic, N. et al., *Phys. Rev. B*, 62, 16279, 2000; Berding, M.A., *Phys. Rev. B*, 60, 8943, 1999; Capper, P. (ed.), *Properties of Narrow Gap Cadmium-Based Compounds*, INSPEC, London, U.K., 1994.

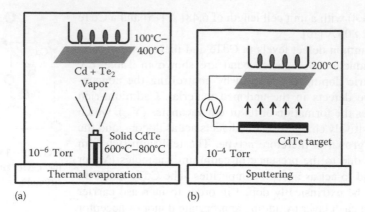

FIGURE 1.5   Schematic diagrams of thermal evaporation (a) and sputtering (b) deposition techniques.

condensation. A schematic of a simple sputtering system is shown in Figure 1.5. The typical deposition rate is 100 nm/min. As-deposited films (2 μm thick) have a grain size of ~300 nm with nearly random grain orientation [26].

1.2.3.3   Close-Spaced Sublimation

Close-spaced sublimation (CSS) is a widely used technique for depositing CdTe and is in fact a modified version of the thermal evaporation method. To date, the most efficient cells use CSS-deposited CdTe [27–30]. This technique is based on reversible dissociation of CdTe at high temperature. The source material is maintained at a higher temperature (e.g., 650°C) than the substrate (e.g., 550°C), a few mm away from it (Figure 1.6). The source dissociates into its elements that recombine on the substrate surface. The deposition occurs through a flowing gas that can be N_2, Ar, He, or O_2. This deposition technique is characterized by a high deposition rate (>1 μm/min), a nearly random orientation of the as-deposited film [24], and a large grain size (several μm) [31].

1.2.3.4   Vapor Transport Deposition

Vapor transport deposition (VTD) is a high-rate deposition technique, also sometimes referred to as modified CSS. This technique was developed by Solar Cells, Inc. [32,33], and deposition occurs by the transfer of Cd and Te vapors from heated CdTe onto a moving and heated substrate

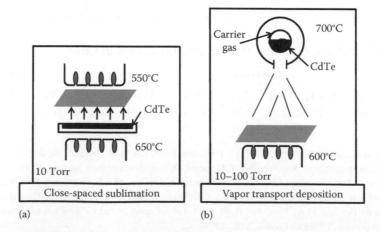

FIGURE 1.6   Schematic diagrams of CSS (a) and VTD (b) deposition techniques.

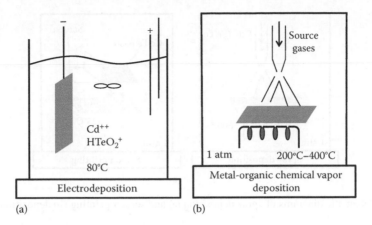

FIGURE 1.7   Schematic diagrams of electrodeposition (a) and MOCVD (b) deposition techniques.

(600°C) by means of a carrier gas (Figure 1.6). As with CSS, the carrier gas can be N_2, Ar, He, or O_2. The deposition rate is usually slower than with CSS. The resultant films consist of large columnar grains.

1.2.3.5   Electrodeposition

The electrodeposition of CdTe has been developed to become a promising method for producing efficient thin film solar cells. It consists of the galvanic reduction of cadmium and tellurium from Cd^{2+} and $HTeO_2^+$ ions in acidic aqueous electrolyte (Figure 1.7). The reduction of these ions utilizes six electrons in the following reactions taking place simultaneously:

$$HTeO_2^+ + 3H^+ + 4e^- \rightarrow Te + 2H_2O, \quad E_0 = +0.559 \text{ V}$$

$$Cd^{2+} + 2e^- \rightarrow Cd, \quad E_0 = -0.4 \text{ V}$$

$$Cd + Te \rightarrow CdTe$$

The thickness deposited and the deposition rate are limited to the ability to maintain the deposition potential over the entire surface of the growing film. The deposition rate is typically ~100 nm/min. As-deposited CdTe films on CdS exhibit strong [111] orientation [34] with columnar grains with a mean lateral diameter of 100–200 nm [35].

1.2.3.6   Metal-Organic Chemical Vapor Deposition

The metal-organic chemical vapor deposition (MOCVD) method usually uses (Figure 1.7) metal-organic precursor gases, such as dimethylcadmium (DMCd) and di-isopropyltelluride (DIPTe), mixed with H_2 at atmospheric pressure. The gases decompose at 290°C and 325°C [36], respectively, allowing the use of lower substrate temperatures compared to the other methods of deposition. The substrates are supported on a heated graphite susceptor. This technique allows the control of [Cd]/[Te] in the deposited film, but also the introduction of a dopant such as P [37] or As [38]. The maximum deposition rate is in the order of 100 nm/min for a substrate temperature of 400°C, yielding films with a strong (111) preferred orientation and a small grained structure [39].

1.2.3.7   Spray Pyrolysis

Pyrolytic spraying is a low-cost technique for the fabrication of large-area CdTe/CdS solar cells. It is a non-vacuum technique for depositing films from a solution (Figure 1.8). Droplets of liquid

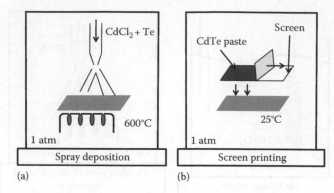

FIGURE 1.8   Schematic diagrams of spray deposition (a) and screen-printing (b) deposition techniques.

are sprayed onto unheated or heated substrates, after which a reaction/recrystallization treatment is performed. CdTe films deposited by this technique usually have large grains and random orientation.

1.2.3.8   Screen Printing

Screen-printing deposition utilizes high-purity elements combined together with a suitable binder into a paste that is applied onto the substrate through a screen (Figure 1.8). A drying step is necessary to remove binder solvents before the layer is heated at elevated temperatures to recrystallize the film and activate the junction. For CdTe films, $CdCl_2$ is combined together with Cd and Te to act as a sintering flux and to avoid the usual post-deposition heat treatment. Films fabricated by this method typically have a thickness of up to $20\,\mu m$ with a large lateral grain dimension (up to ~$5\,\mu m$) and random orientation.

1.2.3.9   Atomic Layer Epitaxy

Atomic layer epitaxy (ALE) consists of the alternate deposition of monolayers of Cd and Te onto a heated substrate by means of a high-temperature carrier gas. In this way, very stoichiometric and pure films can be grown. More details can be found in Section 1.5.2.6. However, this technique yields very low deposition rates, and it is not extensively researched at the present time.

Table 1.4 summarizes the CdTe/CdS thin film solar cells produced to date for each of the techniques described previously. Where possible, this table reports only structures that have both CdS and CdTe grown using the same deposition technique.

TABLE 1.4
CdTe/CdS Thin Film Solar Cells

Growth Technique for the CdTe Absorber Layer	Efficiency (%)	Cell Area (cm²)	Laboratory
CSS	16.5	1.032	NREL [30]
Thermal evaporation	16	0.25	Central Research Laboratory [40]
Electrodeposition	14.2	0.02	BP Solar [41]
RF sputtering	14	—	University of Toledo [42]
ALE	14		Microchemistry, Inc. [43]
Spray pyrolysis	12.7	0.3	Golden Photon, Inc. [37]
Screen printing	12.8	0.78	Matsushita Battery Industrial [44]
MOCVD	13.3	0.25	Centre for Solar Energy Research [45]

Note: Best devices reported for each of the deposition techniques presented in Section 1.2.3.

1.2.4 SOLAR CELL STRUCTURE, FORMATION, AND PROPERTIES

1.2.4.1 Substrate

The choice of an appropriate substrate is very important. It should withstand the cell fabrication process temperature and must not contaminate the layers that are subsequently grown. CdTe/CdS solar cells in the superstrate configuration require a transparent substrate because incident light has to pass through it before reaching the CdS and the CdTe layers; any absorption in the substrate would be detrimental to the current generation in the cell. The general choice is glass because it is transparent and cheap, and withstands relatively high processing temperature. The common types of glass used include soda-lime glass, which is inexpensive, and borosilicate glass. The latter has a higher softening temperature, and for this reason, it is often used for the higher-temperature deposition methods. However, since it is ten times more expensive than soda-lime glass, soda-lime glass is generally preferred for low-cost production. The substrate is usually 2–4 mm thick (to withstand hail impact). It sometimes has an antireflection coating, such as MgF_2, on its surface to minimize reflection losses and a barrier layer to prevent any impurity diffusion from the glass into the cell structure.

Recently, there has been a regained interest in producing efficient solar cells on metal foils and polymer substrates, which offer a substantial weight reduction of the modules compared to glass. Efficiencies up to 8% have been recorded for devices produced on flexible substrates [46].

1.2.4.2 Transparent Conductive Oxide

In general, TCOs are used as the front contact for thin film solar cells. The front contact, or TCO, must be transparent and highly conducting. For high-efficiency cells, it is required that the sheet resistance of the front contact is no more than 10 $\Omega/\square$. The most widely used materials are tin oxide (SnO_2, TO), indium tin oxide (In_2O_3:Sn, ITO), and fluorine-doped tin oxide (SnO_2:F, FTO). The choice between ITO and SnO_2 is primarily determined by the deposition temperature of the CdS and CdTe films [47]. For low-temperature deposition processes, ITO is the material of choice, because it has a high optical transmission for a given sheet resistance. For higher-temperature deposition, SnO_2 is the material preferred, since it is more stable. Recently, the use of cadmium stannate (Cd_2SnO_4, CTO) as front contact has shown improvements in power conversion due to the higher conductivity and higher transmittance of CTO compared to ITO [30]. However, to avoid diffusion of indium from the ITO used in high-temperature deposition techniques, a layer of undoped SnO_2 is often included between the ITO and CdS layers. This intermediate layer between the TCO and the CdS window layer also acts to prevent any shunting through the very thin CdS layer. For this purpose, the intermediate layer needs to be of high resistivity, similar to the role of the ZnO layer in CIGS thin film solar cells (see Section 1.3). Several other materials have been tried (SnO_2, In_2O_3 [40], and Ga_2O_3 [48]), but the most promising is Zn_2SnO_4 [30].

1.2.4.3 Buffer Layer

The energy bandgap of the buffer layer must be sufficiently large so as to maximize the transmission of incident light to the CdTe absorber layer. Also, the window layer should be of relatively high electrical conductivity (1) to ensure that the field region is largely located in the CdTe layer to maximize carrier collection and (2) to minimize resistance losses in the transport of carriers to the external circuit. The polycrystalline CdS is grown n-type and can be deposited by thermal evaporation [49], CSS [50], chemical bath deposition (CBD) [51], magnetron sputtering [52], or MOCVD [27].

As a wide-bandgap semiconductor ($E_g = 2.42$ eV at 300 K), CdS is largely transparent down to a wavelength of around 510 nm. Depending on the thickness of the CdS layer, which is usually ~100 nm, some of the light below the 510 nm wavelength can still pass through to the CdTe, giving additional current in the device [27]. The reduction of layer thickness is then important to allow greater transmission into the CdTe. On the other hand, the uniform coverage of the TCO and the consumption of CdS into the CdTe layer during the annealing treatment require that the thickness is not reduced below a certain limit, otherwise the cell is shunted, resulting in a low open-circuit voltage [53]. An intermediate layer between the CdS and TCO layers can help minimize this problem.

CdS grown by low-temperature deposition techniques (e.g., CBD) generally requires an annealing in air or in chlorine ambient to increase the grain size and reduce the defect density [54]. This treatment has been found to be less influential for layers deposited by methods such as CSS, spray pyrolysis, and screen printing, where a temperature in excess of 500°C is used during the deposition. CBD is the most successful method to deposit CdS, as this is the method employed to produce the most efficient CdTe and CIGS solar cells [9,30]. The CBD CdS process involves the use of aqueous alkaline solutions containing a cadmium salt, a complexing agent (e.g., aqueous ammonia), and a sulfur-containing compound such as thiourea. More details on the deposition technique are given in Section 1.5.2.1. This method results in uniform film deposition even for a very thin layer (50 nm), due mainly to the slow deposition rate, typically <10 nm/min. Lately, there has been some increased interest in using alternative buffer layers for CdTe-based cells, especially those that offer a wider bandgap. The ternary alloy $Cd_{0.9}Zn_{0.1}S$ increases the buffer-layer bandgap to 2.7 eV, making use of the lower part of the light spectrum and increasing the photon-generated current compared to CdS [45].

1.2.4.4 Absorber

CdTe is, in principle, an ideal thin film PV absorber material, as ~99% of the solar radiation is absorbed within a thickness of 2 μm. The polycrystalline CdTe layer should be p-type to form the p–n junction with the n-CdS layer. Since CdTe has a lower carrier concentration than the CdS layer, the depletion region is mostly located within the CdTe layer, and in this region, most of the carrier generation and collection occur.

The conductivity and grain size of the layer depend on the deposition technique and the post-deposition treatment used. Typically, the thickness of this layer is between 2 and 10 μm, with a grain size ranging from 0.5 to 5 μm. In its as-deposited state, the CdTe layer is either n-type or highly resistive p-type. However, a suitable heat treatment in chlorine or oxygen ambient can convert the layer to p-type and also increase its conductivity.

1.2.4.5 CdCl₂ Heat Treatment

Nearly, all the deposition techniques described previously (see Section 1.2.3) require a post-deposition treatment of the CdTe layer in order to produce high-efficiency solar cells. The treatment generally involves a high-temperature processing step with exposure to a chlorine-containing species and/ or oxygen. This treatment changes the electrical and structural properties not only of the absorber layer but also of the junction itself, and increases the efficiency some 2%–5% above that achievable with as-grown material to efficiencies in the range 10%–16% after the treatment. Typically, the heat treatment takes place at a temperature range of 350°C–450°C for 15–30 min depending on the CdTe film thickness.

The treatment steps can be performed in a variety of ways, such as

1. In situ incorporation of chlorine species during CdTe film deposition
2. Dipping the CdTe layer in a $CdCl_2$ aqueous solution, followed by drying in order to precipitate a $CdCl_2$ film, and then annealing [55]
3. Annealing in $CdCl_2$ vapor [56], HCl [57], or Cl_2 gas [58,59]
4. Deposition of a $CdCl_2$ layer onto the absorber followed by annealing [60]

In its as-deposited state, CdTe is usually n-type or highly resistive p-type, making it unsuitable for solar cell applications. The heat treatment modifies the electronic properties of the CdTe layer by "type-converting" the CdTe from n-type to p-type.

The incorporation of chlorine is considered to form an acceptor complex with cadmium vacancies, $[V_{Cd}Cl_{Te}]^-$. This relatively shallow acceptor state has an ionization energy of 0.12 eV

[17] compared to 0.6 eV for $[V_{Cd}]^{2-}$ [20], which makes the complex a more effective dopant than the cadmium vacancies alone. However, excess chlorine can lead to compensating $[Cl_{Te}]^+$ donors. In conjunction with converting the conductivity of the CdTe layer from n-type to p-type, the $CdCl_2$ treatment decreases the sheet resistance of the absorber layer up to three orders of magnitude [61].

The $CdCl_2$ treatment also modifies the structural properties of the CdTe films. The treatment can promote recrystallization and grain growth in small grained films [62]. Levi et al. [63] observed a five-time increase in the initial crystallite size following the treatment of small-grained evaporated films, but not for larger-grained CSS-deposited layers. This was confirmed by Cousins et al., who observed that no grain growth occurs following the $CdCl_2$ heat treatment of 10 µm thick CSS-deposited structures [64]. The predominant recrystallization effect, common to almost all deposition techniques giving small grains, is the randomization of the orientation of the CdTe films.

The post-deposition treatment is usually performed at temperatures greater than 350°C. During this phase, a physical reaction between CdTe and CdS can occur and interdiffusion between the two semiconductors forms a $CdTe_{1-x}S_x$ solid solution in the absorber layer and also a $CdS_{1-x}Te_x$ solid solution in the window layer. These solid solutions have both beneficial and detrimental effects on the performance of the solar cell. The interdiffusion process narrows the energy bandgap of the absorber layer for lower sulfur concentrations ($x < 0.4$), resulting in a higher long-wavelength quantum efficiency (QE) [65]. The intermixing of CdS and CdTe reduces interfacial strain by the reduction of the lattice mismatch at the CdTe/CdS junction [66] and may reduce dark current recombination [67]. Alloy formation also consumes the CdS layer, which can be beneficial for window transmission. However, nonuniform consumption can result in lateral discontinuities at the junction and device shunting [68].

The effects of the $CdCl_2$ heat treatment on the performance of complete CdS/CdTe solar devices are quite astonishing, as shown in Figure 1.9a. All the fundamental parameters of the cells are improved following the $CdCl_2$ heat treatment of the CdTe absorber layer. Figure 1.9b shows the corresponding spectral response curves. When the CdTe absorber layer receives no treatment, the QE remains weak with a CdS/CdTe junction of poor quality, as indicated by the peak QE at ~820 nm, characteristic of a buried junction. Only the $CdCl_2$ treatment allows uniform collection across the visible spectrum; the corresponding QE curve is a testimony of the good-quality junction.

Finally, one of the important effects of the $CdCl_2$ treatment is the passivation of the grain boundaries of the polycrystalline absorber layer. This has been evidenced by Edwards et al. using electron beam–induced current (EBIC) imaging [70,71]. The technique basically consists

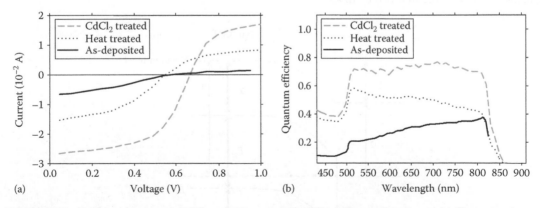

FIGURE 1.9 Effects of heat and $CdCl_2$ treatment on CdS/CdTe device performances. (a) Current–voltage curves for cells made from an as-deposited CdTe film ($\eta = 1.2\%$, FF = 39%), a heat-treated CdTe layer ($\eta = 3.6\%$, FF = 39%), and a $CdCl_2$-treated film ($\eta = 10\%$, FF = 56%); and (b) corresponding QE curves. (From Al-Allak, H.M. et al., *J. Cryst. Growth*, 159, 910, 1996. With permission.)

of mapping the short circuit current collected by the p–n junction after carrier excitation by an electron beam (e-beam). Using the front-wall configuration, i.e., junction irradiated from the front surface after chemical removal of the glass substrate, Edwards et al. showed that the collected images were beam-current dependent and that for CdCl$_2$-treated CSS-grown structures, no image contrast was observed between grains and grain boundaries at low beam current. Further investigations under high beam injection, i.e., higher beam current, gave rise to a newer band structure diagram of grain boundaries, this being consistent with the previously derived model using conductivity experiments indicating an upward band bending at the grain boundaries [72,73].

1.2.4.6 Back Contact

Producing ohmic contacts on most semiconductors is a difficult problem. With p-type material, an ohmic contact is only created when the work function of the conducting layer is larger than that of the semiconductor. When this condition is not satisfied, a Schottky barrier is formed that can reduce device performance. For p-type CdTe, forming an ohmic contact is exceptionally hard due to the high work function of CdTe ($\phi_p \sim 5.7$ eV, electron affinity $\chi_p = 4.3$ eV) [74]. This can be seen in Figure 1.10, which shows a band diagram of a metal contact on p-type CdTe. The Schottky barrier height, defined as the difference of the two work functions, $qV_{bi} = \phi_p - \phi_m$, can be reduced by increasing the carrier concentration in the CdTe in the vicinity of the back contact.

Numerous chemical recipes are employed with more or less success to make a back contact to CdTe/CdS solar cells. One possibility is to prepare a p^+-type layer that can decrease the work function of the top part of the absorber in order to make a quasi-ohmic contact with the metal. Various chemical etchants may be used, which include acidified potassium dichromate solution [75], bromine methanol solution [76], or orthophosphoric acid [77]. The principle of the etching is to leave a tellurium-rich layer at the surface of the CdTe by reduction of the tellurium ions, hence creating a more p-type layer at the surface of the absorber.

A second technique involves direct doping of the top surface of the absorber using a graphite paste with the required dopant, such as Cu$_x$Te in HgTe. Subsequent annealing allows the dopant to diffuse into the CdTe [78]. Doping can also be performed during growth. Both methods increase the acceptor concentration near the surface, and this results in a narrower depletion region when the metal is subsequently applied. This allows thermally assisted tunneling of holes through the barrier.

Finally, the deposition of an intermediate layer of a p-type semiconductor with a lower work function than CdTe, such as ZnTe [79], HgTe [80], or Sb$_2$Te$_3$ [49], can be used sometimes to complement with one of the first two methods.

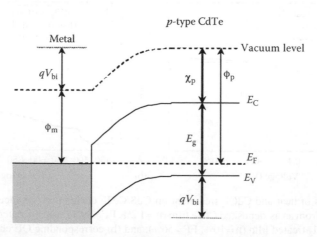

FIGURE 1.10 Band diagram of a metal/p-type CdTe contact at equilibrium.

1.2.5   STATE-OF-THE-ART CdTe PV DEVICES AND MODULES

To date, the most efficient CdTe/CdS thin film solar cell produced on a laboratory scale has an efficiency of 16.5%, as measured under standard air mass 1.5 (AM1.5) illumination conditions [30]. The cell parameters were fill factor, FF=75.5%; short circuit current density, J_{sc}=25.9 mA/cm^2; and open-circuit voltage, V_{oc}=845 mV. The device structure used was glass/CTO/Zn$_2$SnO$_4$/CBD-CdS (0.1 µm)/CSS-CdTe (10 µm)/back contact (mixture of C/HgTe/Cu$_2$Te).

A common feature of CdTe cells, but also of solar cells to which it is difficult to form an ohmic back contact, is a current limiting effect in forward bias, which becomes progressively more important as temperature decreases. This is shown in Figure 1.11, where the current density–voltage (J–V) curve for a CSS-grown solar cell is plotted as a function of temperature [81]. The influence of temperature on the open-circuit voltage and flattening of the J–V curve at high forward bias can be observed; this is referred to as the "rollover effect."

In the particular case of CdTe thin film solar cells, the single diode equation is not sufficient to fully describe the behavior of the device. Stollwerck and Sites [82] described this phenomenon as due to a second diode barrier located at the back contact of the solar cell. The equivalent circuit model corresponding to this is shown in Figure 1.12.

This barrier is also thought to be responsible for the crossover effect between the dark and light J–V curves because of a non-negligible minority carrier current. This feature also demonstrates the poor conductivity of the absorber layer in the dark. However, it has been suggested by Agostinelli

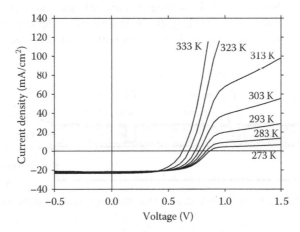

FIGURE 1.11   Current density–voltage characteristics under illumination as a function of temperature for a CSS-grown CdTe/CdS solar cell. (Reproduced from Edwards, P.R., PhD thesis, University of Durham, Durham, England, 1998. With permission.)

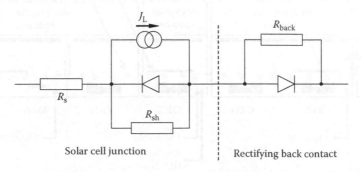

FIGURE 1.12   Equivalent circuit for a solar cell with rectifying back contact. This comprises two diodes associated with the main solar cell and the Schottky junction located at the back contact, respectively.

et al. [83] that the crossover observed between the current–voltage characteristics measured in the dark and under illumination could be associated with a barrier located at the front region of the device due to an increased compensation of donors in the CdS window layer and the presence of a buried junction located deeper in the absorber layer.

CdTe modules are manufactured by interconnecting in series CdTe cells deposited on a single substrate. The principle is shown in Figure 1.13. For manufacturing purposes, the cells are not individually grown. Each layer of the structure is grown over the entire substrate area, and layers are scribed at several points during the process to isolate them between one another, thus avoiding a "dead" module from one small faulty part. Three scribes are performed during the manufacture of the CdTe module. The first scribe, performed by a laser, cuts through the TCO layers to isolate the front contacts. The second scribe, usually done by mechanical scribing, goes through the CdS/CdTe bilayer to provide an electrical path from the front to the back contact between consecutive cells. Finally, a third scribe through the back and CdS/CdTe layers define the final area of the cells. The next processing steps include the contact attachment and finally the lamination. In this way, the individual CdTe cells are connected in series in each module. A schematic of a production line, such as the one employed at Antec GmbH, is shown in Figure 1.14. A complete line can be as long as 150 m as a heat-up and a cool-down chamber need to be included.

Over the past decades, several companies have been involved in the development of large-area modules and production lines using different processes (see also Section 1.1): BP Solarex (electrodeposition), Matsushita Corporation (screen printing), Golden Photon (spray deposition), Antec GmbH (CSS), and First Solar (VTD). First Solar is currently the main commercial producer of CdTe modules, and the company is building up toward gigawatt capacity production, with their aim to achieve cost levels competitive with conventionally produced electricity within the next

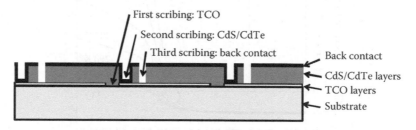

FIGURE 1.13 Schematic of a CdTe module showing the interconnection of cells and the scribing.

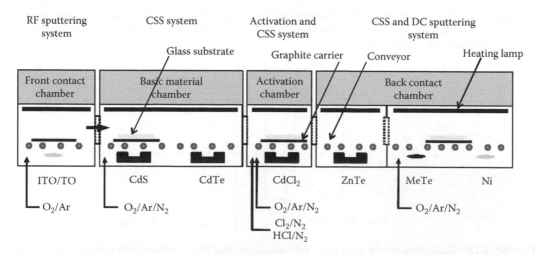

FIGURE 1.14 Schematic representation of a production system used for producing CdS/CdTe modules.

12–24 months. The expansion of sales up to 8% of the entire (and rapidly expanding) PV market during 2008 has been very impressive, making CdTe the leading thin film technology at present. The very high level of success of First Solar has stimulated renewed interest in other companies producing and selling CdTe modules. These include Calyxo in Germany.

1.3   THIN FILM SOLAR CELLS BASED ON THE USE OF CHALCOPYRITE COMPOUNDS

1.3.1   HISTORICAL DEVELOPMENT OF CuInGaSe$_2$ SOLAR CELLS

During the 1970s, the ternary chalcopyrite semiconductors experienced a growth in interest, particularly at Bell Laboratories in the United States. The research into these semiconductors focused on their potential as broadband near-infrared PV detectors and light-emitting diodes (LEDs) [84–86]. This work provided the underpinning for developing PV solar cells using chalcopyrite materials [87]. In particular, it was realized that copper indium diselenide (CIS) had an energy bandgap of 1.04 eV, i.e., in the range needed for making solar cells; that the energy bandgap was direct, leading to high values of the optical absorption coefficient for photons with energies greater than the energy bandgap; and that the material was amphoteric, giving flexibility in device design. The work at Bell Laboratories used annealing in selenium to control the composition and, hence, the p-type conductivity of single-crystal CIS. Using this single-crystal material and vacuum-evaporated CdS, small-area CIS solar cells were produced with efficiencies up to 12% [88]. For larger-area cells, micro-cracks in the crystals were observed, and these were believed to limit the device performance. Devices based on the use of single-crystal CIS were however impractical as commercial solar cells due to the very high cost needed to produce good-quality single-crystal material.

During the 1970s, significant developments were also reported by workers at the University of Maine. The group produced solar cells using single-crystal absorbers and reported thin film CdS/CIS and CdZnS/CIS heterojunctions as well as CdS/CuInS$_2$ devices. They also produced p–n homojunctions of thin film CuInS$_2$ and CIS, and compared their performance with heterojunction solar cells made using thin film and single-crystal CIS. The thin film materials were produced by dual-source thermal evaporation of the compound (CIS or CuInS$_2$) and the chalcogen (Se or S) by Kazmerski et al. [89–92]. The best thin film CdS/CIS solar cells had efficiencies of 6.6%, and the best CdS/CuInS$_2$ solar cell, an efficiency of 3.25% (measured using 100 mW/cm^2 tungsten halogen illumination).

In the early 1980s, Boeing produced an all thin film CIS-based solar cell with a world-record efficiency of 9.4% [93,94]. By 1983, the device was further developed to have an efficiency >10% [95].

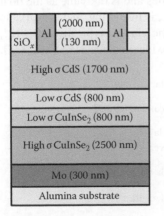

FIGURE 1.15 Schematic diagram indicating the structure of the high-performance Boeing CuInSe$_2$ solar cell.

A cross-sectional view of a Boeing cell is shown in Figure 1.15. The Boeing cell structure consists of two layers of CIS deposited onto molybdenum: one Cu-rich adjacent to the Mo, followed by a layer that was In-rich next to the device interface. Two layers of CdS were also deposited: an undoped layer next to the CIS, followed by a high-conductivity (σ) In-doped CdS layer, adjacent to the top contact. This design resulted in high-conductivity materials next to the contacts, minimizing the contact resistivity, with relatively low resistivity CdS and CIS to form the junction.

Further developments in the United States were mainly made by Boeing, the Institute of Energy Conversion (IEC) at the University of Delaware, and SERI (the Solar Energy Research Institute, Denver, now the National Renewable Energy Laboratory [NREL]) [96]. They investigated heat treatment of the back contact using high-resolution EBIC. Their work showed that heating in oxygen to 200°C for 2–3 h seemed to activate the heterojunction response in the thin film devices [97,98].

In 1987, Boeing also reported the production of CIGS thin film PV solar cells (Ga/(In+Ga) ratio of 0.23) with efficiencies >10%, wherein the CIGS was produced by the co-evaporation of the elements [99]. This work and that in Europe demonstrated that the energy bandgap value of the thin film absorber was increased by introducing Ga [100].

During this period, the first steps to industrialization of CIS-based technology were made by Boeing and by ARCO Solar in California. The methods used to form the CIS were based on two distinctly different processing techniques. Boeing concentrated on developing the co-evaporation route, whereas ARCO began the development of the so-called two-stage process. The latter involved the deposition of metallic precursors followed by conversion to selenide. Their process was described by a number of patents [101–103]. Initially, ARCO sputter-deposited copper and indium layers and followed by selenization using hydrogen selenide (H_2Se) at temperatures between 400°C and 450°C. Subsequently, the company reduced the dependence on toxic H_2Se by using an evaporated layer of Se as a chalcogen source; the patents also indicate the planned addition of gallium and sulfur to their absorber layers. While sputtering was the basis of the ARCO process, their patents also considered the deposition of the precursors by other methods ranging from e-beam evaporation to electrodeposition.

By the late 1980s, ARCO Solar were producing large-area modules with efficiencies of 11.1% for a module area of $938\,cm^2$ and 9% for a module area of $3900\,cm^2$ [104,105]. The unencapsulated larger module yielded an output power of over 35.8 W equivalent to 75% of the output of crystalline silicon modules available, at that time of comparable area.

1.3.2 Choice of Device Configuration, Substrate, and Back-Contact Material

While both superstrate and substrate configurations were investigated during the 1970s and 1980s, the highest performances were associated with the substrate configuration devices. A diagram of the most commonly used device structure is shown in Figure 1.16. In this configuration, the substrate provides mechanical support to the film and the back-contact metal is the first layer to be deposited. The early devices used a variety of substrates, and the Boeing research originally used alumina substrates. Borosilicate glass was also used. The choice of substrate affects the subsequent processing steps and material growth due to the heating necessary during processing and the influence of surface and chemical properties on the growing thin films. The need to heat the substrates also brings into consideration the need to minimize the different thermal expansion coefficients between the cell layers and that of the substrate. The highest efficiency devices have all used soda-lime glass as the substrate. This material is available in large quantities and is the same as the float glass used for making windows in buildings. It has a smooth surface, is insulating, and is available at relatively low cost. The thermal expansion coefficient of soda-lime glass is typically ~$10^{-6}\,K^{-1}$, a good thermal match to CIS thin films [106]. The alkali metal oxides include Na_2O and K_2O.

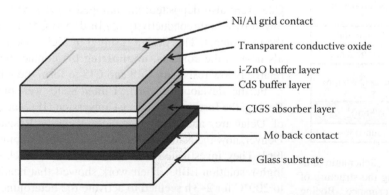

FIGURE 1.16 Schematic diagram of a CIGS-based solar cell in the substrate configuration.

Sodium was found to have a significant effect on the CIGS films and crystallographic properties there being a higher degree of $\langle 112 \rangle$ texturing for CIGS films grown on soda-lime glass substrates. This work showed that both stress and the presence of sodium were important in establishing the texture and structure of films, and contributed to increased performance (16.9% active area) compared to earlier devices [107,108]. Sodium diffuses from the soda-lime glass through the back contact to the growing CIGS film, and therefore the back contact plays an important role in the growth of the absorber layer.

Molybdenum (Mo) is used as a back-contact layer in CIGS-based solar cells because of its low contact resistance, and it also exhibits insignificant diffusion into the absorber layer during device processing. A variety of metals such as Pt, Au, Ni, W, Ta, Cr, Nb, V, Mn, and Ti have also been tested by research workers [109,110]. Of these metals, Pt, Au, Ni, Cr, V, Ti, and Mn showed significant diffusion into the CIGS-based absorber layer with annealing at higher temperature or reaction with Se. There is little literature on the use of W, Ta, Nb, and V as the back contact for CIGS-based solar cells. During processing, the Mo back-contact layer is required to keep its properties such as conductivity and adherence to the substrate. In general, Mo has been used as a refractory metal for protective coatings and high-temperature applications due to its outstanding mechanical properties, such as high melting point, high mechanical hardness, and high thermal stability [111].

Various techniques have been used to deposit Mo layers, including direct current (DC) or RF magnetron sputtering, chemical vapor deposition, ion beam–assisted deposition, e-beam evaporation, and laser ablation deposition [110,112]. However, in general, for CIGS-based thin film solar cells, the Mo layer has been deposited using e-beam evaporation or magnetron sputtering onto soda-lime glass substrates. Back contacts deposited using these techniques yield comparatively inexpensive, inert layers that are conformal to the substrate even at temperatures in the range 500°C–600°C. The use of a Mo back contact also results in the formation of a $MoSe_2$ interfacial layer [113], and some researchers suggest that the $MoSe_2$ layer acts as a wide-bandgap back-surface field layer. The control of the morphology of the deposited Mo film has been associated with a change in orientation in the CIGS film and higher performance levels from record cells.

1.3.3 THIN FILM DEPOSITION USING THE CO-EVAPORATION METHOD

In 1987, Chen et al. described the production of CIGS thin films and, subsequently, thin film CdS/CIGS solar cells [99]. The CIGS material was produced by the simultaneous vacuum evaporation of the elements Cu, In, Ga, and Se to form the quaternary compound. A schematic of the deposition is shown in Figure 1.17. They followed a process similar to the two-layer "Boeing" process that had been used to deposit $CuInSe_2$ thin films. The first Cu-rich layer ($2\,\mu m$ thick) was grown using a substrate temperature in the range 400°C–450°C. A second Cu-poor layer ($1\,\mu m$) was then deposited with a slight increase in temperature to 550°C. Different amounts of Ga were incorporated into the films by changing the evaporation rates of the appropriate sources. A range of cells $1\,cm^2$ in area were made by varying Ga/(In + Ga) ratio from 1.0 ($CuGaSe_2$) to 0.25 (25% Ga content in CIGS or $CuIn_{0.75}Ga_{0.25}Se_2$) and depositing a $Zn_{0.12}Cd_{0.88}S$ window layer. The fill factor (FF) was significantly reduced with the increase in Ga content. A Ga/(In + Ga) ratio of 1 gave an open-circuit voltage of 0.68 V with an efficiency of 2.71%. In contrast, only 25% Ga content gave an open-circuit voltage of 0.51 V, FF of 66%, but a conversion efficiency greater than 10%. Similar work on CIGS was carried out by Dimmler et al. at the University of Stuttgart, Germany, and they yielded cell efficiencies up to 5.8% for $CuGaSe_2$, 9.3% for CIS, and 3% for $CuIn_{0.56}Ga_{0.44}Se_2$ thin films [100].

The work described above used constant-rate evaporation, while the technique that has yielded the highest performances has three separate and distinct stages. This technique is referred to as "the three-stage process." Figure 1.18 summarizes the deposition rate of the elements and the substrate temperature for constant rate and for the three-stage process. The first stage involves the deposition of a Cu-free selenide layer, which is followed by an In- and Ga-free Cu-Se layer, and finally a second Cu-free layer terminates the structure and adjusts the final composition. Kessler [114] was

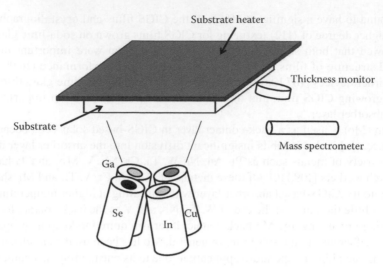

FIGURE 1.17   Multisource co-evaporation technique for CIGS deposition.

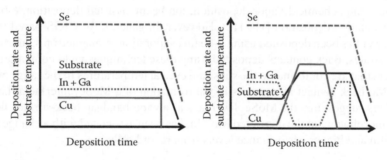

FIGURE 1.18   Deposition rate and substrate–temperature profile for constant-rate evaporation (a) and the three-stage process (b).

the first to propose this process with subsequent modifications suggested by Gabor et al. [115]. The higher performances are suggested to be the result of grading the energy bandgap from a higher value (higher Ga concentration) at the Mo/CIGS interface to a lower value toward the junction. This is thought to increase the carrier collection and, thus, increase the current generation in the device. A higher value of the energy bandgap at the free surface is associated with achieving a higher open-circuit voltage and FF. This process yields high-performance cells, but requires careful monitoring and close control over the deposition parameters. The design of the source is important to ensure a controllable evaporation rate. A range of techniques is used to monitor the film composition during growth [95,116–118]. Either thermocouple or optical pyrometer monitoring of the substrate temperature has also been associated with the production of the highest-performance devices [119,120]. The highest efficiency reported for a CIGS-based cell is 19.9% (0.419 cm² active area). This was based on an absorber layer produced using the three-stage co-evaporation process [9].

1.3.4   Thin Film Synthesis Using Selenization (or Sulfidization) of Pre-Deposited Precursor Layers

The two-stage process of CIGS-based absorber-layer formation involves the deposition of metallic precursor layers by magnetron sputtering or evaporation, followed by selenization/sulfidization of the precursor layers at an elevated temperature in a Se/S atmosphere created by H_2Se/H_2S gas or elemental Se/S. A schematic of this process is shown in Figure 1.19.

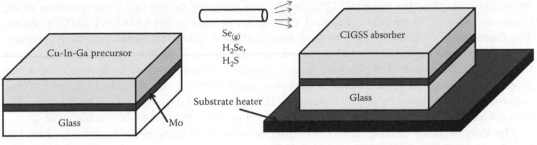

Stage 1: Precursor preparation   Stage 2: Selenization/sulfidization

FIGURE 1.19 Schematic of the two-stage synthesis of CIGS cells.

The process of sputtering is a well-established deposition technique used in semiconductor and thin-film-coating industries. There are several distinguishing features of magnetron sputtering, such as uniform and cost-effective deposition of thin films over large areas [121]. In addition, the advantage of the sputtering technique is its higher degree of material utilization (75%) as compared to that achieved using the thermal evaporation method (35%) [122].

Showa Shell in Japan (using a process licensed from Siemens Solar Industry) have fabricated $CuInGa(S,Se)_2$ (CIGSS, copper indium gallium diselenide disulfide) thin film modules by a two-stage method using In/Cu-Ga/Mo stacked precursors and H_2Se gas. The Cu-In-Ga precursor layers were deposited by magnetron sputtering followed by selenization in H_2Se gas. A sulfurization process was then used to introduce sulfur in the absorber layer. The deposition process and the process control technologies were improved to enable the processing of substrate sizes of $3600\,cm^2$ or larger. The highest efficiency reported was 13.4% for a $Zn(O,S,OH)_x/Cu(In,Ga)(S,Se)_2$ module area of $3600\,cm^2$ [123].

The CIGSS-based module fabrication process of Avancis GmbH, Germany (a joint venture between Shell Solar GmbH and Saint-Gobain, the glass manufacturing company), avoided the use of highly toxic H_2Se gas in the selenization process [124]. The CIGSS production method involves

1. The deposition of the Cu-In-Ga precursor layers using magnetron sputtering
2. The deposition of a Se layer on top of the precursor layer using thermal evaporation
3. Rapid thermal processing (RTP) of the Se capping precursor layers in an environment containing both Se and S

Avancis have produced CIGSS modules with an efficiency of 13.5% for a $30\times30\,cm^2$ module and an efficiency of 13.1% for a $60\times90\,cm^2$ module [125]. The main advantages of this method are its suitability for a large-area deposition and improved environmental safety by avoiding the use of H_2Se.

1.3.5 DEVELOPMENT OF $CuInS_2$ SOLAR CELLS

The energy bandgap of $CuInS_2$ is close to the optimum for PV solar energy conversion (1.5 eV), and as with CIS, the direct energy bandgap results in high values of optical absorption coefficients for photons with energies larger than the energy bandgap. The first $CuInS_2$ devices of note were produced by Kazmerski et al. during the late 1970s [89–92]. These devices were produced using single-crystal $CuInS_2$ and thin film $CuInS_2$, as described in Section 1.3.1. The thin film materials were produced by dual-source thermal evaporation of the compound ($CuInS_2$) and the chalcogen (S). The efficiency of the best thin film $CdS/CuInS_2$ solar cells produced was 3.25%.

During the 1990s, the University of Stuttgart extended their work on CIGS to $CuInS_2$. The achievement of device efficiencies of over 10% [126] indicated the viability of $CuInS_2$ as an important solar absorber material [129]. $CuInS_2$ was prepared by thermal co-evaporation of the elements and was formed into a typical CIS device structure, $ZnO:Al/CdS/CuInS_2/Mo/glass$. The limitations of the device performance were attributed to the moderate open-circuit voltage. Efficiencies were improved to >12% by improving the absorber–buffer interface and increasing the V_{oc} values to over 800 mV. Etching using a cyanide solution was used to remove copper-rich phases that segregated to the surface of the layers, and the device structures included an intrinsic zinc oxide layer between the highly conductive TCO layer (ZnO:Al) and the CdS buffer layer. The workers also investigated the use of incorporating Zn into the absorber at the interface [127].

The Hahn-Meitner-Institute (HMI) in Berlin continued the development of solar cells based on $CuInS_2$, and in 1998, the development of an 11.1% cell began the effort that led to the formation of Sulfurcell Solartechnik GmbH [128]. These cells were made using a two-stage process: the deposition of Cu/In precursor layers using DC sputtering followed by sulfidization using elemental sulfur.

1.3.6   STATE-OF-THE-ART CHALCOPYRITE PV DEVICES AND MODULES

1.3.6.1   Würth Solar Process

The Würth Solar process was developed from the co-evaporation technology route developed at the University of Stuttgart. Pilot scale production began during the period 1999–2006 with a production capacity of 1.5 MWp (Watt peak). During this phase, the main aim was to demonstrate that high-quality CIGS could be produced with high productivity. This was established, and in 2005–2006 a large-scale production facility was built. By 2007, the capacity was 15 MWp, and from July 2008, this was doubled to 30 MWp. The production at Würth Solar is a fully integrated, continuous in-line operation with a high level of automation from the raw material to the completed module. The modules have less energy payback time of 18 months (if operated in a middle-European climate). The modules have been tested and have a proven long-term stability performance under different climatic conditions.

1.3.6.2   Avancis GmbH & Co. KG

Avancis was formed in 2006 as a joint venture between Shell Solar and the glass manufacturer, Saint-Gobain. Shell Solar bought Siemens Solar that had in turn bought ARCO Solar. Avancis is therefore the direct descendent of the first CIGS production process. The technology is based on depositing copper-indium-gallium precursor layers onto Mo-coated glass substrates, depositing a selenium capping layer onto the precursor layers, and then performing rapid thermal annealing in a H_2S atmosphere to create a $CuInGa(S,Se)_2$ absorber layer. The structures then have the buffer layer (ZnO/CdS) and the transparent front contact (ZnO:Al) deposited. In October 2008, Avancis opened a new 20 MWp facility in Germany to produce 11% thin film CIS modules. The new facility will be supported by the experience gained from their existing 3 MWp facility in Camerillo, California.

1.3.6.3   Showa Shell

Showa Shell licensed the Siemens Solar two-stage process and began commercial production in 2007 with an annual production capacity of 20 MWp at their first production facility in Miyazaki, Japan. A second plant with an annual production capacity of 60 MWp is due to start commercial production in 2009. By 2011, the company is planning to start a new production facility that will have an annual production capacity of 1 GWp. Showa Shell holds the record performance for a cadmium-free $CuInGa(Se,S)_2$-based module with an efficiency of 13.5% for a 3459 cm^2 aperture area [129].

1.3.6.4   Solarion AG/Photovoltaics

Solarion uses ion beam–assisted deposition of CIGS onto molybdenum-coated polyimide flexible substrates. This reel-to-reel process deposits either CIS or CIGS absorber layers using the ion beam–assisted deposition to reduce the deposition temperature needed. The substrate uses a buffer layer deposited via a wet chemical deposition process, followed by the deposition of the zinc oxide–based TCO front contact. The light, flexible solar cells are particularly suitable for space applications. The roll-to-roll process enables significant cost reductions compared to conventional production processing techniques based on rigid glass substrates.

1.3.6.5 Nanosolar, Inc.

Nanosolar, Inc. use proprietary "nanostructured" inks and a printing process to deposit the semiconductor absorber. The deposition method has been designed to ensure uniform distribution of the four elements in the desired atomic ratios. This is to ensure that the composition remains uniform across any areas that are printed. The benefits of the process are that it is a non-vacuum, low-cost, simple to maintain, and high-yield process that also allows the material to be printed where required and thus ensures high material utilization. The process involves deposition onto a flexible metal foil that is 20 times more conductive than stainless steel that is commonly used for roll-to-roll processes. This reduces the cost of the bottom electrode.

The printing technique utilizes equipment with, relatively, a very small footprint and is claimed to give a very high throughput by enabling substrate rolls of meters wide and miles long to be processed efficiently at minimal capital cost. Other key advantages of roll-to-roll processing are identified as the significantly greater uniformity of the process and minimized edge effects compared to processing glass sheets or wafers. The company also claims to be using a low-cost, transparent top electrode material that has an order of magnitude higher current-carrying capacity than other solar cells.

The Nanosolar process is designed to allow matching of cells to ensure a consistent efficiency distribution in modules, which is not possible with conventional thin film technologies that use large-area rigid substrates. It is claimed that this cell-sorting reduces the yield-loss costs by a factor of hundred compared to monolithic cell integration techniques.

The technology is claimed to be able to deliver high-power solar modules with current values up to 10 times higher than those using conventional thin film technologies. It is also claimed that the energy payback time for the Nanosolar panels would be less than 1 month compared to 1.7 years for vacuum-based thin film technologies on glass and 3 years for silicon wafer–based technologies.

1.3.6.6 HelioVolt Corporation

The HelioVolt Corporation uses their award-winning manufacturing process, FASST, to produce $CuInGaSe_2$ thin films and a printing process that are claimed to be between 10 and 100 times faster than conventional thin film $CuInGaSe_2$ processes. The company's 25 MWp facility will start producing its first products in 2009. The FASST process uses electrostatic clamping and high temperature to print the CIGS onto glass or other suitable substrate. Device conversion efficiencies of up to 12.2% have been reported by the company [130].

1.3.6.7 Solyndra, Inc.

Solyndra, Inc. is producing cylindrical CIGS PV modules. The proprietary modules are designed to enable improved cooling and light collection compared to conventional flat-plate modules. The company is producing their modules to have conversion efficiencies between 12% and 14%. They claim that the modules are faster to install and more resistant to moisture, and that it is possible to fit twice the number onto a roof, compared with conventional modules. Solyndra's panels employ a cylindrical geometry to enable the conversion of direct, diffuse, and reflected sunlight into electricity. The current capacity is 110 MWp with a further 420 MWp planned to open before the end of 2009.

1.3.6.8 Honda Soltec Co., Ltd.

Honda began commercial sales of its CIGS modules in Japan in 2007. The 27.5 MWp production facility was developed as part of the company's efforts to reduce its CO_2 emissions. The facility followed with over 15 years of research and development at the Honda Fundamental Technology

Research Centre and the achievement of laboratory efficiencies of over 18% (0.25 cm² area) in 1999. 125 and 115 W (1.4 × 0.79 m area) modules have average conversion efficiencies of 11.1% and are produced by Honda Engineering. The company is aiming to reach a maximum value of 12.2% for the pilot production line. The three technical features of the CIGS modules are as follows: (1) A sodium solution is sprayed onto low alkali glass using a paint-spraying system. (2) This type of glass enables the heating above 500°C (which is difficult for soda-lime glass) and selenization at a higher temperature to improve material crystallinity. (3) The company also uses an InS buffer layer to avoid the use of CdS.

1.3.6.9 Solibro GmbH

Solibro GmbH is a German company that has been established to commercialize CIGS PV. The company is a joint venture between Q-Cells SE (Germany) and Solibro AB (Sweden). The details of the processing used by Solibro are commercially sensitive, but it is based on the research and development activity at the Uppsala University. The current world-record CIGS mini-module with a 16.6% efficiency [129] is held by researchers at an Uppsala spin-off company, Solibro AB, for a mini-module with an aperture area of 16 cm². A 25 MWp production facility was planned to begin and ramp up production during the second half of 2008 with further expansion to follow.

1.3.6.10 Sulfurcell Solartechnik GmbH

Sulfurcell Solartechnik GmbH had its origins in the work on $CuInS_2$ carried out by the HMI, Berlin (now The Helmholtz Centre Berlin, for Materials and Energy). Sulfurcell began developing and manufacturing thin film solar modules based on $CuInS_2$ chalcopyrite-type semiconductors in 2003. The pilot line production capacity exceeded 1 MWp in 2007 when the median module efficiency for their 125 × 65 cm modules reached 55 Wp [131]. In February 2009, work began on the building of a new factory that will take the production capacity from 3 to 75 MWp. Sulfurcell uses sputtering to deposit the molybdenum back contact and the copper and indium precursor layers. Elemental sulfur vapor at about 500°C reacts with the precursors to form the $CuInS_2$ absorber layer. A wet chemical process is used to deposit the CdS buffer layer followed by the deposition of a front contact. The module is contacted and then encapsulated to protect the cell from moisture.

1.3.6.11 Odersun AG

Odersun AG has two production facilities with a planned annual output over 35 MWp. Odersun AG in addition to Sulphurcell is the only manufacturer of $CuInS_2$-based solar modules. The company uses their patented CISCuT process to manufacture the solar cells. This involves electroplating a copper tape with indium, and then heating the tape to 600°C in a sulfidizing environment at atmospheric pressure to form $CuInS_2$ [132]. The tapes are 1 cm wide and 0.1 mm thick and act as a substrate for the devices that are produced using a "reel-to-reel" process. Solar cells are then cut from the completed reels to form strips that overlap slightly, and interconnected using conductive glue to form "SuperCells."

1.3.6.12 Global Solar Energy, Inc.

The CIGS is deposited onto a narrow flexible foil that allows roll-to-roll production, and the relatively small area enables good control over uniformity. The company has a planned increase in efficiency of 1% per year, and has manufacturing plants of 40 MWp (plus 100 MWp planned) in Tuscon, Arizona, in the United States and 35 MWp in Berlin, Germany. A total capacity of 170 MWp production in the United States and Germany is planned by the end of 2010. In December 2008, the SOLON corporation, a subsidiary of SOLON AG in Berlin, installed the world's largest array of CIGS-based PV modules based on Global Solar's technology. The 750 kWp array uses 6600 solar modules covering a 28,800 m² area and was designed to provide power to Global Solar. The system is ground-mounted and is estimated to offset 27,200 tonnes of CO_2 during its predicted 25-year lifetime by generating over 1.1 million kilowatt-hour (kW h) of renewable electricity per year.

1.3.6.13 CIS Solartechnik GmbH & Co. KG

CIS Solartechnik GmbH & Co. KG is a producer of $CuInSe_2$ solar cells using a reel-to-reel process and deposition onto a metal foil substrate. The company has a 30 MWp solar module production facility in Bremerhaven, Germany.

1.4 NOVEL ABSORBER LAYERS

1.4.1 CuInAlSe₂

Current state-of-the-art CIGS solar devices can be made with efficiencies up to 19.9% [9]. However, this absorbing material has a relatively low bandgap ($E_g = 1.2$ eV for Ga/(In+Ga) = 0.3) compared to the 1.5 eV bandgap required for optimum solar energy conversion. CIGS-based devices fabricated with $E_g > 1.3$ eV (Ga/(In+Ga) > 0.5) are found to have substantially reduced efficiencies [133] compared to those with lower bandgap. This is attributed to the FF and open-circuit voltage losses arising from an increased defect density and stronger interfacial recombination when the Ga content is increased. The need for higher bandgap material ($E_g > 1.5$ eV) is driven by the desire to produce (1) higher-efficiency devices and (2) wider-energy-bandgap cells, for use as the upper cells in multi-junction devices. Substituting Ga by Al makes it possible to produce $CuInAlSe_2$ (CIAS, copper indium aluminum deselenide) material with the same energy bandgap as CIGS but using less Al compared to Ga. This is because the variation of E_g with Al content is up to 2.7 eV for $CuAlSe_2$ (CAS, copper aluminum diselenide) compared to 1.7 eV for $CuGaSe_2$ (CGS, copper gallium diselenide). This is shown in Figure 1.20, where the variations of energy bandgap versus composition are plotted for CIGS, CIAS, and $CuIn(S,Se)_2$ (CISS, copper indium disulfide diselenide) absorber. Another advantage is that for the equivalent energy bandgap, the lattice structure in CIAS is less distorted from the optimum CIS structure compared to CIGS. This is explicitly shown in Figure 1.21.

Niemi and Stolt first reported the production of CAS thin films for application in solar cells in 1988 [136]. The fabrication process used the co-evaporation of Se, Cu, and an Al/Cu alloy (Al/Cu = 9) using resistively heated boats, onto Corning 7059 glass substrates. The substrate temperature was maintained at 450°C during the deposition, and excess Se was supplied during the deposition to compensate for any Se loss in the thin films deposited. All the CAS thin films were found to be p-type. A substantial effort was made between 1997 and 2003, by Marsillac and coworkers, at

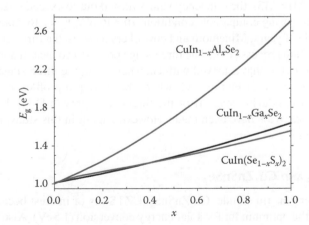

FIGURE 1.20 Variations of the energy bandgap, E_g, with composition x ($x = X/(In+X)$, where X = Al, Ga, or S). (Data from Kessler, J. et al., Progress in low-cost electrodeposition of $Cu(In,Ga)(S,Se)_2$: The CISEL project, in *Proceedings of the 20th European Photovoltaic Solar Energy Conference*, Barcelona, Spain, p. 1704, 2005; Paulson, P.D. et al., *J. Appl. Phys.*, 91 10153, 2002.)

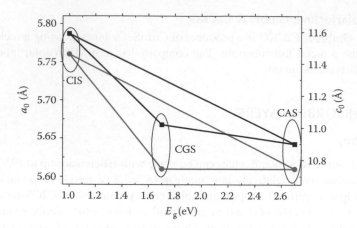

FIGURE 1.21 Variation of lattice parameters a_0 (circles) and c_0 (squares) when alloying CIS with Ga and Al.

understanding the properties of CAS films formed by selenization of Cu/Al stacks, with and without layers of selenium included in the stack, followed by an anneal in selenium atmosphere [137,138]. Although some good-quality layers were produced, most were found to have poor crystallinity. There was also difficulty in reproducibly producing better-quality layers. This group of researchers also investigated the use of CAS as an n-type buffer material due to the large bandgap of CAS (2.7 eV) [139,140]. Reddy and Raja deposited $CuAlSe_2$ films using the co-evaporation method and investigated how changing the stoichiometry of the films altered the optical and electrical properties [141,142].

Several groups have looked at the properties of thin film CIAS with varying composition, deposited by co-evaporation [143,144], vacuum evaporation followed by selenization [145–148], and precursor sputtering followed by selenization [149,150]. Efficient devices have been reported by three groups, ITN Energy Systems, Inc. [151], IEC, Delaware [152], and Ritsumeikan University [153]. ITN focused their research on wide-energy-bandgap material ($E_g > 1.5$ and $\eta > 10\%$) for application in high-altitude airship and spacecraft, by developing both single- and multiple-junction devices. The group at Delaware used multisource evaporation to produce CIAS thin films. Using this technique to produce the CIAS, in substrate configuration devices, they have produced solar cells with efficiencies up to 16.9% [152]. Although the energy bandgap in these devices is relatively low ($E_g = 1.16$ eV, Al/(In + Al) = 0.13), the efficiency values turned out to exceed those achieved based on CIGS devices deposited using comparable conditions ($\eta = 16.6\%$, $E_g = 1.16$, Ga/(Ga + In) = 0.26).

Over the last couple of years, Minemoto and coworkers have used a three-stage molecular beam epitaxy (MBE) deposition process [153]. The three-stage process produces a double grading of Al in the absorber layer (similar to that achieved with Ga when using the three-stage process to produce CIGS). An example is shown in Figure 1.22, where the Al depth profiles are plotted for a range of Cu/(Al + In) ratios used during stage II of the three-stage growth [154]. Using this technique, devices with efficiencies >12% have been made. However, again in this study, low-energy-bandgap CIAS layers were used.

1.4.2 Cu_2ZnSnS_4 AND $Cu_2ZnSnSe_4$

The compound copper zinc tin sulfide, Cu_2ZnSnS_4 (CZTS), is of interest because it has an optical energy bandgap near the optimum for PV solar energy conversion (1.5 eV). Also because of its direct energy bandgap, it has an optical absorption coefficient, $\alpha > 10^4$ cm^{-1} [155], resulting in the need for only a few microns of material to absorb most of the incident light. All of the constituent elements needed to make CZTS, i.e., copper, zinc, tin, and sulfur, are highly abundant and therefore cheap, and all are environmentally acceptable.

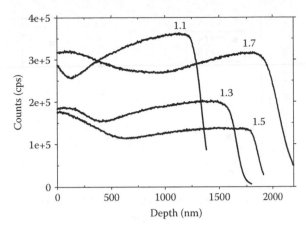

FIGURE 1.22 Al depth profiles of CIAS thin films grown by MBE for different Cu/(Al + In) ratios. Samples are provided by T. Minemoto, Ritsumeikan University, Japan, and measurements by G. Zoppi.

The PV effect was first observed in this material in 1988 by Ito and Nakazawa [155]. Almost a decade later (1997), Friedlmeier et al., at the University of Stuttgart, produced CZTS films using thermal evaporation of the elements and binary chalcogenides [156]. They were able to use these layers to make solar cells with efficiencies up to 2.3%, using substrate configuration device structures similar to those used with CIGS, i.e., glass, magnetron sputtered Mo, CZTS absorber layer, CBD CdS buffer layer, and aluminum-doped zinc oxide TCO layer. Both groups showed that the presence of secondary Zn and Sn phase plays an important role in affecting the electrical activity of the absorber layer.

From 1997 onward, Katagiri et al., at the University of Nagaoka, Japan, have investigated the production of thin films of CZTS, primarily using a two-stage method [157–161]. This method consists of firstly the deposition of the metallic precursor layers and secondly their conversion into CZTS by a sulfidization process. In their earlier work, the Cu/Sn/ZnS precursor layers were produced using e-beam evaporation and then sulfidized in an environment containing H_2S and N_2 gases [160]. Devices were made by synthesizing the CZTS on Mo-coated glass substrates, followed by the deposition of CBD CdS, ZnO:Al, and a top contact grid to complete the device. The devices were found to have efficiencies up to 2.62%. Two years later, a 5.45% efficient cell was produced by using a different Cd source in the CBD CdS process, CdI_2 instead of $CdSO_4$, and also by controlling the amount of Na present at the Mo/substrate interface [158]. In 2005, the efficiency of CZTS devices was increased to 5.74%. This was achieved by producing the CZTS by co-sputtering SnS, Cu, and ZnS simultaneously, followed by an H_2S anneal without breaking vacuum [162]. This method has now been developed further to produce devices with efficiencies >6.77% [159]. The latest increase in efficiency is thought to be due to "preferential deionized water etching" of the absorber surface, prior to depositing the CdS buffer layer. Most of the absorber layers used in these devices were Zn-rich with Cu/(Zn + Sn) ~ 0.9, Zn/Sn ~ 1.2, and S/(Cu + Zn + Sn) ~ 1.1.

In their latest work, the same group has investigated forming the CZTS by sulfidizing the metallic precursors using a mixture of sulfur vapor with nitrogen, as sulfur vapor is much less toxic than H_2S. However, so far, the efficiencies achieved (up to 1.79%) are not as high as those achieved with H_2S [163]. Two groups have also reported the production of a CZTS absorber by an electrochemical route. Scragg et al. electrodeposited a stack of Zn/Sn/Cu onto Mo-coated glass substrates [164]. The precursors were then annealed at 550°C in a sulfur ambient for 2 h. Ennaoui et al. produced a mixed Cu/Zn/Sn precursor layer, and then annealed it in an Ar/H_2S atmosphere. In the latter case, device efficiencies up to 3.4% were produced [165].

The work on CZTS has also stimulated work on the closely related compound copper zinc tin selenide, $Cu_2ZnSnSe_4$ (CZTSe). This material also has a direct energy bandgap, resulting in a high

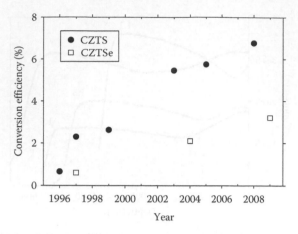

FIGURE 1.23 Evolution of solar cell efficiency using CZTS and CZTSe absorbers.

optical absorption coefficient, $\alpha > 10^4$ cm^{-1}. However, in this case, the energy bandgap is around 0.9 eV, much narrower than that obtained with CZTS. Frieldmeier et al. produced the first CZTSe device with an efficiency of 0.6% in 1997 [156]. Altosaar and coworkers at the Tallim University of Technology are synthesizing CZTSe by mixing together CuSe, ZnSe, SnSe, and elemental Se, and annealing them together in a sealed quartz ampoule, and have used the CZTSe films produced to fabricate solar cells with efficiencies up to 2.1%[166,167]. At the Northumbria University, we have followed a more conventional approach in which we have produced multilayers of Cu/Zn/Sn using RF magnetron sputtering and converted the precursors to CZTSe by annealing in an Ar/Se atmosphere [168]. Devices made using the substrate configuration were made with efficiencies up to 4.0%. The evolution of the conversion efficiency versus time is shown in Figure 1.23 for both CZTS and CZTSe devices.

Given the promising success with both CZTS and CZTSe, alloying these materials offers the possibility of grading the absorber layer for more effective light capture and minority carrier transport to the junction and more flexibility in lattice and electron affinity matching to the buffer layers available, especially those that are Cd free. There is also the possibility of using CZTSe devices in tandem with CZTS devices to more fully utilize the absorption of the solar spectrum while minimizing thermalization losses.

1.4.3 SnS

SnS has a direct optical energy bandgap (E_g) of 1.3–1.5 eV with a high optical absorption coefficient (>10^4 cm^{-1}) [169,170]. It exhibits p-type conductivity, and the electrical properties can be easily controlled by doping with suitable elements such as Ag, Al, N, and Cl [171,172]. Its major advantages are however that the constituent elements of this semiconductor are abundant and nontoxic and that mass production methods for producing tin and for sulfidizing metals are well established. These properties make SnS suitable as an absorber layer for PV applications using CdS, ZnS, and other II–VI compounds as window layers in similar heterojunction device structures to those developed for CdTe and CIGS. Although Loferski [10] diagrams predicted a solar conversion efficiency up to 25% for SnS in 1956, there has been no significant effort made to develop SnS solar cells until the last two decades. SnS thin films have now been deposited using CBD [173], plasma-enhanced chemical vapor deposition (PECVD) [174], electrochemical deposition [175], chemical spray pyrolysis [176,177], close-spaced vapor transport [178], RF magnetron sputtering [179], sulfidization of tin layers [180,181], and vacuum evaporation [182,183]. The films deposited usually indicate the presence of different phases of Sn and S, such as SnS$_2$, SnS, and Sn$_2$S$_3$, with a varied surface morphology that depends on the growth conditions, particularly the deposition temperature. The SnS films grown by most of the techniques have the orthorhombic crystal structure with x-ray diffraction indicating a predominant (111) orientation, although a few studies give the (040) planes as giving rise to the dominant reflection [184].

The optical energy bandgap of the films is most commonly found to be direct with a value in the range 1.3–1.5 eV, the precise value depending on the growth technique used. The electrical resistivity of the SnS films can be changed over a wide range from 10 to 10^5 Ω cm by altering the deposition conditions [178,185,186]. The films display good photoconductivity from the near-infrared to ultraviolet region with the thinner layers showing higher photoconductivity [187].

SnS films grown by chemical techniques have shown PV activity in photoelectrochemical cells. The use of vacuum-evaporated SnS layers as an absorber in PV cells was started in 1994. The best cells were found to have an open-circuit voltage (V_{oc}) of 120 mV and a short circuit current density (J_{sc}) of 7 mA/cm^2 with a conversion efficiency $\eta = 0.29\%$ for n-CdS/p-SnS solar cells [188]. Reddy et al. reported improved V_{oc} and J_{sc} of 140 mV and 8.4 mA/cm^2, respectively, for an Al/CdS/SnS/Ag structure device. This device produced using sprayed SnS layers had an efficiency of 0.5% [189]. Ristov et al. developed CdO/SnS and Cd$_2$SnO$_4$/SnS junctions using chemically deposited SnS layers with $V_{oc} > 200$ mV [190]. However, the observed J_{sc} values were very low in these junctions. Electrodeposited SnS layers were used by Subramanian et al. to develop photoelectrochemical cells with the configuration p-SnS|Fe^{3+}, Fe^{2+}|Pt; these devices had a higher open-circuit voltage of 320 mV but a reduced short circuit current density, <1 mA/cm^2 [191]. Thin film SnS$_2$/SnS heterojunctions, developed using PECVD-grown SnS layers, have been made with $V_{oc} > 350$ mV. However, the device efficiency was limited to <1% due to the low value of J_{sc} (1.5 mA/cm^2) [192]. Reddy et al. have produced SnS-based cells with efficiencies >1.3%. These superstrate configuration (glass/SnO$_2$:F/CdS/SnS/In) cells had a much improved short circuit current density of 9.6 mA/cm^2 with an open-circuit voltage of 260 mV [193]. In these cells, the thickness of the SnS used was 0.6 μm, i.e., they were far too thin to absorb most of the incident light and too thin to support the depletion region. Increasing the film thickness to >1.5 μm should in principle improve the short circuit current, simply due to the layers now being thick enough to absorb most of the solar spectrum and recombination losses minimized due to the larger grain size; the open-circuit voltage should also be improved as the distance between the back contact and the metallurgical junction will now be greater than the depletion region width.

Electrochemically deposited SnS layers were used by Gunasekaran and Ichimura to develop CdS/SnS and CdZnS/SnS heterojunctions [194]. The latter junction showed a $V_{oc} > 288$ mV and $J_{sc} > 9.2$ mA/cm^2 leading to an efficiency of 0.7%. Although the open-circuit voltage and short circuit current density obtained in this work are comparable with the best reported values, the efficiency is limited by a low value for the FF (30%). Avellaneda et al. fabricated SnS-based PV cells with the configuration glass/SnO$_2$:F/CdS/SnS/(CuS)/Ag using all chemically deposited layers. Although they produced cells with a V_{oc} up to 400 mV, the J_{sc} was <1 mA/cm^2. A heat treatment in air at 423 K improved J_{sc}, but correspondingly V_{oc} decreased. The most efficient cells had an open-circuit voltage of 340 mV with a short circuit current density of 6 mA/cm^2 [195]. ZnO/SnS heterostructures have also been developed using all electrodeposited layers. These cells showed PV behavior, but the device efficiency was very low [196]. An open-circuit voltage as high as 480 mV and a short circuit current density of 8.76 mA/cm^2 were obtained by Li et al. However, the overall conversion efficiency was <0.7% due to a poor FF. These devices were formed by depositing p-type SnS onto n-type silicon substrates using CBD [197]. Li et al. also developed n-CdS/p-SnS and n-ZnO/p-SnS junctions. The former solar cells were produced with a $J_{sc} > 13.2$ mA/cm^2. However, the conversion efficiency obtained in these junctions was very low due to a poor V_{oc} and poor FF. The best solar cell parameters (in separate devices) are $V_{oc} = 480$ mV, $J_{sc} = 13.2$ mA/cm^2, and FF = 60%. These values indicate that efficiencies >10% are probably achievable with further research efforts.

1.5 NOVEL BUFFER LAYERS

1.5.1 BUFFER LAYER REQUIREMENTS

The "buffer" or "window" layer acts as the *n*-type partner to the *p*-type absorber material. The buffer layer used should normally possess the following characteristics:

1. It should be *n*-conductivity type so that it forms an anisotype heterojunction with the *p*-type absorber layer.
2. Its energy bandgap should be >2.0 eV to allow light to be transmitted to the absorber layer [198].
3. It should have minimal lattice mismatch with the absorber-layer material, otherwise misfit dislocations will form at the junction, and these will act as recombination centers in the junction region.
4. It should have an electron affinity match with the absorber-layer material so that "spikes" and "notchs" are avoided at the device interface.
5. It should relax mechanical strain at the interface.

The buffer materials that have been used until today in the development of thin film solar cells mainly involve the compounds that contain metals such as cadmium, zinc, indium, and tin and nonmetals such as sulfur, selenium, and oxygen. Table 1.5 gives the various metal chalcogen and metal oxide buffers that have been used in thin film solar cells.

Table 1.6 gives the most successful buffer-layer materials and the deposition methods that have been used to fabricate efficient solar cells, mainly using the chalcopyrite compounds and CdTe as the absorber layer. It should be noted that while analyzing the performance of the various cells, one must keep in mind that the quality of the chalcopyrite absorber layer varies significantly from one device structure to the other.

Table 1.7 lists the highest efficiencies realized using different buffer layers with chalcopyrite-based solar modules. Among all the buffer materials that have been investigated, only two materials, ZnS and In_2S_3, have been successful in developing $30 \times 30 \, cm^2$ area modules with efficiencies higher than those achieved with CdS buffer layers. The remaining modules consist of monolithically interconnected prototype sub-modules of 5×5 or $10 \times 10 \, cm^2$ area. Showa Shell has developed $900 \, cm^2$ area PV modules using CIGSS films synthesized by sulfurization/selenization of metal precursors and CBD ZnS as the buffer layer, which have demonstrated a world-record efficiency of 14.3%. In_2S_3 buffer layers (deposited using ALE by ZSW) have also been used with evaporated CIGS layers, to fabricate $900 \, cm^2$ area modules with efficiencies >12.9%.

TABLE 1.5
Buffer Materials Reviewed in This Work

	Cd		Zn		In	Sn
Chalcogenides						
S	CdS	CdZnS	ZnS		In_2S_3	
Se			ZnSe	$ZnIn_2Se_4$	In_xSe_y	
Oxides						
O			ZnO	ZnMgO		SnO_2
			Zn		Mg	

TABLE 1.6
Performance of Polycrystalline Thin Film Solar Cells with Various Buffer Layers

Absorber Material	Type of Material	Buffer Layer	Growth Technique	Efficiency (%)	Reference
Chalcopyrite	Cd based	CdS	CBD	19.9	[9]
		CdZnS	CBD	19.5	[199]
	In based	In_2S_3	ALE	16.4	[200]
			CBD	15.7	[201]
			Evaporation	15.2	[202]
			ILGAR	14.7	[203]
			Sputtering	13.3	[204]
			Spray pyrolysis	12.4	[205]
			MOCVD	12.3	[206]
		InS	PVD	14.8	[207]
		$In(OH)_3$	CBD	14.0	[208]
		In_xSe_y	Evaporation	13.0	[209]
	Zn based	ZnS	CBD	18.6	[210]
			ALE	16.0	[211]
			ILGAR	14.2	[212]
			Evaporation	9.1	[213]
		ZnSe	CBD	15.7	[214]
			MOVPE	12.6	[215]
			ALE	11.6	[216]
			Evaporation	9.0	[213]
		ZnO	Sputtering	15.0	[204]
			ILGAR	15.0	[217]
			CBD	14.3	[218]
			ALE	13.9	[219]
			CVD	13.9	[220]
			Electrodeposition	11.4	[221]
		ZnMgO	ALE	18.1	[222]
			Sputtering	16.2	[223]
		ZnInSe	Evaporation	15.3	[224]
	Sn-based	SnO_2	CBD	12.2	[225]
CdTe	Cd-based	CdS	CBD	16.5	[30]

TABLE 1.7
Efficiencies of Chalcopyrite-Based Solar Modules with Different Buffer Layers

Buffer Layer	Deposition	Module Area (cm^2)	Module Efficiency (%)	Institute/Company	Reference
CdS	CBD	6500	13.0	Würth Solar	[226]
ZnS	CBD	900	14.3	Showa Shell	[123]
ZnS	CBD	3659	13.2	Showa Shell	[123]
ZnSe	CBD	20	11.7	HMI/Siemens	[218]
ZnO	ILGAR	20	10.9	HMI/Siemens	[212]
In_2S_3	ALE	13	13.4	ZSW	[227]
In_2S_3	ALE	900	12.9	ZSW	[228]
In_2S_3	CBD	717	9.7	ZSW	[229]

1.5.2　Deposition Techniques

A large number of deposition methods have been used to grow the buffer layers, resulting in high-quality layers and high-efficiency cells on the one hand and economic production of modules on the other. These include CBD, physical vapor deposition (PVD), sputtering, spray pyrolysis, and electrodeposition. CBD is the most popular technique for obtaining highest-efficiency devices. With the development of other alternate buffer layers, other deposition methods that have also been effectively used include ALE, MOCVD, and ion layer gas reaction (ILGAR).

1.5.2.1　Chemical Bath Deposition

The CBD technique is a simple process used to synthesize thin films of a wide range of materials over large areas [230,231]. Although this technique has mainly been used for the deposition of II–VI compounds, such as CdS and ZnS, the deposition mechanism is largely the same for all inorganic materials [232–234]. The technique is based on the kinetically controlled reaction between the constituent ions. It involves the controlled precipitation from the solution of a compound onto a suitable substrate. Film growth occurs when the concentrations of the ions in the solution are sufficiently high such that their ionic product exceeds the solubility product for the compound to be deposited. Conformal and pinhole-free CdS layers can be obtained using this process to make these films for buffer-layer applications. Figure 1.24 shows a typical CBD system used for the synthesis of CdS films.

A soluble salt of the required metal is dissolved in an aqueous solution, to release cations. The nonmetallic element is provided by a suitable precursor compound, which decomposes in the presence of hydroxide ions, releasing the anions. The anions and cations then react to form the compound. Generally, CdS deposition in CBD is carried out in an alkaline medium, pH > 9, using the following constituents: (1) a cadmium precursor: $CdSO_4$, $CdCl_2$, or $Cd(CH_3COO)_2$; (2) a sulfur precursor: thiourea or thioacetamide; and (3) a complexing agent, which is commonly ammonia. The concentrations of the precursor solutions vary over a wide range, and each laboratory uses its own specific recipe.

For example, in a typical CdS deposition process, cadmium sulfate, thiourea, and ammonia solutions of 1.5×10^{-3} M, 0.15 M, and 1 M concentrations can be used as the recipe to deposit CdS layers. The reaction for this process is given below.

The decomposition of cadmium sulfate is given by

$$CdSO_4 \rightarrow Cd^{2+} + SO_4{}^{2-}$$

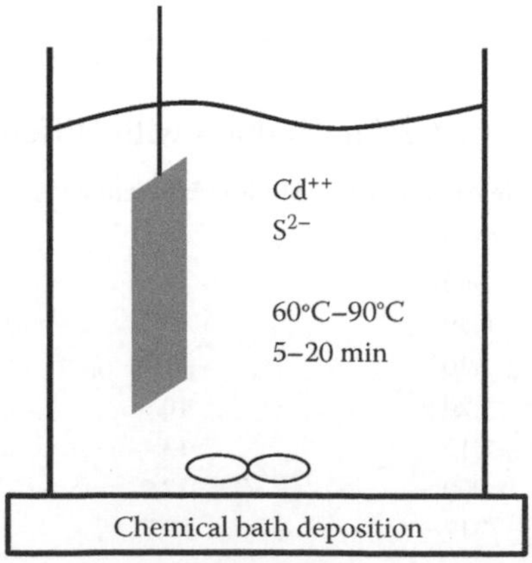

Cd^{++}

S^{2-}

60°C–90°C

5–20 min

Chemical bath deposition

FIGURE 1.24　Schematic of a typical CBD system.

The decomposition of thiourea is given by

$$CS(NH_2)_2 + OH^- \rightarrow SH^- + CH_2N_2 + H_2O$$

$$SH^- + OH^- \leftrightarrow S^{2-} + H_2O$$

The cation and anion then react to form CdS in the solution as follows:

$$Cd^{2+} + S^{2-} \rightarrow CdS$$

After immersing the substrates in the solution, the bath is heated to the desired temperature, usually 60°C–80°C. This is to improve the surface morphology of the deposited layer. Due to its low solubility, the CdS produced in the solution precipitates onto the exposed surfaces. Films produced in this process have a rough topology, resulting in low optical transmittance. In order to minimize this, a complexing agent is used to form complex ions with the metal ions. The most widely used complexing agent is ammonia. This exists in equilibrium with ammonium hydroxide, which also provides the hydroxide ions for the decomposition of thiourea. The ammonia used in most recipes is considered to be crucial for the cleaning of the absorber surface and the removal of oxides and impurities. The chemical reaction is given by

$$NH_4 + OH^- \leftrightarrow NH_3 + H_2O$$

$$Cd^{2+} + 4NH_3 \leftrightarrow Cd(NH_3)_4^{2+}$$

The Cd-complex ion and the sulfide ion migrate to the substrate surface, where they react to form CdS:

$$Cd(NH_3)_4^{2+} + S^{2-} \rightarrow CdS + 4NH_3$$

The growth of the film in CBD occurs by an ion-by-ion reaction or by the clustering of colloidal particles on the active sites. CBD-grown films are usually sulfur deficient, and contain large amounts of oxygen-related impurities either in the form of O or OH$^-$ [235,236] in addition to traces of carbon, hydrogen, and nitrogen, which originate from the deposition bath [237]. The concentration of these impurities has been correlated to a decrease of the energy bandgap in CdS layers [238]. A compositional deviation from stoichiometry is generally observed in this process. Layers formed by this process are dense, homogeneous, and uniform over the substrate. The growth parameters, such as film thickness and deposition rate, can be controlled by varying the solution pH, reagent concentration, and temperature.

1.5.2.2 Thermal Evaporation Method

In this method, the layers are deposited in high vacuum from a heated evaporation source or an effusion cell. The deposition rate and uniformity of the vapor species arriving at the substrate are controlled by the source temperature, the geometry of the source, the source–substrate distance, and the vacuum in the chamber. A schematic has been presented in Section 1.2 (Figure 1.5). Typical substrate temperatures used for the preparation of CdS layers were in the range 150°C–200°C with a source temperature of 750°C–900°C [239,240]. The layers grown at this temperature are crystalline, conformal, homogeneous, and free from pinholes.

1.5.2.3 Sputtering

Sputtering is a rapid deposition process, compatible with the vacuum deposition of the absorber and TCO window layers, which can be easily scaled up for industrial production, particularly CIGS solar cells. For example, In_2S_3 films have been deposited by RF magnetron sputtering either using the compound target or elemental indium in a H_2S ambient with Ar as the sputter gas [241]. In the former case, the mass transfer of In and S occurs via sputtering of the In_2S_3 target, diffusion of these atoms onto the substrate surface, followed by condensation. The energetic ionized sputter gas atoms form the plasma that strike the target and remove the material atoms. These atoms consequently form the film on the substrate, which is placed on the counter electrode facing the target. A schematic of a typical sputtering system has been given earlier (Figure 1.5).

Generally, the deposition is carried out using substrate temperatures of 200°C–350°C and at a sputter pressure of $\sim 1 \times 10^{-2}$ mbar. The layers were deposited using a typical deposition rate of ~20 nm/min and a power density of 1 W/cm^2 using pure Ar as the sputter gas [242].

1.5.2.4 Spray Pyrolysis

This is a non-vacuum technique that can be used for industrial production by using a linear array of nozzles in an inline system. An aerosol of water containing droplets that contain heat-decomposable compounds of the constituent elements is sprayed onto heated substrates. The droplets on reaching the substrate surface undergo pyrolytic decomposition followed by reaction among the ions leading to the formation of the required compound in solid form. Figure 1.8 shows the schematic of the spray deposition system.

This technique has been used effectively for the growth of oxides and chalcogens of various metals [243–248]. For example, to deposit ZnS layers using this process, the Zn precursor, such as $ZnSO_4$/$ZnCl_2$/Zn acetate, is mixed in solution form with thiourea (a commonly used S precursor) and the mixture sprayed using N_2 as the carrier gas. The concentration of salts in the solution, the gas flow rate, the source–nozzle distance, and the droplet size are the important parameters that control the quality of the layers. The layers grown by this technique adhere well to the substrate surface and are uniform and pinhole free.

1.5.2.5 Electrodeposition

The electrodeposition of II–VI compounds has been widely studied, mostly for the synthesis of selenides, sulfides, and tellurides. The electrodeposition of ZnO has been intensively studied by the group at ENCL, Paris [249,250], and the Weizmann Institute, Israel [251], particularly for solar cell applications. The first successful attempt to use electrodeposited ZnO as a window layer for CIGS solar cells was reported to produce devices with efficiencies up to 10% [252]. The ZnO films were prepared in this process from a dimethylsulfoxide (DMSO) solution composed of dissolved O_2 and 0.05 M zinc perchlorate with 0.1 M lithium perchlorate used as the supporting electrolyte, and the bath was maintained at 150°C. Evaporated CIGS films on Mo-coated glass were used as the cathode, and a platinum coil as the anode. Electrodeposition was carried out potentiostatically with a saturated calomel electrode as the reference electrode connected via a salt bridge. Figure 1.7 showed the typical electrodeposition system used for the growth of thin films.

1.5.2.6 Atomic Layer Epitaxy

ALE is a simple and self-limiting process used to grow thin films, one atomic layer at a time, with a precise control over film deposition down to the atomic scale. The principle of ALE is based on sequential pulsing of precursor vapors, each of which forms about one monolayer per pulse. Figure 1.25 is a schematic of an ALE system.

The reaction sequence is as follows: the first precursor gas is admitted into the reaction chamber so that a monolayer of gas is produced on the substrate surface. The second precursor gas is then introduced into the process chamber so that it reacts with the first precursor gas, thereby producing

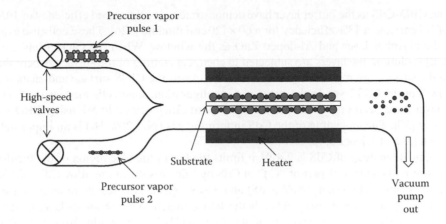

FIGURE 1.25 Schematic of an ALE system.

a monolayer of the film on the substrate surface. The process is then repeated. The layer composition and thickness is controlled by the pulsing sequence of the precursor gases and the number of pulsing cycles. Generally, in the ALE process, the growth temperatures vary from 150°C to 350°C depending on the type of material to be deposited [253,254]. A wide variety of thin films can be deposited with this technique using gas, liquid, or solid precursors. The precursor gases used in the process should be stable and have a high vapor pressure enough to fill the reaction chamber so that the monolayer growth takes place in a reasonable length of time.

1.5.2.7 MOCVD

In the MOCVD process, thin film deposition occurs by the pyrolytic decomposition of the precursor gases onto a heated substrate surface (Figure 1.7). The substrate temperature and the precursor gas flow rate are the most important deposition conditions in this method. The substrates are kept on graphite susceptors that can be heated radiatively or using an RF generator. Diethylzinc (DEZn) and O_2 gas/H_2O vapor (high purity) are generally used as the sources of Zn and O, respectively. These are introduced into the growth chamber through separate injectors to avoid any pre-reaction of the precursors. Ultrahigh purity Ar is used as the carrier gas for DEZn. The flow rates of argon carrier gas and oxygen are set each at 10 sccm, in which the O/Zn elemental ratio is fixed as 73.3. The temperature of DEZn is kept at 10°C in a coolant water bath. Additional argon gas with a flow rate of 100 sccm is supplied to the MOCVD system to reduce the parasitic reaction of the reevaporated ZnO from the reactor and to retain a highly reactive gas flow; this improves the growth rate of ZnO films at low temperatures [255].

1.5.2.8 ILGAR

ILGAR is a new deposition process developed by the HMI, Germany, for the deposition of oxides and sulfides that are used as buffer layers in solar cells [256,257]. It is a cyclic process that involves the formation of a metal precursor on the substrate surface using a dipping or spraying process, drying the deposited substrate, followed by the reaction of the metal precursor with gaseous hydrogen chalcogenides to form the corresponding metal chalcogen film. The resulting films are conformal and adhere well to the substrate. In this method, the film thickness is easily controlled by the number of deposition cycles. This method has an advantage over CBD in that it produces less waste.

1.5.3 CADMIUM-BASED MATERIALS

Cadmium sulfide (CdS) is a *n*-type semiconductor with an optical energy bandgap of 2.4 eV. It is the most widely used buffer-layer material to make polycrystalline thin film solar cells. Solar cells made using CBD-CdS as the buffer layer have demonstrated a world-record efficiency of 19.9% over an area of $0.4\,cm^2$ and a 13% efficiency for a $60\times120\,cm^2$ module [226]. These cells use evaporated CIGS as the absorber layer and Al-doped ZnO as the window. When the CIGS films are dipped into the CBD solution, the layers are subjected to chemical etching of the surface, where the native oxides on the surface are removed by ammonia. This cleans the CIGS surface and enables the epitaxial growth of the CdS layer. In the superstrate configuration, solar cells made using CdTe as the absorber layer have been produced with a world-record efficiency of 16.5% using CBD-CdS as a window layer [30]. The annealing of the CdS in forming gas ($N_2+20\%$ H_2) is an important step to produce such high efficiencies.

As the optical bandgap of CdS is 2.4 eV, it limits the QE in the blue region of the incident solar spectrum. The replacement of part of "Cd" in CdS by "Zn" results in the alloy $Cd_{1-x}Zn_xS$, which has an energy bandgap between 2.4 eV (CdS) and 3.8 eV (ZnS). This results in an increase of blue response of the solar cell and also yields a better lattice match to the absorber layer depending on the Zn/Cd ratio in the film. The replacement of CdS with $Cd_xZn_{1-x}S$ should also lead to a favorable conduction band offset at the heterojunction interface [258].

$Cd_xZn_{1-x}S/CuGaSe_2$ solar cells have showed an enhanced open-circuit voltage and short circuit current, leading to a higher conversion efficiency compared to $CdS/CuGaSe_2$ devices [259]. Although $Cd_xZn_{1-x}S$ films can be deposited using a variety of physical and chemical techniques, CBD is the most commonly used process to deposit the buffer layer in thin film solar cells. By varying the Zn/(Zn+Cd) ratio in the range 15%–20% in $Cd_xZn_{1-x}S$ films (grown using the CBD method), Devaney et al. developed solar cells with conversion efficiencies >12% using co-evaporated CIS [260]. Recently, the influence of CBD-grown $Cd_xZn_{1-x}S$ layer thickness on the performance of CIGS solar cells was reported by Song et al., who reported solar conversion efficiencies up to 13%, for a $Cd_xZn_{1-x}S$ thickness of 40 nm [260]. A real breakthrough in achieving an efficiency of 19.5% with $Cd_{1-x}Zn_xS/CIGS$ devices was made by Bhattacharya et al. [199].

Although CdS and its alloy $Cd_xZn_{1-x}S$ have been successfully used to produce high-efficiency devices, "Cd" is a heavy metal that has been perceived to be a "toxic." The environmental and health hazard issues related to this element have been investigated by several independent groups around the world [261,262]. In addition, there are practical problems related to Cd-containing dust in the production process and the subsequent disposal of waste material. It can be observed from Tables 1.5 through 1.7 that there are several materials that can be used to develop highly efficient PV cells without using cadmium. All the listed materials include one or more of the elements In, Zn, S, Se, and O. These materials are deposited either by CBD or by alternative methods, discussed in earlier sections. An important advantage with all these novel materials is that they have higher energy bandgaps than CdS, which minimizes the absorption losses in the buffer layer and improves the spectral response of the final device in the blue region.

1.5.4 INDIUM-BASED MATERIALS

Indium sulfide (In_2S_3) is an attractive alternative buffer to CdS among this group of materials, particularly in CIGS-based solar cells. It has an energy bandgap of 2.7 eV with an optical transmittance >80%. Although thin layers of this material can be deposited by different physical and chemical techniques such as sputtering, evaporation, CBD, and ALE, the record solar conversion efficiency was demonstrated for the films grown by atomic layer epitaxy (ALE) using CIGS as the absorber-layer material.

The development of ALE-In_2S_3 as a buffer for solar cell fabrication was initiated at the ENSCP, France, using indium acetylacetonate and H_2S as the precursor materials [263]. These layers in combination with CIGS as the absorber layer and ALE-ZnO as the TCO yielded an efficiency of 13.5%. The optimum deposition temperature was found to be in the range 200°C–220°C with a layer thickness of 30–50 nm. Optimizing the post-deposition annealing temperature and time were found to be crucial in producing the best efficiencies. ZSW and Würth Solar have used the ALE-deposited In_2S_3

to produce In_2S_3/CIGS solar cells with efficiencies up to 16.4% [200] and $30 \times 30 \, cm^2$ modules with efficiencies of 12.9% [228]. CIGS devices with an ALE-In_2S_3 buffer do not show any significant degradation in the performance and are quite stable even under hot, damp conditions [264].

Initial work on $In_x(OH,S)_y$ was started at the University of Stuttgart that developed a 9.5% efficient cell, using a CBD process based on the use of $InCl_3$ and thiourea [265]. Depending on the type of sulfur precursor used in the CBD process, there is a possibility of oxide/hydroxide introduction in the In_2S_3 layer that helps in enhancing the device efficiency. Hariskos et al. [201] deposited $In_x(OH,S)_y$ films on a co-evaporated CIGS absorber by the CBD method using $InCl_3$ and thioacetamide at 70°C. The post-deposition of the structure at 200°C and light soaking resulted in an active-area efficiency of 15.7%. Compared to the standard CdS buffer in these devices, the $In_x(OH,S)_y$ buffer-layer devices resulted in an improved open-circuit voltage and comparable FF, but a slightly reduced short circuit current density [266]. Detailed studies on the growth mechanisms of the In_2S_3-based buffers were investigated by the CIEMAT research group [267]. Small-area cells with efficiencies up to 15.7% and $30 \times 30 \, cm^2$ modules with efficiencies up to 9.7% have been produced for these In_2S_3/CIGS devices [201].

Evaporated In_2S_3 layers have also been used with device efficiencies up to 11.2% [268]. These devices used In_2S_3 layers deposited by the co-evaporation of indium and sulfur. Solar cell efficiencies >12.4% have been achieved at the University of Nantes also using co-evaporated In_2S_3. They found that optimizing the substrate temperature plays a crucial role in achieving higher efficiencies [269]. At lower substrate temperatures (<130°C), the diffusion of Cu from the absorber layer into the In_2S_3 buffer leads to the formation of a $CuIn_5S_8$ phase on the surface of the absorber layer. Using higher deposition temperatures (>200°C), there is a degradation in the device efficiency. In order to avoid Cu diffusion into the buffer, sodium fluoride was evaporated along with "In" and "S" so that "Na" can play a role similar to "Cu" by forming a $NaIn_5S_8$ phase [270]. In_2S_3 can also be directly evaporated onto CIGS to produce devices with efficiencies up to 15.2% [202].

In_2S_3 buffer layers have also been deposited by ultrasonic spray pyrolysis on CIGSS absorbers using a substrate temperature of 200°C with an In/S ratio of 1:4; devices with efficiencies up to 12.4% were produced [205]. Magnetron-sputtered In_2S_3 layers can also be used as a buffer layer for CIGS-based solar cells. Solar cell efficiencies of 12.2% were realized by Hariskos et al. [241] using an In_2S_3 target and also by sputtering metallic indium in an Ar/H_2S ambient.

CIGS-based solar cells have also been developed using an InS buffer deposited by PVD, wherein the In_2S_3 powder is sublimed onto the CIGS using a source temperature of 725°C. Devices made with a buffer-layer thickness of 90 nm, formed at a substrate temperature of 120°C, subsequently annealed in air at 200°C for 24 min, had efficiencies up to 14.8% [207].

A novel buffer layer of $In(OH)_3$:Zn^{2+} was prepared on CIGS using the CBD method. The solution contained $ZnCl_2$, $InCl_3$, and thiourea at a temperature of 60°C [208]. After deposition, the $In(OH)_3$:Zn^{2+}/CIGS structure was subjected to heat treatment at 200°C for 15 min to improve the quality of the hetero-interface. The TCO used was ZnO:B deposited by MOCVD. It was found that Zn doping in the CIGS layer produced a buried junction in the CIGS layer. The resulting cells had efficiencies of up to 14%.

Although indium sulfide and other related buffers are very promising materials for efficient solar cell development, an uninterrupted supply of highly refined indium metal is critical for the accelerating success of In-based solar cell technology. At the present time, there is a shortage of indium, and this will most likely prohibit the large-scale manufacture of solar modules using this technology.

1.5.5 ZINC-BASED MATERIALS

ZnS, with an energy bandgap of 3.7 eV, is an obvious choice as an alternate buffer material to CdS. ZnS films have been grown by CBD using a thioacetamide- and ammonia-free bath on CIS films to produce solar cells. These have been found to have higher short circuit currents than the CdS/CIS cells fabricated using similar conditions. However, the overall efficiencies are lower (<10%) because of a lower open-circuit voltage. Most ZnS layers grown by CBD contain zinc hydroxide, which breaks down into zinc oxide after annealing at higher temperatures. The development of CIGS solar cells using CBD-grown ZnS started in 1992 with efficiencies in the range 9%–10% reported by Kessler et al. [271] and Ortega Borges et al. [272]. The precursors used for synthesis of the ZnS used in this work were zinc acetate, thiourea, and ammonia. Later in 1996, Kushiya et al. deposited a 50 nm thick Zn(S,O,OH) buffer layer onto CIGS to produce $30 \times 30 \, cm^2$ modules with efficiencies up to 12.9% [273]. Light soaking was found to improve the device performance remarkably. The best Zn(S,O,OH)/CIGS cell made to date has an efficiency of 18.6%. This device used the CIGS layers produced by NREL with a CBD Zn(S,O,OH) buffer layer produced at Aoyama Gakuin University [210]. It is suggested by these workers that the Zn(S,O,OH) buffer layer increases the positive conduction band offset (CBO) at the CIGS/buffer interface. It also has an energy bandgap of 3.4 eV (after air annealing for 15 min at 200°C), which is much higher than that of CdS, which hence enhances the blue response of the cells [274]. Zn(S,O) buffer layers have also been used in "sulfur cells," with $CuInS_2$ as the absorber layer. A $10 \times 10 \, cm^2$ area Zn(S,O)/CIGSS module has also been produced with an efficiency of 12.3% [275].

ALE has also been used to produce the ZnS used in ZnS/CIGS solar cells, which have efficiencies up to 16% [211]. In these cells, ZnS was grown using diethylzinc mixed with 20% H_2S in a reactor at a temperature of 120°C. Although these devices had high efficiencies, the devices could not be reproducibly produced, due to buffer-layer thickness variations. Improvements in the ALE process using longer pulses improved the reproducibility, but with an efficiency reduced to 12.1%. ZnS films prepared by the ILGAR process on Siemens Solar CIGSS absorber layers were used to produce devices with efficiencies up to 14.2% [212]. This efficiency was the same as that achieved using an ILGAR CdS buffer layer. The cells with ZnS showed an improved V_{oc} and FF but lower J_{sc} than those made with the CdS buffer layer. Vacuum-evaporated ZnS layers (50 nm thick) have been deposited using substrate temperatures <150°C onto evaporated CIGS films. These devices had an efficiency of 5.4% that was increased to 9.1% after light soaking for 5 h [213].

With an energy bandgap of 2.7 eV, ZnSe is another potential candidate to replace CdS in solar cells. Although thin layers of this material have been grown using many different techniques, CBD-grown ZnSe has been used to produce the best efficiencies with CIGS absorber layers. CBD-grown Zn(Se, OH) layers (8 nm thick) have been deposited from selenourea, ammonia, and hydrazine hydrate onto CIGSS films. The resulting devices had active-area efficiencies up to 15.7% [214]. This was despite large energy band offsets ($\Delta E_V = 0.6 \, eV$ and $\Delta E_C = 1.26 \, eV$) at the interface. QE studies confirmed the expected improvement in short-wavelength response in these cells. The buffer-layer thickness is crucial in controlling the efficiency. Layers of higher thickness (>8 nm) decrease the efficiency to 10.4%, due to a drop in the FF, probably due to the higher resistivity of the films. ZnSe films deposited by CBD have also been used as buffer layers in $CuInS_2$ solar cells. These cells had higher short circuit currents and open-circuit voltages but lower FFs compared to CdS-based solar cells. The overall efficiency of these devices was 10.1% [276]. The bath temperature and composition were found to be the crucial parameters that control the open-circuit voltage, the FF, and device stability.

ZnSe has been grown on CIGS using MOCVD with ditertiary-butylselenide and dimethylzinc-triethylamine as precursors and H_2 as the carrier gas. The best cells demonstrated a conversion efficiency of 12.6% [215], while conventional cells with a CdS buffer had efficiencies of 15.4%. The Tokyo Institute of Technology (TIT), Japan, has developed ZnSe layers using the ALE technique and made solar cells with CIGS films. For a (10 nm thick) ZnSe buffer layer, the device efficiency was found to be 11.6% [216]. Evaporated ZnSe layers (50 nm thick) have also been used by Romeo et al. on evaporated CIGS layers to produce solar cells with efficiencies up to 9% [213]. A high defect concentration at the interface and the formation of an oxide layer on the surface of CIGS (as observed from the x-ray photoelectron spectroscopy measurements) were considered to be responsible for the observed low efficiency.

ZnS_xSe_{1-x} is another material that has been studied recently as a novel buffer material for solar cells [193]. Tailoring the optical bandgap and lattice constants by altering the S/Se ratio in the layers made this material attractive because it will minimize the lattice mismatch and possibly the electron affinity mismatch, improving the device interface [277,278].

Although ZnO is also normally used as the TCO layer in chalcopyrite solar cells, it has also been used as a buffer layer in CIGS-based solar cells. These studies were first initiated at the HMI in Germany, where a CBD ZnO layer was used as a buffer in $CuInS_2$-based solar cells [279]. ZnO has also been synthesized using a novel CBD process. In this different method, layers of $Zn(OH)_2$ are grown by CBD using $ZnSO_4$ and ammonia solution; this is followed by annealing at temperatures >120°C to convert the $Zn(OH)_2$ into ZnO. Although the use of such layers as a buffer in CIGS solar cells showed a lower conversion efficiency of 3.8% compared to 8.6% reported with a CdS buffer, the short circuit current is approximately the same in both the cells. ZnO buffers grown by electrodeposition on CIGS absorbers with sputtered ZnO as the TCO were used to produce devices with efficiencies up to 11.4%, with a better FF than the comparable CBD CdS/CIGS devices [221].

Various non-wet techniques, such as sputtering, ILGAR, ALE, and chemical vapor deposition (CVD), have also been used to deposit ZnO for buffer-layer application. ILGAR ZnO layers grown on CIGSS have an efficiency of 15% [217], which is higher than the reference cell fabricated using CBD CdS. The research group working at TIT, Japan, have fabricated CIGS-based solar cells using ALE-grown ZnO and reported an efficiency of 13.9% [219]. ZnO layers were grown on evaporated CIGS substrates from NREL using a CVD process, by reacting a zinc adduct with tetrahydrofuran, with hydrogen as the carrier gas. The resulting devices had efficiencies up to 13.9% [220].

The addition of a small amount of "Mg" to ZnO, resulting in ZnMgO, has been found to widen the energy bandgap of ZnO, thereby improving the blue response of the device and enhancing the conversion efficiency. $Zn_{1-x}Mg_xO$ films (150 nm thick) have been grown as buffer layers on co-evaporated CIGS films. The ALE process uses nitrogen as the carrier gas, and deionized water, diethyl zinc, and bis-cyclopentadienyl magnesium as source materials. The deposition temperature was 120°C with 1000 pulsing cycles used. The best cells had an efficiency of 18.1% [223], compared to that of CBD ZnS/CIGS devices (18.6% efficiency). Sputtered $Zn_{1-x}Mg_xO$ ($x=0.1$) has also been used as a buffer on evaporated CIGS films to produce devices with efficiencies up to 16.2%. This was considered to be due to the adjustment of the conduction band minima of both the layers [224].

Evaporated In_xSe_y and $ZnIn_xSe_y$ films were also used as buffer layers in CIGS-based solar cells by the research groups working at TIT and Energy Photovoltaics. The optimum substrate temperature for the buffer-layer deposition was found to be 300°C, since at higher temperatures, the diffusion of Cu from the CIGS absorber layer into the buffer layer was noticed. Solar cells fabricated with such layers had efficiencies of 13% when made using In_xSe_y [209] and 15.3% when made using $ZnIn_xSe_y$ buffer layers [225].

1.6 TRANSPARENT CONDUCTING OXIDES

1.6.1 INTRODUCTION

TCOs are used in both substrate and superstrate configuration, thin film solar cell device structures. Their purpose is to reduce the lateral resistance of the device, by being highly conductive and, yet, highly transmissive of incident light. They are usually used in the region in between the buffer layer and the top contact grid in substrate configuration devices, or the buffer layer and the glass substrate in superstrate configuration devices.

TCOs are used in other applications, e.g., as transparent contacts in liquid crystal and LED displays, as transparent conductive layers on glass for electromagnetic screening, and as transparent strip heaters for defrosting glass.

Most TCOs are binary or ternary compounds, which have energy bandgaps >3 eV and which can be doped to produce material with resistivities as low as 10^{-4} Ω cm and with an optical extinction coefficient of 0.0001 in the visible range of wavelengths. A 100 nm thick layer of ITO with these properties will have a transmittance >90% and a sheet resistance < 10 $\Omega/\square$. The TCO should also be highly adherent to the surface onto which it is deposited, and be stable in hostile environments, such as acidic and alkaline environments, and in highly oxidizing or reducing atmospheres. It should also be able to withstand the elevated temperatures used in device processing.

The most developed TCO material is indium tin oxide (In_2SnO_4), which is commonly referred to as ITO. ITO layers have in fact been produced with resistivities as low as 7.7×10^{-5} Ω cm corresponding to an electron concentration of 2.5×10^{25} cm^{-3} [1]. This material is also found to be environmentally very stable. A wide range of methods have been used to produce ITO, including physical vapor deposition methods (usually magnetron sputtering) and CVD methods (usually, atmospheric pressure chemical vapor deposition, APCVD) [280]. The latter method is compatible with the float glass production process, ensuring a fresh surface and exploitation of the high temperature of the glass in the process. The magnetron sputtering method gives better-quality layers and is more flexible for controlling the quality and properties of the layers deposited, but it is also more expensive. In commercial production, the glass sheets are passed through long modular vacuum systems (40–160 m in length), in which there are typically 20–60 rotary targets. Multilayer stacks of ITO are deposited as the panels pass underneath the targets, traveling with speeds of typically 1 m/s. Most manufacturers (>75%) produce ITO as the TCO of choice on soda-lime glass substrates. Other manufacturers supply FTO (SnO_2:F), indium-doped zinc oxide (ZnO:In), or aluminum-doped zinc oxide (ZnO:Al) on glass.

Despite the achievements with ITO, the lack of abundance and, consequently, high cost of indium has motivated research into developing other materials that avoid the use of indium. Foremost among these materials are FTO (SnO_2:F) and aluminum-doped zinc oxide (ZnO:Al). The former material (or ITO) is used in superstrate configuration, CdTe-based solar cells, while the latter material is commonly used in substrate configuration, CIGS, and $CuInS_2$-based solar cells. A list of commonly used TCO layers and typical values, and their resistivity and transmittance are given in Table 1.8.

Much research efforts have been made over the past few decades to incorporate novel dopants into SnO_2 and ZnO to try to improve the "figures of merit" and stability of these materials, into developing novel materials, e.g., CdO, Cd_2SnO_4, and Zn_2SnO_4, and into developing p-type TCOs for application in blue-green lasers. These efforts are discussed in the following sections. Novel dopants that have been used with some of the more common TCO materials are given in Table 1.9.

TABLE 1.8
Properties of Some of the Most Common TCOs
and Buffer Layers

Material	Resistivity (Ω cm)	Transmittance in the 400–900 nm Region (%)
SnO_2:F	8×10^{-4}	80
In_2O_3:Sn	2×10^{-4}	80
In_2O_3:Ga	2×10^{-4}	85
In_2O_3:F	2.5×10^{-4}	85
Cd_2SnO_4	2×10^{-4}	85
Zn_2SnO_4	1×10^{-2}	90
ZnO:Al	8×10^{-4}	85
ZnO:B	2×10^{-3}	85
ZnO:In	8×10^{-4}	85

Source: Morales-Acevedo, A., Solar Energy, 80, 675, 2006.

TABLE 1.9
**List of the Dopants That Have Been Used in the More
Commonly Used TCO Materials**

TCO	Dopants
SnO_2	Sb, F, As, Nb, Ta
ZnO	Al, Ga, B, In, Y, Sc, F, V, Si, Ge, Ti, Zr, Hf, Mg, As, H
In_2O_3	Sn, Mo, Ta, W, Zr, F, Ge, Nb, Hf, Mg
CdO	In, Sn
$GaInO_3$	Sn, Ge
$CdSb_2O_3$	Y

Source: Minami, T., *Semicond. Sci. Technol.*, 20, S35, 2005.

1.6.2 TIN OXIDE

SnO_2 has the advantage over ITO that it does not contain indium. With a direct energy bandgap of 3 eV, it is also a low-toxicity material, with good thermal and chemical stabilities. The lowest resistivities have been achieved for SnO_2 doped with fluorine or doped with antimony, and these are therefore the commonest dopants used in large-scale production [283,284]. Other dopants that have been used to dope SnO_2 include As, Nb, and Ta; however, the resistivities achieved with these materials are not as low as those achieved with F or Sb [282].

1.6.3 ZINC OXIDE

The main advantages of using ZnO include its high abundance (and therefore low cost of the constituent elements), low toxicity, and high thermal and chemical stabilities, compared to ITO. Its direct energy bandgap equal to 3.37 eV is wide enough to transmit the entire solar spectrum. It can also be easily doped, most commonly with Al or Ga, to achieve resistivities approaching those of ITO [280]. Other materials can also be used to dope ZnO, but not with resistivities as low as those achieved with Al and Ga; these alternative dopants include Y, Sc, F, V, Si, Ge, Ti, Zr, Hf, Mg, As, and H.

A wide variety of methods can be used to deposit the doped ZnO layers including magnetron sputtering, chemical spray pyrolysis, pulsed laser deposition, chemical vapor deposition, sol-gel processing, and MBE [285–289]. Of these methods, magnetron sputtering is usually used because of the ease with which the sputtering parameters can be controlled [290].

To deposit ZnO:Al, the sputtering target used is usually a mixture of ZnO with Al_2O_3, containing approximately 3 wt% of the Al_2O_3 in the ZnO. The source–substrate distance is typically 70 mm, and the deposition is evacuated to a base pressure better than 5×10^{-4} Pa prior to admitting the sputtering gases. The sputtering gas is a mixture of argon and oxygen, introduced into the chamber by mass controllers, such that the total flow rate is 15 sccm and the working pressure is 1 Pa. To deposit Al-doped ZnO, the flow rate of oxygen is typically 1% of that of argon, and for a constant sputtering power of 150 W maintained throughout the deposition, the film thickness deposited is approximately 500 nm for a deposition time of 15 min. A resistivity of 8.5×10^{-5} Ω cm has been reported by Agura et al. [291], which is close to the lowest resistivity value ever reported for ITO, 7.7×10^{-5} Ω cm.

Some workers, when producing substrate configuration devices, deposit a layer of undoped ZnO prior to depositing the doped layer of ZnO. This is said by some to improve the junction properties, possibly by blocking shunting paths through the very thin buffers often used. To deposit undoped ZnO, the target is pure ZnO (99.99%), the oxygen flow rate increased to 25%, and the deposition time extended to 1 h, to produce a 500 nm thick film (for a sputtering power of 150 W) [290]. The chemical and physical properties of the layers deposited depend on the sputtering parameters, the most important of which are the partial pressure of oxygen used in the chamber, the sputtering power, the substrate temperature, and the deposition rate.

1.6.4 Cadmium Oxide

Historically, CdO was the first TCO to be identified [16]. The energy bandgap of CdO is typically in the range 2.40–2.45 eV, shifting significantly due to the Moss–Burstein effect. Despite a wide direct energy bandgap, it can easily be produced with a resistivity as low as 10^{-4} Ω cm, using a wide range of methods, such as thermal evaporation or sputtering of the compound, or by reactive deposition by thermally evaporating or sputtering Cd in an oxygen-containing environment [292–294]. It has also been used as a partner material to CdTe as an alternative to CdS with some success, with devices having efficiencies >10% produced by Sravani et al. [293]. The main drawback of this material is that CdO is more toxic than Cd. This restricts its use in PV solar cell devices and modules.

1.6.5 Cadmium Stannate and Zinc Stannate

CTO (Cd_2SnO_4) and zinc stannate (Zn_2SnO_4) are relatively novel TCO materials. They have however been used as the TCO in CdS/CdTe solar cell devices by workers at NREL to manufacture the highest-efficiency CdTe-based devices produced to date (16.5%) [21]. The improvement in device efficiency is attributed to the Cd_2SnO_4 with higher transmittance and lower resistivity than all other TCOs at present; the Zn_2SnO_4 is incorporated into the best devices to block shunting paths through the junction that would reduce cell performance. For the CdTe devices, both Cd_2SnO_4 and Zn_2SnO_4 were deposited using RF magnetron sputtering using hot-pressed targets of the compounds. A further advantage of using these layers is that the Cd_2SnO_4, the Zn_2SnO_4, and the CdS buffer layer can all be deposited using RF magnetron sputtering at room temperature, minimizing the time for warm-up and cool-down cycles, consequently increasing the throughput of the production process. During the deposition of the CdTe using the "close-spaced sublimation" (CSS) process at 570°C–625°C, the layers previously deposited are recrystallized and the interdiffusion of the layers promoted to reduce stress at the interface.

1.6.6 *p*-Type Transparent Conducting Oxides

A major stimulus for developing *p*-doped TCOs is to develop "transparent electronic devices." In addition to giving flexibility to device design, such devices would be very useful for making the interconnections between solar cells in "tandem solar cell structures" [280]. However, fabricating *p*-type TCOs has proved to be extremely difficult, with *p*-doping achieved only for a limited number of materials. One of the first materials to be made was *p*-type $CuAlO_2$ [295]. This early work has led to the synthesis of *p*-type TCO compounds, such as $CuMO_2$ and $AgMO_2$ (where M is a trivalent metal) [296]. The resistivities of these materials are however not as good as those achieved with *n*-type TCOs, as the transmittances are <80% and the resistivities >1 Ω cm.

The most successful *p*-type TCO produced to date is *p*-type ZnO [280]. Yamamoto and Yoshida proposed that co-doping using donor and acceptor dopants (e.g., Ga and N) might lead to *p*-type doping on theoretical grounds [297]. This led Joseph et al. to dope ZnO using N (an acceptor) with Ga (a donor), using an acceptor concentration equal to twice the donor concentration [298]. The *p*-type layers produced had an optical transmittance >85%, but the resistivity was approximately 1 Ω cm. *p*-type ZnO has also been produced using Sb as a dopant [299]. These layers also have transmittances >85% and resistivities of approximately 2 Ω cm, with a net hole density of 4×10^{16} cm^{-3} and mobilities in the range 9–20 cm^2/V s. The challenge remains to reduce the resistivity to <10^{-3} Ω cm to make the *p*-type layers useful in solar cell devices.

1.7   FUTURE ISSUES AND PROSPECTS OF THIN FILM PHOTOVOLTAICS

Thin film solar cells based on the use of compound semiconductors are now commercially available, and their use is expanding rapidly due to the low cost of manufacture, particularly in the case of the CdTe solar cell technology developed by First Solar. The market share of CdTe-based cells has in fact risen from about 3% during 2006 to 8% during 2008, accounting for an annual production rate >500 MW in 2008. The challenges now for CdTe are, as with other PV materials, to achieve higher efficiencies at the module level while continuing to drive down production costs.

CIGS solar cells have also been developed with considerable success. These do not have the problem of lack of environmental acceptability, due to the development of Cd-free cells. The most promising novel Cd-free buffer layers have been discussed in this chapter and include ZnS, In_2S_3, and $Mg_{1-x}Zn_xO$. However, the lack of abundance of indium and gallium is limiting how cheaply these cells can be produced and may act as a bottleneck for production in the medium/long term.

There are many alternative materials for developing thin film solar cells in the medium-/long-term future. These materials should consist of environmentally acceptable, abundant elements. Grain boundary passivation in these materials will be essential, and the materials will also need to be stable under illumination. The properties of some promising candidates, e.g., Cu_2SnZnS_4 and SnS, have been discussed in this chapter.

It is also possible that in order to minimize thermalization and transmission losses, multijunction devices based on the use of thin film solar cells will be developed. It remains to be established whether or not the best approach will be to produce two-terminal cells with the problem of loss of matching during parts of the day or to develop four-terminal cells without this problem but with a higher balance of system costs. These devices will be more efficient than single-junction devices, but without too high an increase in materials cost (given that thin films are used throughout the device structure).

ACKNOWLEDGMENT

One of the coauthors, Prof. K.T.R. Reddy, would like to acknowledge the British Commonwealth for financial support to work at the Northumbria University, Newcastle upon Tyne, United Kingdom.

ABBREVIATIONS

ALE	Atomic layer epitaxy
AM1.5	Air mass 1.5
BIPV	Building-integrated photovoltaics
CAS	Copper aluminum diselenide, $CuAlSe_2$
CBD	Chemical bath deposition
CGS	Copper gallium diselenide, $CuGaSe_2$
CIAS	Copper indium aluminum diselenide, $CuInAlSe_2$
CIGS	Copper indium gallium diselenide, $CuInGaSe_2$
CIGSS	Copper indium gallium diselenide disulfide, $CuInGa(S,Se)_2$
CIS	Copper indium diselenide, $CuInSe_2$
CISS	Copper indium disulfide diselenide, $CuIn(S,Se)_2$
CSS	Close-spaced sublimation
CTO	Cadmium stannate
CVD	Chemical vapor deposition
CZTS	Copper zinc tin sulfide, Cu_2ZnSnS_4
CZTSe	Copper zinc tin selenide, $Cu_2ZnSnSe_4$
DC	Direct current
DEZn	Diethylzinc

DIPTe	Di-isopropyltelluride
DMCd	Dimethylcadmium
E-beam	Electron beam
EBIC	Electron beam–induced current
HMI	Hahn-Meitner-Institute
ILGAR	Ion layer gas reaction
ITO	Indium tin oxide
J–V	Current density–voltage
LED	Light-emitting diode
MBE	Molecular beam epitaxy
MOCVD	Metal-organic chemical vapor deposition
MOVPE	Metal-organic vapor phase epitaxy
NREL	National Renewable Energy Laboratory
PV	Photovoltaic
PECVD	Plasma-enhanced chemical vapor deposition
PVD	Physical vapor deposition
QE	Quantum efficiency
RF	Radio-frequency
RTP	Rapid thermal processing
TCO	Transparent conductive oxide
TIT	Tokyo Institute of Technology
TO	Tin oxide
VTD	Vapor transport deposition

SYMBOLS

a_0, c_0	Lattice parameters (Å)
E_g	Energy bandgap (eV)
FF	Fill factor (%)
J_{sc}	Short circuit current density (mA/cm^2)
q	Electronic charge (C)
qV_{bi}	Barrier height (eV)
R_s	Series resistance (Ω/cm^2)
R_{sh}	Shunt resistance (Ω/cm^2)
V_{bi}	Built-in potential (V)
V_{oc}	Open-circuit voltage (V)
α	Absorption coefficient (cm^{-1})
$\Delta E_V, \Delta E_C$	Energy band offset (eV)
η	Efficiency (%)
ϕ_p, ϕ_m	Work function of semiconductor, metal (eV)
σ	Conductivity (S/cm)
χ_ρ	Electron affinity (eV)

REFERENCES

1. Navigant Consulting, Inc., *Solar Outlook*. Palo Alto, CA (2009).
2. Partain LD, *Solar Cells and Their Applications*, K. Chang, Ed. John Wiley & Sons, New York, 1995.
3. Miles RW, Zoppi G, and Forbes I, Inorganic photovoltaic cells. *Materials Today* **10** (2007) 20.
4. Miles RW, Hynes KM, and Forbes I, Photovoltaic solar cells: An overview of state-of-the-art cell development and environmental issues. *Progress in Crystal Growth and Characterization of Materials* **51** (2005) 1–42.

5. Luque A and Andreev V, *Concentrator Photovoltaics*. Springer Series in Optical Sciences, Springer, Berlin, Germany, 2007.
6. Archer M and Hill R, Eds., *Clean Electricity from Photovoltaics*. Imperial College Press, London, U.K., 1998.
7. Wronski CR and Carlson DE, Amorphous silicon solar cells, in *Clean Electricity from Photovoltaics*, M.D. Archer and R. Hill, Eds. Imperial College Press, London, U.K., pp. 199–236, 2001.
8. Rentzing S, Going for slim. *New Energy* **3** (2008) 58–68.
9. Repins I, Contreras MA, Egaas B et al., 19.9%-efficient ZnO/CdS/CuInGaSe$_2$ solar cell with 81.2% fill factor. *Progress in Photovoltaics: Research and Applications* **16** (2008) 235–239.
10. Loferski JJ, Theoretical considerations governing the choice of the optimum semiconductor for photovoltaic solar energy conversion. *Journal of Applied Physics* **27** (1956) 777–784.
11. Zanio K, *Cadmium Telluride*, R.K. Willardson and A.C. Beer, Eds. Academic Press, New York (1978).
12. Cusano DA, CdTe solar cells and photovoltaic heterojunctions in II–VI compounds. *Solid-State Electronics* **6** (1963) 217–218.
13. Adirovich EI, Yuabov YM, and Yagudaev GR, Photoelectric effects in film diodes with CdS-CdTe heterojunctions. *Fizika Tekhnika Poluprovodnikov* **3** (1969) 81–85.
14. Bonnet D and Rabenhorst H, New results on the development of a thin-film *p*-CdTe *n*-CdS heterojunction solar cell, in *Proceedings of the 9th IEEE Photovoltaic Specialists Conference*, Silver Springs, MD, 1972, pp. 129–132.
15. *International Committee for Diffraction Data*, Card number 15-770.
16. Mahnke HE, Haas H, Holub-Krappe E et al., Lattice distortion around impurity atoms as dopants in CdTe. *Thin Solid Films* **480–481** (2005) 279–282.
17. Hoschl P, Grill R, Franc J et al., Native defect equilibrium in semi-insulating CdTe(Cl). *Materials Science and Engineering B* **16** (1993) 215–218.
18. Abulfotuh FA, Balcioglu A, Wangensteen T et al., Study of the defects levels, electrooptics and interface properties of polycrystalline CdTe and CdS thin films and their junction, in *Proceedings of the 26th IEEE Photovoltaic Specialists Conference*, Anaheim, CA, 1997, p. 451.
19. Emanuelsson P, Omling P, Meyer BK et al., Identification of the cadmium vacancy in CdTe by electron paramagnetic resonance. *Physical Review B* **47** (1993) 15578–15580.
20. Wienecke M, Schenk M, and Berger H, Native point defects in Te-rich p-type Hg$_{1-x}$Cd$_x$Te. *Semiconductor Science and Technology* **8** (1993) 299–302.
21. Krsmanovic N, Lynn KG, Weber MH et al., Electrical compensation in CdTe and Cd$_{0.9}$Zn$_{0.1}$Te by intrinsic defects. *Physical Review B* **62** (2000) 16279–16282.
22. Berding MA, Native defects in CdTe. *Physical Review B* **60** (1999) 8943–8950.
23. Capper P, Ed., *Properties of Narrow Gap Cadmium-Based Compounds*. INSPEC, London, U.K., 1994.
24. Moutinho H, Hasoon FS, Abulfotuh FA et al., Investigation of polycrystalline CdTe thin films deposited by physical vapor deposition, close-spaced sublimation, and sputtering. *Journal of Vacuum Science and Technology A* **13** (1995) 2877–2883.
25. Birkmire RW and Phillips JE, *Processing and Modelling Issues for Thin Film Solar Cell Devices*. Institute of Energy Conversion, Newark, DE, 1997.
26. Wendt R, Fischer A, Grecu D et al., Improvement of CdTe solar cell performance with discharge control during film deposition by magnetron sputtering. *Journal of Applied Physics* **84** (1998) 2920–2925.
27. Aramoto T, Kumazawa S, Higuchi H et al., 16.0% efficient thin film CdS/CdTe solar cells. *Japanese Journal of Applied Physics* **36** (1997) 6304–6305.
28. Ferekides CS, Marinskiy D, Viswanathan V et al., High efficiency CSS CdTe solar cells. *Thin Solid Films* **361–362** (2000) 520–526.
29. Ohyama H, Aramoto T, Kumazawa S et al., 16.0% efficient thin-film CdS/CdTe solar cells, in *Proceedings of the 26th IEEE Photovoltaic Specialists Conference*, Anaheim, CA, 1997, pp. 343–346.
30. Wu X, Keane JC, Dhere R et al., 16.5%-efficient CdS/CdTe polycrystalline thin-film solar cell, in *Proceedings of the 17th European Photovoltaic Solar Energy Conference*, Munich, Germany, 2001, pp. 995–1000.
31. Al-Jassim MM, Dhere RG, Jones KM et al., The morphology, microstructure, and luminescent properties of CdS/CdTe films, in *Proceedings of the Second World Conference on Photovoltaic Solar Energy Conversion*, Vienna, Austria, 1998, pp. 1063–1066.
32. Meyers PV, Zhou T, Powell RC et al., Elemental vapor-deposited polycrystalline CdTe thin film photovoltaic modules, in *Proceedings of the 23rd IEEE Photovoltaics Specialists Conference*, Louisville, KY, 1993, pp. 400–404.

33. Sandwisch DW, Development of CdTe module manufacturing, in *Proceedings of the First World Conference on Photovoltaic Solar Energy Conversion*, Waikoloa, HI, 1994, pp. 836–839.

34. Kim D, Pozder S, Qi B et al., Thin film CdS/CdTe solar cells fabricated by electrodeposition. *AIP Conference Proceedings* **306** (1994) 320–328.

35. Johnson DR, Microstructure of electrodeposited CdS/CdTe cells. *Thin Solid Films* **361–362** (2000) 321–326.

36. Irvine SJC, Metal-organic vapour phase epitaxy, in *Narrow-Gap II–VI Compounds for Optoelectronic and Electromagnetic Applications*, P. Capper, Ed. Chapman & Hall, London, U.K., pp. 71–96, 1997.

37. Chu TL and Chu SS, Thin-film II–VI photovoltaics. *Solid-State Electronics* **38** (1995) 533–549.

38. Barrioz V, Rowlands RL, Jones EW et al., A comparison of in situ As doping with ex situ CdCl$_2$ treatment of CdTe solar cells. *Materials Research Society Symposium Proceedings* **865** (2005) F3.4.

39. Zoppi G, Durose K, Irvine SJC et al., Grain and crystal texture properties of absorber layers in MOCVD-grown CdTe/CdS solar cells. *Semiconductor Science and Technology* **21** (2006) 763.

40. Takamoto T, Agui T, Kurita H et al., Improved junction formation procedure for low temperature deposited CdS/CdTe solar cells. *Solar Energy Materials and Solar Cells* **49** (1997) 219–225.

41. Woodcock JM, Turner AK, Ozsan ME et al., Thin-film solar-cells based on electrodeposited CdTe, in *Conference Record of the Twenty Second IEEE Photovoltaic Specialists Conference*, Las Vegas, NV, 1991, pp. 842–847.

42. Gupta A and Compaan AD, All-sputtered 14% CdS/CdTe thin-film solar cell with ZnO:Al transparent conducting oxide. *Applied Physics Letters* **85** (2004) 684–686.

43. Skarp J, Koskinen Y, Lindfors S et al., Development and evaluation of CdS/CdTe thin film PV cells, in *Proceedings of the 10th European Photovoltaic Solar Energy Conference*, Lisbon, Portugal, 1991, pp. 567–569.

44. Ikegami S, CdS/CdTe solar cells by the screen-printing-sintering technique: Fabrication, photovoltaic properties and applications. *Solar Cells* **23** (1988) 89–105.

45. Irvine SJC, Barrioz V, Lamb D et al., MOCVD of thin film photovoltaic solar cells—Next-generation production technology? *Journal of Crystal Growth* **310** (2008) 5198–5203.

46. Mathew X, Enriquez JP, Romeo A et al., CdTe/CdS solar cells on flexible substrates. *Solar Energy* **77** (2004) 831–838.

47. Birkmire RW and Eser E, Polycrystalline thin film solar cells: Present status and future potential. *Annual Review of Materials Research* **27** (1997) 625–653.

48. McCandless BE and Birkmire RW, Influence of window and absorber layer processing on device operation in superstrate thin film CdTe solar cells, in *28th IEEE Photovoltaic Specialists Conference*, Anchorage, AK, 2000, p. 491.

49. Romeo A, Tiwari AN, Zogg H et al., Influence of transparent conducting oxides on the properties of CdTe/CdS solar cells, in *Proceedings of the Second World Conference on Photovoltaic Solar Energy Conversion*, Vienna, Austria, 1998, pp. 1105–1108.

50. Bonnet D, Manufacturing of CSS CdTe solar cells. *Thin Solid Films* **361–362** (2000) 547–552.

51. Turner AK, Woodcock JM, Ozsan ME et al., BP solar thin film CdTe photovoltaic technology. *Solar Energy Materials and Solar Cells* **35** (1994) 263–270.

52. Romeo N, Bosio A, Tedeschi R et al., Cadmium and zinc chloride treatments of CdS films for the preparation of high efficiency CdTe/CdS thin films solar cells, in *Proceedings of the 14th European Photovoltaic Solar Energy Conference*, Barcelona, Spain, 1997, pp. 2351–2353.

53. McCandless BE and Dobson KD, Processing options for CdTe thin film solar cells. *Solar Energy* **77** (2004) 839–856.

54. Ferekides CS, Dugan K, Ceekala V et al., The effects of CdS processing and glass substrates in the performance of CdTe solar cells, in *Proceedings of the First World Conference on Photovoltaic Energy Conversion*, Waikoloa, HI, 1994, pp. 99–102.

55. McCandless BE and Birkmire RW, Analysis of post deposition processing for CdTe/CdS thin film solar cells. *Solar Cells* **31** (1991) 527–535.

56. McCandless BE, Hichri H, Hanket GM et al., Vapor phase treatment of CdTe/CdS thin films with CdCl$_2$:O$_2$, in *25th IEEE Photovoltaic Specialists Conference*, Washington, DC, 1996, pp. 781–784.

57. Qu Y, Meyers PV, and McCandless BE, HCl vapor post-deposition heat treatment of CdTe/CdS films, in *Proceedings of the 25th IEEE Photovoltaic Specialists Conference*, Washington, DC, 1996, pp. 1013–1016.

58. Edwards PR, Durose K, Beier J et al., A comparative EBIC study of CdTe solar cell activation using CdCl$_2$ and Cl$_2$. *Thin Solid Films* **387** (2001) 189–191.

59. Zhou TX, Reiter N, Powell RC et al., Vapor chloride treatment of polycrystalline CdTe/CdS films, in *Proceedings of the First World Conference on Photovoltaic Solar Energy Conversion*, Waikoloa, HI, 1994, pp. 103–106.

60. Zoppi G, Durose K, Irvine SJC et al., Towards MOCVD-grown CdTe/CdS solar cells, in *Proceedings of the 19th European Photovoltaic Solar Energy Conference*, Paris, France, 2004, pp. 1921–1924.

61. Bayhan H and Erçelebi C, Electrical characterization of vacuum-deposited n-CdS/p-CdTe heterojunction devices. *Semiconductor Science and Technology* **12** (1997) 600.

62. McCandless BE, Moulton LV, and Birkmire RW, Recrystallization and sulfur diffusion in $CdCl_2$-treated CdTe/CdS thin films. *Progress in Photovoltaics: Research and Applications* **5** (1997) 249–260.

63. Levi DH, Moutinho HR, Hasoon FS et al., Micro through nanostructure investigations of polycrystalline CdTe: Correlations with processing and electronic structures. *Solar Energy Materials and Solar Cells* **41–42** (1996) 381–393.

64. Cousins MA and Durose K, Grain structure of CdTe in CSS-deposited CdTe/CdS solar cells. *Thin Solid Films* **361–362** (2000) 253–257.

65. Jensen DG, McCandless BE, and Birkmire RW, Thin film cadmium telluride-cadmium sulfide alloys and devices, in *Proceedings of the 25th IEEE Photovoltaic Specialists Conference*, Washington, DC, 1996, pp. 773–776.

66. Suyama N, Arita T, Nishiyama Y et al., CdS/CdTe solar cells by the screen-printing-sintering technique, in *Proceedings of the 21st IEEE Photovoltaic Specialists Conference*, Kissimmee, FL, 1990, pp. 498–503.

67. Oman DM, Dugan KM, Killian JL et al., Reduction of recombination current in CdTe/CdS solar cells. *Applied Physics Letters* **67** (1995) 1896–1898.

68. Conibeer GJ, Lane DW, Painter JD et al., Diffusion in CdTe and CdS polycrystalline thin films and single crystal CdS, in *Proceedings of the Second World Conference on Photovoltaic Solar Energy Conversion*, Vienna, Austria, 1998, pp. 1097–1100.

69. Al-Allak HM, Brinkman AW, Richter H et al., Dependence of CdS/CdTe thin film solar cell characteristics on the processing conditions. *Journal of Crystal Growth* **159** (1996) 910–915.

70. Edwards PR, Durose K, Galloway SA et al., Front-wall electron beam-induced current studies on thin film CdS/CdTe solar cells, in *Proceedings of the Second World Conference on Photovoltaic Solar Energy Conversion*, Vienna, Austria, 1998, pp. 472–476.

71. Edwards PR, Halliday DP, Durose K et al., The influence of $CdCl_2$ treatment and interdiffusion on grain boundary passivation in CdTe/CdS solar cells, in *Proceedings of the 14th European Photovoltaic Solar Energy Conference*, Barcelona, Spain, 1997, pp. 2083–2086.

72. Woods LM, Levi DH, Kaydanov V et al., Electrical characterization of CdTe grain boundary properties from As processed CdTe/CdS solar cells, in *Proceedings of the Second World Conference on Photovoltaic Solar Energy Conversion*, Vienna, Austria, 1998, pp. 1043–1046.

73. Woods LM and Robinson GY, The effects of $CdCl_2$ on CdTe electrical properties using a new theory for grain boundary conduction, in *Proceedings of the 28th IEEE Photovoltaic Specialists Conference*, Anchorage, AK, 2000, pp. 603–606.

74. Streetman BG and Banerjee S, *Solid State Electronic Devices*, 5th edn. Prentice Hall, Upper Saddle River, NJ, 2000.

75. Jäger H and Seipp E, Transition resistances of ohmic contacts to p-type CdTe and their time-dependent variation. *Journal of Electronic Materials* **10** (1981) 605–618.

76. Kotina IM, Tukhkonen LM, Patsekina GV et al., Study of CdTe etching process in alcoholic solutions of bromine. *Semiconductor Science and Technology* **13** (1998) 890.

77. Sarlund J, Ritala M, Leskela M et al., Characterization of etching procedure in preparation of CdTe solar cells. *Solar Energy Materials and Solar Cells* **44** (1996) 177–190.

78. Dhere R, Rose D, Albin D et al., Influence of CdS/CdTe interface properties on the device properties, in *Proceedings of the 26th IEEE Photovoltaic Specialists Conference*, Anaheim, CA, 1997, pp. 435–438.

79. Meyers PE, Design of a thin film CdTe solar cell. *Solar Cells* **23** (1988) 59–67.

80. Janik E and Triboulet R, Ohmic contacts to p-type cadmium telluride and cadmium mercury telluride. *Journal of Physics D: Applied Physics* **16** (1983) 2333–2340.

81. Edwards PR, Beam-induced current studies of CdTe/CdS solar cells PhD thesis, University of Durham, Durham, England, 1998.

82. Stollwerck G and Sites JR, Analysis of CdTe back-contact barriers, in *Proceedings of the 13th European Photovoltaic Solar Energy Conference*, Nice, France, 1995, pp. 2020–2023.

83. Agostinelli G, Batzner DL, and Burgelman M, An alternative model for V, G and T dependence of CdTe solar cells IV characteristics, in *Proceedings of the 29th IEEE Photovoltaic Specialists Conference*, New Orleans, LA, 2002.

84. Wagner S, Shay JL, Migliorato P et al., CuInSe$_2$/CdS heterojunction photovoltaic detectors. *Applied Physics Letters* **25** (1974) 434–435.

85. Migliorato P, Tell B, Shay JL et al., Junction electroluminescence in CuInSe$_2$. *Applied Physics Letters* **24** (1974) 227–228.

86. Wagner S, in *Device Applications of Ternary Compounds*, vol. 35, G.D. Holah, Ed. Institute of Physics, London, U.K., p. 205, 1977.

87. Shay JL and Wernick JH, *Ternary Chalcopyrite Semiconductors: Growth, Electronic Properties and Applications.* Pergamon Press, Oxford, U.K., 1975.

88. Shay JL, Wagner S, and Kasper HM, Efficient CuInSe$_2$/CdS solar cells. *Applied Physics Letters* **27** (1975) 89–90.

89. Kazmerski LL, Ayyagari MS, and Sanborn GA, CuInS$_2$ thin films: Preparation and properties. *Journal of Applied Physics* **46** (1975) 4865–4869.

90. Kazmerski LL, White FR, and Morgan GK, Thin-film CuInSe$_2$/CdS heterojunction solar cells. *Applied Physics Letters* **29** (1976) 268–270.

91. Kazmerski LL and Sanborn GA, CuInS$_2$ thin-film homojunction solar cells. *Journal of Applied Physics* **48** (1977) 3178–3180.

92. Kazmerski LL and Juang YJ, Vacuum-deposited CuInTe$_2$ thin films—Growth, structural, and electrical properties. *Journal of Vacuum Science & Technology* **14** (1977) 769–776.

93. Mickelsen RA and Chen WS, High photocurrent polycrystalline thin-film CdS/CuInSe$_2$ solar cell. *Applied Physics Letters* **36** (1980) 371–373.

94. Mickelsen RA and Chen WS, Development of a 9.4% efficient thin-film CuInSe$_2$/CdS solar cell, in *Proceedings of the 15th IEEE Photovoltaic Specialists Conference*, Orlando, FL, 1981, p. 800.

95. Mickelsen RA and Chen WS, Polycrystalline thin film CuInSe$_2$ solar cells, in *Proceedings of the 16th IEEE Photovoltaic Specialists Conference*, San Diego, CA, 1982, p. 781.

96. Meakin JD, Status of CuInSe$_2$ solar cells, in *Proceedings of the 14th Critical Reviews of Technology Conference*, Arlington, VA, 1985, pp. 108–118.

97. Russell PE, Jamjoum O, Ahrenkiel RK et al., Properties of the Mo–CuInSe$_2$ interface. *Applied Physics Letters* **40** (1982) 995–997.

98. Kazmerski LL, Russel PE, Jamjoum O et al., Initial formation and development of CdS/CuInSe$_2$ solar cell interfaces, in *Proceedings of the 16th IEEE Photovoltaic Specialists Conference*, San Diego, CA, 1982, p. 786.

99. Chen WS, Stewart JM, Stanbery BJ et al., Development of polycrystalline CuIn$_{1-x}$Ga$_x$Se$_2$ solar cells, in *Proceedings of the 19th IEEE Photovoltaic Specialists Conference*, New Orleans, LA, 1987, p. 1445.

100. Dimmler B, Dittrich H, Menner R et al., Performance and optimization of heterojunctions based on Cu(Ga,In)Se$_2$, in *Proceedings of the 19th IEEE Photovoltaic Specialists Conference*, New Orleans, LA, 1987, pp. 1454–1460.

101. Love RB and Choudary UV, Method for forming PV cells employing multinary semiconductor films. U.S. Patent Number 4,465,575, 1984.

102. Emer JH and Love RB, Method for forming CuInSe films. U.S. Patent Number 4,798,660, 1989.

103. Eberspacher C, Ermer JH, and Mitchell K, Process for making thin film solar cell. European Patent Application Number 0318315A2, 1989.

104. Mitchell KW, Eberspacher C, Ermer JH et al., Single and tandem junction CuInSe$_2$ technology, in *Proceedings of the Fourth International PV Science and Engineering Conference*, Sydney, Australia, 1989, p. 889.

105. Ermer JH, Frederic C, Pauls K et al., Recent progress in large area CuInSe$_2$ submodules, in *Proceedings of the Fourth International PV Science and Engineering Conference*, Sydney, Australia, 1989, p. 475.

106. Bodegard M, Stolt L, and Hedstrom J, The influence of sodium on the grain structure of CuInSe$_2$ films for photovoltaic applications, in *Proceedings of the 12th European Photovoltaic Solar Energy Conference*, Amsterdam, the Netherlands, 1994, p. 1743.

107. Bodegård M, Granath K, Stolt L et al., The behaviour of Na implanted into Mo thin films during annealing. *Solar Energy Materials and Solar Cells* **58** (1999) 199–208.

108. Hedstrom J, Ohlsen H, Bodegard M et al., ZnO/CdS/Cu(In,Ga)Se$_2$ thin film solar cells with improved performance, in *Proceedings of the 23rd IEEE Photovoltaic Specialists Conference*, Louisville, KY, 1993, pp. 364–371.

109. Scofield JH, Duda A, Albin D et al., Sputtered molybdenum bilayer back contact for copper indium diselenide-based polycrystalline thin-film solar cells. *Thin Solid Films* **260** (1995) 26–31.

110. Schmid U and Seidel H, Effect of substrate properties and thermal annealing on the resistivity of molybdenum thin films. *Thin Solid Films* **489** (2005) 310–319.

111. Orgassa K, Schock HW, and Werner JH, Alternative back contact materials for thin film $Cu(In,Ga)Se_2$ solar cells. *Thin Solid Films* **431–432** (2003) 387–391.

112. Malikov IV and Mikhailov GM, Electrical resistivity of epitaxial molybdenum films grown by laser ablation deposition. *Journal of Applied Physics* **82** (1997) 5555–5559.

113. Wada T, Kohara N, Nishiwaki S et al., Characterization of the $Cu(In,Ga)Se_2$/Mo interface in CIGS solar cells. *Thin Solid Films* **387** (2001) 118–122.

114. Kessler J, $CuInSe_2$ film formation from sequential depositions of In(Se):Cu:Se, in *Proceedings of the 12th European Photovoltaic Solar Energy Conference*, Amsterdam, the Netherlands, 1994, pp. 648–652.

115. Gabor AM, Tuttle JR, Albin DS et al., High-efficiency $CuIn_xGa_{1-x}Se_2$ solar cells made from $(In_x,Ga_{1-x})_2Se_3$ precursor films. *Applied Physics Letters* **65** (1994) 198–200.

116. Eisgruber IL, Treece RE, Hollingsworth RE et al., In-situ measurements of $Cu(In,Ga)Se_2$ composition by x-ray fluorescence, in *Proceedings of the 28th IEEE Photovoltaic Specialists Conference*, Anchorage, AK, 2000, pp. 505–508.

117. Stolt L, Hedstrom J, and Sigurd D, Coevaporation with a rate control-system based on a quadrupole mass-spectrometer. *Journal of Vacuum Science & Technology A—Vacuum Surfaces and Films* **3** (1985) 403–407.

118. Powalla M, Voorwinden G, and Dimmler B, Continuous $Cu(In,Ga)Se_2$ deposition with improved process control, in *Proceedings of the 14th European Photovoltaic Solar Energy Conference*, Barcelona, Spain, 1997, p. 1270.

119. Nishitani M, Negami T, and Wada T, Composition monitoring method in $CuInSe_2$ thin film preparation. *Thin Solid Films* **258** (1995) 313–316.

120. Negami T, Nishitani M, Kohara N et al., Real time composition monitoring methods in physical vapor deposition of $Cu(In,Ga)Se$-2 thin films, in *Symposium on Thin Films for Photovoltaic and Related Device Applications, at the 1996 MRS Spring Meeting*, San Francisco, CA, 1996, pp. 267–278.

121. Müller J, Nowoczin J, and Schmitt H, Composition, structure and optical properties of sputtered thin films of $CuInSe_2$. *Thin Solid Films* **496** (2006) 364–370.

122. Kapur VK and Basol BM, Key issues and cost estimates for the fabrication of $CuInSe_2$ (CIS) PV modules by the two-stage process, in *Proceedings of the 21st IEEE Photovoltaic Specialists Conference*, Kissimmee, FL, 1990, p. 467.

123. Kushiya K, Progress in large-area $Cu(InGa)Se_2$-based thin-film modules with the efficiency of over 13%, in *Proceedings of the Third World Conference on Photovoltaic Energy Conversion*, Osaka, Japan, 2003, pp. 319–324.

124. Probst V, Stetter W, Palm J et al., CIGSSE module pilot processing: From fundamental investigations to advanced performance, in *Proceedings of the Third World Conference on Photovoltaic Solar Energy Conversion*, Osaka, Japan, 2003, pp. 329–334.

125. Palm J, Visbeck S, Stetter W et al., CIS process for commercial power module production, in *Proceedings of the 21st European Photovoltaic Solar Energy Conference*, Dresden, Germany, 2006, pp. 1796–1800.

126. Scheer R, Walter T, Schock HW et al., $CuInS_2$ based thin film solar cell with 10.2% efficiency. *Applied Physics Letters* **63** (1993) 3294.

127. Braunger D, Hariskos D, Walter T et al., An 11.4% efficient polycrystalline thin film solar cell based on $CuInS_2$ with a Cd-free buffer layer. *Solar Energy Materials and Solar Cells* **40** (1996) 97–102.

128. Klaer J, Bruns J, Henninger R et al., Efficient $CuInS_2$ thin-film solar cells prepared by a sequential process. *Semiconductor Science and Technology* **13** (1998) 1456.

129. Green MA, Emery K, Hishikawa Y et al., Solar cell efficiency tables (version 33). *Progress in Photovoltaics: Research and Applications* **17** (2009) 85–94.

130. Stanbery BJ, Entrepreneurship on the road from science to sales, in *Proceedings of the 33rd IEEE Photovoltaic Specialists Conference*, San Diego, CA, 2008.

131. Meeder A, Neisser A, Rühle U et al., Manufacturing the first MW of large-area $CuInS_2$-based solar module—Recent experiences and progress, in *Proceedings of the 22nd European Photovoltaic Solar Energy Conference*, Milan, Italy, 2007.

132. Penndorf J, Winkler M, Tober O et al., $CuInS_2$ thin film formation on a Cu tape substrate for photovoltaic applications. *Solar Energy Materials and Solar Cells* **53** (1998) 285–298.

133. Ramanathan K, Hasoon F, Smith S et al., Properties of Cd and Zn partial electrolyte treated CIGS solar cells, in *Proceedings of the 29th IEEE Photovoltaic Specialists Conference*, New Orleans, LA, 2002, pp. 523–526.

134. Kessler J, Sicx-Kurdi J, Naghavi N et al., Progress in low-cost electrodeposition of $Cu(In,Ga)(S,Se)_2$: The CISEL project, in *Proceedings of the 20th European Photovoltaic Solar Energy Conference*, Barcelona, Spain, 2005, pp. 1704–1708.

135. Paulson PD, Haimbodi MW, Marsillac S et al., $CuIn_{1-x}Al_xSe_2$ thin films and solar cells. *Journal of Applied Physics* **91** (2002) 10153–10156.

136. Niemi E and Stolt L, $CuAlSe_2$ for thin film solar cells, in *Proceedings of the Eighth International European Photovoltaic Solar Energy Conference*, Florence, Italy, 1988, pp. 1070–1074.

137. Marsillac S, Bernede JC, El Moctar C et al., Physico-chemical characterization of $CuAlSe_2$ films obtained by reaction, induced by annealing, between Se vapour and Al/Cu/Al…Cu/Al/Cu thin films sequentially deposited. *Materials Science and Engineering B* **45** (1997) 69–75.

138. Bernède JC, Marsillac S, Moctar CE et al., Optical and electrical properties of $CuAlSe_2$ thin films obtained by selenization of Cu/Al/Cu…Al/Cu layers sequentially deposited. *Physica Status Solidi (a)* **161** (1997) 185–192.

139. Barkat L, Morsli M, Amory C et al., Study on the fabrication of n-type $CuAlSe_2$ thin films. *Thin Solid Films* **431–432** (2003) 99–104.

140. Benchouk K, El Moetar O, Marsillac S et al., New ternary compounds $CuAlX_2$ (X = Te, Se) buffer layers: Large band gap. *Synthetic Metals* **103** (1999) 2644–2645.

141. Reddy YBK and Raja VS, Preparation and characterization of $CuAlSe_2$ thin films prepared by co-evaporation, in *Twenty-Ninth IEEE Photovoltaic Specialists Conference*, New Orleans, LA, 2002, pp. 664–667.

142. Reddy YBK and Raja VS, Effect of Cu/Al ratio on the properties of $CuAlSe_2$ thin films prepared by co-evaporation. *Materials Chemistry and Physics* **100** (2006) 152–157.

143. Dhananjay, Nagaraju J, and Krupanidhi SB, Structural and optical properties of $CuIn_{1-x}Al_xSe_2$ thin films prepared by four-source elemental evaporation. *Solid State Communications* **127** (2003) 243–246.

144. Reddy YBK, Raja VS, and Sreedhar B, Growth and characterization of $CuIn_{1-x}Al_xSe_2$ thin films deposited by co-evaporation. *Journal of Physics D: Applied Physics* **39** (2006) 5124.

145. Itoh F, Saitoh O, Kita M et al., Growth and characterization of $Cu(InAl)Se_2$ by vacuum evaporation. *Solar Energy Materials and Solar Cells* **50** (1998) 119–125.

146. Halgand E, Bernede JC, Marsillac S et al., Physico-chemical characterisation of $Cu(In,Al)Se_2$ thin film for solar cells obtained by a selenisation process. *Thin Solid Films* **480–481** (2005) 443–446.

147. López-García J and Guillén C, $CuIn_{1-x}Al_xSe_2$ thin films obtained by selenization of evaporated metallic precursor layers. *Thin Solid Films* **517** (2009) 2240–2243.

148. Sugiyama M, Umezawa A, Yasuniwa T et al., Growth of single-phase $Cu(In,Al)Se_2$ photoabsorbing films by selenization using diethylselenide. *Thin Solid Films* **517** (2009) 2175–2177.

149. Yun JH, Chalapathy RBV, Lee JC et al., Formation of $CuIn(1-x)Al(x)Se(2)$ thin films by selenization of metallic precursors in Se vapor. *Solid State Phenomena* **124–126** (2007) 975–978.

150. Zoppi G, Forbes I, Nasikkar P et al., Characterisation of thin films $CuIn_{1-x}Al_xSe_2$ prepared by selenisation of magnetron sputtered metallic precursor, in *Thin-Film Compound Semiconductor Photovoltaics*, T. Gessert, S. Marsillac, T. Wada, K. Durose, and C. Heske, Eds. (*Mater. Res. Soc. Symp. Proc.*), San Francisco, CA, 2007, p. Y12-02.

151. Woods LM, Kalla A, Gonzalez D et al., Wide-bandgap CIAS thin-film photovoltaics with transparent back contacts for next generation single and multi-junction devices. *Materials Science and Engineering B* **116** (2005) 297–302.

152. Marsillac S, Paulson PD, Haimbodi MW et al., High-efficiency solar cells based on $Cu(InAl)Se_2$ thin films. *Applied Physics Letters* **81** (2002) 1350–1352.

153. Minemoto T, Hayashi T, Araki T et al., Electronic properties of $Cu(In,Al)Se_2$ Solar cells prepared by three-stage evaporation, in ENERGEX 2007, *The 12th International Energy Conference & Exhibition*, Singapore, 2007.

154. Hayashi T, Minemoto T, Zoppi G et al., Effect of composition gradient in $Cu(In,Al)Se_2$ solar cells. *Solar Energy Materials and Solar Cells* **93** (2007) 922–925.

155. Ito K and Nakazawa T, Electrical and optical-properties of stannite-type quaternary semiconductor thin-films. *Japanese Journal of Applied Physics* **27** (1988) 2094–2097.

156. Friedlmeier TM, Wieser N, Walter T et al., Heterojunctions based on Cu_2ZnSnS_4 and $Cu_2ZnSnSe_4$ thin films, in *Proceedings of the 14th European Photovoltaic Specialists Conference*, Barcelona, Spain, 1997, pp. 1242–1245.

157. Katagiri H, Cu_2ZnSnS_4 thin film solar cells. *Thin Solid Films* **480–481** (2005) 426–432.

158. Katagiri H, Jimbo K, Moriya K et al., Solar cell without environmental pollution by using CZTS thin film, in *Third World IEEE PVSC*, Osaka, Japan, 2003, pp. 2874–2877.

159. Katagiri H, Jimbo K, Yamada S et al., Enhanced conversion efficiencies of Cu_2ZnSnS_4-based thin film solar cells by using preferential etching technique. *Applied Physics Express* **1** (2008) 041201.

160. Katagiri H, Saitoh K, Washio T et al., Development of thin film solar cell based on Cu_2ZnSnS_4 thin films. *Solar Energy Materials and Solar Cells* **65** (2001) 141–148.

161. Katagiri H, Sasaguchi N, Hando S et al., Preparation and evaluation of Cu_2ZnSnS_4 thin films by sulfurization of E—B evaporated precursors. *Solar Energy Materials and Solar Cells* **49** (1997) 407–414.

162. Jimbo K, Kimura R, Kamimura T et al., Cu_2ZnSnS_4-type thin film solar cells using abundant materials. *Thin Solid Films* **515** (2007) 5997.

163. Araki H, Mikaduki A, Kubo Y et al., Preparation of Cu_2ZnSnS_4 thin films by sulfurization of stacked metallic layers. *Thin Solid Films* **517** (2009) 1457–1460.

164. Scragg JJ, Dale PJ, Peter LM et al., New routes to sustainable photovoltaics: evaluation of Cu_2ZnSnS_4 as an alternative absorber material. *Physica Status Solidi (b)* **245** (2008) 1772–1778.

165. Ennaoui A, Lux-Steiner M, Weber A et al., Cu_2ZnSnS_4 thin film solar cells from electroplated precursors: Novel low-cost perspective. *Thin Solid Films* **517** (2009) 2511–2514.

166. Altosaar M, Raudoja J, Timmo K et al., $Cu_2ZnSnSe_4$ monograin powders for solar cell application, in *Proceedings of the IEEE Fourth World Conference on Photovoltaic Energy Conversion*, Waikoloa, HI, 2006, pp. 468–470.

167. Mellikov E, Meissner D, Varema T et al., Monograin materials for solar cells. *Solar Energy Materials and Solar Cells* **93** (2008) 65–68.

168. Zoppi G, Forbes I, Miles RW et al., $Cu_2ZnSnSe_4$ thin film solar cells produced by selenisation of magnetron sputtered precursors. *Progress in Photovoltaics: Research and Applications*, **17** (2009) 315–319.

169. Sharon M and Basavaswaran K, Photoelectrochemical behaviour of tin monosulphide. *Solar Cells* **25** (1988) 97–107.

170. Pramanik P, Basu PK, and Biswas S, Preparation and characterization of chemically deposited tin(II) sulphide thin films. *Thin Solid Films* **150** (1987) 269–276.

171. Devika M, Reddy NK, Ramesh K et al., Low resistive micrometer-thick SnS:Ag films for optoelectronic applications. *Journal of the Electrochemical Society* **153** (2006) G727–G733.

172. Albers W, Haas C, Vink HJ et al., Investigations on SnS. *Journal of Applied Physics* **32** (1961) 2220–2225.

173. Tanusevski A, Optical and photoelectric properties of SnS thin films prepared by chemical bath deposition. *Semiconductor Science and Technology* **18** (2003) 501.

174. Ortiz A, Alonso JC, Garcia M et al., Tin sulphide films deposited by plasma-enhanced chemical vapour deposition. *Semiconductor Science and Technology* **11** (1996) 243.

175. Ichimura M, Takeuchi K, Ono Y et al., Electrochemical deposition of SnS thin films. *Thin Solid Films* **361–362** (2000) 98–101.

176. Lopez S and Ortiz A, Spray pyrolysis deposition of Sn_xS_y thin films. *Semiconductor Science and Technology* **9** (1994) 2130–2133.

177. Koteswara Reddy N and Ramakrishna Reddy KT, Growth of polycrystalline SnS films by spray pyrolysis. *Thin Solid Films* **325** (1998) 4–6.

178. Yanuar Y, Guastavino F, Llinares C et al., SnS thin films grown by close-spaced vapor transport. *Journal of Materials Science Letters* **19** (2000) 2135–2137.

179. Guang-Pu W, Zhi-Lin Z, Wei-Ming Z et al., SnS films by rf sputtering for photovoltaic application, in *Proceedings of the First World Conference on Photovoltaic Energy Conversion*, Waikoloa, HI, 1994, pp. 365–368.

180. Ramakrishna Reddy KT, Purandhara Reddy P, Datta PK et al., Formation of polycrystalline SnS layers by a two-step process. *Thin Solid Films* **403–404** (2002) 116–119.

181. Sugiyama M, Miyauci K, Minemura T et al., Preparation of SnS films by sulfurization of Sn sheet. *Japanese Journal of Applied Physics* **47** (2008) 4494–4495.

182. Devika M, Reddy KTR, Reddy NK et al., Microstructure dependent physical properties of evaporated tin sulfide films. *Journal of Applied Physics* **100** (2006) 023518.

183. Ogah OE, Zoppi G, Forbes I et al., Thin films of tin sulphide for use in thin film solar cell devices. *Thin Solid Films* **517** (2009) 2485–2488.

184. Yue GH, Peng DL, Yan PX et al., Structure and optical properties of SnS thin film prepared by pulse electrodeposition. *Journal of Alloys and Compounds* **468** (2009) 254–257.

185. Koteswara Reddy N and Ramakrishna Reddy KT, Electrical properties of spray pyrolytic tin sulfide films. *Solid-State Electronics* **49** (2005) 902–906.

186. Ristov M, Sinadinovski G, Grozdanov I et al., Chemical deposition of TIN(II) sulphide thin films. *Thin Solid Films* **173** (1989) 53–58.

187. Johnson JB, Jones H, Latham BS et al., Optimization of photoconductivity in vacuum-evaporated tin sulfide thin films. *Semiconductor Science and Technology* **14** (1999) 501.

188. Noguchi H, Setiyadi A, Tanamura H et al., Characterization of vacuum-evaporated tin sulfide film for solar cell materials. *Solar Energy Materials and Solar Cells* **35** (1994) 325–331.

189. Koteeswara Reddy N and Ramakrishna Reddy KT, Tin sulphide films for solar cell application, in *Proceedings of the 26th IEEE Photovoltaic Specialists Conference*, Anaheim, CA, 1997, pp. 515–518.

190. Ristov M, Sinadinovski G, Mitreski M et al., Photovoltaic cells based on chemically deposited p-type SnS. *Solar Energy Materials and Solar Cells* **69** (2001) 17–24.

191. Subramanian B, Sanjeeviraja C, and Jayachandran M, Cathodic electrodeposition and analysis of SnS films for photoelectrochemical cells. *Materials Chemistry and Physics* **71** (2001) 40–46.

192. Sánchez-Juárez A, Tiburcio-Silver A, and Ortiz A, Fabrication of SnS_2/SnS heterojunction thin film diodes by plasma-enhanced chemical vapor deposition. *Thin Solid Films* **480–481** (2005) 452–456.

193. Ramakrishna Reddy KT, Koteswara Reddy N, and Miles RW, Photovoltaic properties of SnS based solar cells. *Solar Energy Materials and Solar Cells, 14th International Photovoltaic Science and Engineering Conference* **90** (2006) 3041–3046.

194. Gunasekaran M and Ichimura M, Photovoltaic cells based on pulsed electrochemically deposited SnS and photochemically deposited CdS and $Cd_{1-x}Zn_xS$. *Solar Energy Materials and Solar Cells* **91** (2007) 774–778.

195. Avellaneda D, Delgado G, Nair MTS et al., Structural and chemical transformations in SnS thin films used in chemically deposited photovoltaic cells. *Thin Solid Films* **515** (2007) 5771–5776.

196. Ichimura M and Takagi H, Electrodeposited ZnO/SnS heterostructures for solar cell application. *Japanese Journal of Applied Physics* **47** (2008) 7845–7847.

197. Li W, Wei-Ming S, Juan Q et al., The investigations of heterojunction solar cells based on the p-SnS thin films, in *Technical Digest of the 18th Photovoltaic Science and Engineering Conference*, Kolkata, India, 2009.

198. Fenske F, Kliefoth K, Elstner L et al., ZnO/c-Si heterojunction interface tuning by interlayers, in *Symposium on Thin Films for Photovoltaic and Related Device Applications, at the 1996 MRS Spring Meeting*, San Francisco, CA, 1996, pp. 135–140.

199. Bhattacharya RN, Contreras MA, Egaas B et al., High efficiency thin-film $CuIn_{1-x}Ga_xSe_2$ photovoltaic cells using a $Cd_{1-x}Zn_xS$ buffer layer. *Applied Physics Letters* **89** (2006) 253503.

200. Naghavi N, Spiering S, Powalla M et al., High-efficiency copper indium gallium diselenide (CIGS) solar cells with indium sulfide buffer layers deposited by atomic layer chemical vapor deposition (ALCVD). *Progress in Photovoltaics: Research and Applications* **11** (2003) 437–443.

201. Hariskos D, Ruckh M, Rühle U et al., A novel cadmium free buffer layer for $Cu(In,Ga)Se_2$ based solar cells. *Solar Energy Materials and Solar Cells* **41–42** (1996) 345–353.

202. Pistor P, Caballero R, Hariskos D et al., Quality and stability of compound indium sulphide as source material for buffer layers in $Cu(In,Ga)Se_2$ solar cells. *Solar Energy Materials and Solar Cells* **93** (2009) 148–152.

203. Allsop NA, Schönmann A, Muffler HJ et al., Spray-ILGAR indium sulfide buffers for $Cu(In,Ga)(S,Se)_2$ solar cells. *Progress in Photovoltaics: Research and Applications* **13** (2005) 607–616.

204. Hariskos D, Spiering S, and Powalla M, Buffer layers in $Cu(In,Ga)Se_2$ solar cells and modules. *Thin Solid Films* **480–481** (2005) 99–109.

205. Buecheler S, Corica D, Guettler D et al., Ultrasonically sprayed indium sulfide buffer layers for $Cu(In,Ga)$ $(S,Se)_2$ thin-film solar cells. *Thin Solid Films* **517** (2009) 2312–2315.

206. Spiering S, Bürkert L, Hariskos D et al., MOCVD indium sulphide for application as a buffer layer in CIGS solar cells. *Thin Solid Films* **517** (2009) 2328–2331.

207. Strohm A, Schlotzer T, Nguyen Q et al., New approaches for the fabrication of Cd-free $Cu(In,Ga)Se_2$ heterojunctions, in *Proceedings of the 19th European Photovoltaic Solar Energy Conference*, Paris, France, 2004, pp. 1741–1744.

208. Tokita Y, Chaisitsak S, Yamada A et al., High-efficiency $Cu(In,Ga)Se_2$ thin-film solar cells with a novel $In(OH)_3$:Zn^{2+} buffer layer. *Solar Energy Materials and Solar Cells* **75** (2003) 9–15.

209. Konagai M, Ohtake Y, and Okamoto T, Development of $Cu(InGa)Se_2$ thin film solar cells with Cd-free buffer layers, in *Symposium on Thin Films for Photovoltaic and Related Device Applications, at the 1996 MRS Spring Meeting*, San Francisco, CA, 1996, pp. 153–163.

210. Contreras MA, Nakada T, Hongo M et al., ZnO/ZnS(O,OH)/$Cu(In,Ga)Se_2$/Mo solar cell with 18.6% efficiency, in *Proceedings of the Third World Conference on Photovoltaic Energy Conversion*, Osaka, Japan, 2003, 2LN-C-08.

211. Platzer-Bjorkman C, Kessler J, and Stolt L, Atomic layer deposition of Zn(O,S) buffer layers for high efficiency $Cu(In,Ga)Se_2$ solar cells, in *Proceedings of the Third World Conference on Photovoltaic Energy Conversion*, Osaka, Japan, 2003, pp. 461–464.

212. Muffler H, Bar M, Fischer C-H et al., Sulfidic buffer layers for Cu(InGa)(S,Se)$_2$ solar cells prepared by ion layer gas reaction (ILGAR), in *Proceedings of the 28th IEEE Photovoltaic Specialists Conference*, Anchorage, AK, 2000, pp. 610–613.

213. Romeo A, Abou-Ras D, Gysel R et al., Properties of CIGS solar cells developed with evaporated II–VI buffer layers, in *Technical Digest of the 14th International Photovoltaic Science and Engineering Conference*, Bangkok, Thailand, 2004, pp. 705–706.

214. Ennaoui A, Siebentritt S, Lux-Steiner MC et al., High-efficiency Cd-free CIGSS thin-film solar cells with solution grown zinc compound buffer layers. *Solar Energy Materials and Solar Cells* **67** (2001) 31–40.

215. Munzel M, Deibel C, Dyakonov V et al., Electrical characterization of defects in Cu(In,Ga)Se$_2$ solar cells containing a ZnSe or a CdS buffer layer. *Thin Solid Films* **387** (2001) 231–234.

216. Ohtake Y, Kushiya K, Ichikawa M et al., Polycrystalline Cu(InGa)Se$_2$ thin-film solar cells with ZnSe buffer layers. *Japanese Journal of Applied Physics* **34** (1995) 5949–5955.

217. Bar M, Fischer CH, Muffler HJ et al., High efficiency chalcopyrite solar cells with ILGAR-ZnO WEL-device characteristics subject to the WEL composition, in *Proceedings of the 29th IEEE Photovoltaic Specialists Conference*, New Orleans, LA, 2002, pp. 636–639.

218. Mikami R, Miyazaki H, Abe T et al., Chemical bath deposited (CBD)-ZnO buffer layer for CIGS solar cells, in *Proceedings of the Third World Conference on Photovoltaic Energy Conversion*, Osaka, Japan, 2003, pp. 519–522.

219. Chaisitsak S, Yamada A, Konagai M et al., Improvement in performances of ZnO:B/i-ZnO/Cu(InGa) Se$_2$ solar cells by surface treatments for Cu(InGa)Se$_2$. *Japanese Journal of Applied Physics* **39** (2000) 1660–1664.

220. Olsen LC, Lei W, and Addis FW, High efficiency CIGS and CIS cells with CVD ZnO buffer layers, in *Proceedings of the 26th IEEE Photovoltaic Specialists Conference*, Anaheim, CA, 1997, pp. 363–366.

221. Gal D, Hodes G, Lincot D et al., Electrochemical deposition of zinc oxide films from non-aqueous solution: a new buffer/window process for thin film solar cells. *Thin Solid Films* **361–362** (2000) 79–83.

222. Hultqvist A, Platzer-Bjorkman C, Torndahl T et al., Optimization of i-ZnO window layers for Cu(In,Ga) Se$_2$ solar cells with ALD buffers, in *Proceedings of the 22nd European Photovoltaic Solar Energy Conference*, Milan, Italy, 2007, pp. 2381–2384.

223. Negami T, Aoyagi T, Satoh T et al., Cd free CIGS solar cells fabricated by dry processes, in *Proceedings of the 29th IEEE Photovoltaic Specialists Conference*, New Orleans, LA, 2002, pp. 656–659.

224. Chaisitsak S, Tokita Y, Miyazaki H et al., in *Proceedings of the 17th European Photovoltaic Solar Energy Conference*, Munich, Germany, 2001, pp. 1011–1015.

225. Hariskos D, Heberholtz R, Ruckh M et al., in *Proceedings of the 13th European Photovoltaic Solar Energy Conference*, Nice, France, 1995, pp. 1995–1998.

226. Powalla M, Dimmler B, Schaffler R et al., in *Proceedings of the 19th European Photovoltaic Solar Energy Conference*, Paris, France, 2004.

227. Spiering S, Hariskos D, Powalla M et al., CD-free Cu(In,Ga)Se$_2$ thin-film solar modules with In$_2$S$_3$ buffer layer by ALCVD. *Thin Solid Films* **431–432** (2003) 359–363.

228. Spiering S, Eicke A, Hariskos D et al., Large-area Cd-free CIGS solar modules with In$_2$S$_3$ buffer layer deposited by ALCVD, in *Thin Solid Films Proceedings of Symposium D on Thin Film and Nano-Structured Materials for Photovoltaics, of the E-MRS 2003 Spring Conference*, Strasbourg, France, vol. 451–452, 2004, pp. 562–566.

229. Dimmler B, Gross E, Hariskos D et al., *Proceedings of the Second World Conference on Photovoltaic Solar Energy Conversion*, Vienna, Austria, 1998, pp. 419–422.

230. Nair PK, Nair MTS, García VM et al., Semiconductor thin films by chemical bath deposition for solar energy related applications. *Solar Energy Materials and Solar Cells* **52** (1998) 313–344.

231. Savadogo O, Chemically and electrochemically deposited thin films for solar energy materials. *Solar Energy Materials and Solar Cells* **52** (1998) 361–388.

232. Cashman RJ, New photoconductive cells. *Journal of the Optical Society of America* **36** (1946) 356.

233. Kainthla RC, Pandya DK, and Chopra KL, Solution growth of CdSe and PbSe films. *Journal of The Electrochemical Society* **127** (1980) 277–283.

234. Kitaev G, Uritskaya A, and Mokrushin S, *Russian Journal of Physical Chemistry* **39** (1965) 1101.

235. Kylner A, Rockett A, and Stolt L, Oxygen in solution grown CdS films for thin film solar cells. *Solid State Phenomena* **51–52** (1996) 533–539.

236. Hashimoto Y, Kohara N, Negami T et al., Chemical bath deposition of Cds buffer layer for GIGS solar cells. *Solar Energy Materials and Solar Cells* **50** (1998) 71–77.

237. Kylner A, Lindgren J, and Stolt L, Impurities in chemical bath deposited CdS films for Cu(In,Ga)Se$_2$ solar cells and their stability. *Journal of the Electrochemical Society* **143** (1996) 2662–2669.

238. Kilner A and Nieme E, Chemical bath deposited CdS films with different impurity concentrations—film characterisation and Cu(In,Ga)Se$_2$ solar cell results, in *Proceedings of the 14th European Photovoltaic Solar Energy Conference*, Barcelona, Spain, 1997, pp. 1326–1329.

239. Kim SY, Kim DS, Ahn BT et al., Electrical and optical properties of vacuum-evaporated CdS films. *Journal of Materials Science: Materials in Electronics* **4** (1993) 178–182.

240. Ray S, Banerjee R, and Barua AK, Properties of vacuum-evaporated CdS thin-films. *Japanese Journal of Applied Physics* **19** (1980) 1889–1895.

241. Hariskos D, Menner R, Spiering S et al., In$_2$S$_3$ buffer layer deposited by magnetron sputtering for Cu(InGa)Se$_2$ solar cells, in *Proceedings of the 19th European Photovoltaic Solar Energy Conference*, Paris, France, 2004, p. 1894.

242. Abou-Ras D, Kostorz G, Hariskos D et al., Structural and chemical analyses of sputtered In$_x$S$_y$ buffer layers in Cu(In,Ga)Se$_2$ thin-film solar cells. *Thin Solid Films* **517** (2009) 2792–2798.

243. Raja Mohan Reddy L, Prathap P, and Ramakrishna Reddy KT, Influence of substrate temperature on physical properties of sprayed Zn$_{0.85}$Mn$_{0.15}$O films. *Current Applied Physics* **9** (2009) 667–672.

244. Bouzouita H, Bouguila N, and Dhouib A, Spray pyrolysis of CuInS$_2$. *Renewable Energy* **17** (1999) 85–93.

245. Mooney JB and Lamoreaux RH, Spray pyrolysis of CuInSe$_2$. *Solar Cells* **16** (1986) 211–220.

246. Ramakrishna Reddy KT, Subbaiah YV, Reddy TBS et al., Pyrolytic spray deposition of ZnS$_x$Se$_{1-x}$ layers for photovoltaic applications. *Thin Solid Films* **431–432** (2003) 340–343.

247. Hernández-Fenollosa MA, López MC, Donderis V et al., Role of precursors on morphology and optical properties of ZnS thin films prepared by chemical spray pyrolysis. *Thin Solid Films* **516** (2008) 1622–1625.

248. Prathap P, Devi GG, Subbaiah YPV et al., Preparation and characterization of sprayed In$_2$O$_3$:Mo films. *Physica Status Solidi (a)* **205** (2008) 1947–1951.

249. Pauporte T and Lincot D, Heteroepitaxial electrodeposition of zinc oxide films on gallium nitride. *Applied Physics Letters* **75** (1999) 3817–3819.

250. Peulon S and Lincot D, Mechanistic study of cathodic electrodeposition of zinc oxide and zinc hydroxychloride films from oxygenated aqueous zinc chloride solutions. *Journal of the Electrochemical Society* **145** (1998) 864–874.

251. Hodes G, *Chemical Solution Deposition of Semiconductor Films*. Marcel Dekker, Inc., New York, 2001.

252. Lincot D, Canava B, and Quenet S, in *Proceedings of the 14th European Photovoltaic Solar Energy Conference*, Barcelona, Spain, 1997.

253. Törndahl T, Platzer-Björkman C, Kessler J et al., Atomic layer deposition of Zn$_{1-x}$Mg$_x$O buffer layers for Cu(In,Ga)Se$_2$ solar cells. *Progress in Photovoltaics: Research and Applications* **15** (2007) 225–235.

254. Kim YS and Yun SJ, Studies on polycrystalline ZnS thin films grown by atomic layer deposition for electroluminescent applications. *Applied Surface Science* **229** (2004) 105–111.

255. Kim DC, Kong BH, Jun SO et al., Pressure dependence and micro-hillock formation of ZnO thin films grown at low temperature by MOCVD. *Thin Solid Films* **516** (2008) 5562–5566.

256. Bär M, Fischer CH, Muffler HJ et al., Replacement of the CBD-CdS buffer and the sputtered i-ZnO layer by an ILGAR-ZnO WEL: Optimization of the WEL deposition. *Solar Energy Materials and Solar Cells* **75** (2003) 101–107.

257. Muffler HJ, Fischer CH, Diesner K et al., ILGAR—A novel thin-film technology for sulfides. *Solar Energy Materials and Solar Cells* **67** (2001) 121–127.

258. Reddy KTR and Reddy PJ, Studies of Zn$_x$Cd$_{1-x}$S films and Zn$_x$Cd$_{1-x}$S/CuGaSe$_2$ heterojunction solar cells. *Journal of Physics D: Applied Physics* (1992) 1345.

259. Ramakrishna Reddy KT, Gopalaswamy H, and Jayarama Reddy P, Polycrystalline CuGaSe$_2$ thin film solar cells. *Vacuum* **43** (1992) 811–815.

260. Song J, Li SS, Chen L et al., Investigations of CdZnS buffer layers on the performance of CuInGaSe$_2$ and CuGaSe$_2$ solar cells, in *Proceedings of the Fourth World Conference on Photovoltaic Energy Conversion*, Waikoloa, HI, 2006, pp. 534–537.

261. Hynes KM and Newham J, An investigation of the environmental implications of the chemical bath deposition of CdS through environmental risk assessment, in *Proceedings of the 16th European Photovoltaic Solar Energy Conference*, Glasgow, U.K., 2000, p. 2297.

262. Hartmut S, Health, safety and environmental risks from the operation of CdTe and CIS thin-film modules. *Progress in Photovoltaics: Research and Applications* **6** (1998) 99–103.

263. Yousfi EB, Weinberger B, Donsanti F et al., Atomic layer deposition of zinc oxide and indium sulfide layers for Cu(In,Ga)Se$_2$ thin-film solar cells. *Thin Solid Films* **387** (2001) 29–32.

264. Spiering S, Hariskos D, Schröder S et al., Stability behaviour of Cd-free Cu(In,Ga)Se$_2$ solar modules with In$_2$S$_3$ buffer layer prepared by atomic layer deposition. *Thin Solid Films* **480–481** (2005) 195–198.

265. Velthaus KO, Kessler J, Ruckh M et al., Novel buffer layers for the CuInSe$_2$/buffer/ZnO devices, in *Proceedings of the 11th European Photovoltaic Solar Energy Conference*, Montreux, Switzerland, 1992, p. 842.

266. Huang CH, Li SS, Shafarman WN et al., Study of Cd-free buffer layers using In$_x$(OH,S)$_y$ on CIGS solar cells. *Solar Energy Materials and Solar Cells* **69** (2001) 131–137.

267. Bayón R and Herrero J, Reaction mechanism and kinetics for the chemical bath deposition of In(OH)$_x$S$_y$ thin films. *Thin Solid Films* **387** (2001) 111–114.

268. Karg F, Aulich HA, and Riedl W, CIS-module development within the FORSOL program: Structure and first results, in *Proceedings of the 14th European Photovoltaic Specialists Conference*, Barcelona, Spain, 1997, p. 2012.

269. Gall S, Barreau N, Harel S et al., Material analysis of PVD-grown indium sulphide buffer layers for Cu(In,Ga)Se$_2$-based solar cells. *Thin Solid Films* **480–481** (2005) 138–141.

270. Barreau N, Kessler F, Neghavi N et al., Investigation of CuInS$_2$/In$_2$S$_3$ interface and related cells, in *Proceedings of the 22nd European Photovoltaic Solar Energy Conference*, Milan, Italy, 2007, p. 1915.

271. Kessler J, Ruckh M, Hariskos D et al., Interface engineering between CuInSe$_2$ and ZnO, in *Proceedings of the 23rd IEEE Photovoltaic Specialists Conference*, Louisville, KY, 1993, pp. 447–452.

272. Ortega Borges R, Lincot D, and Vedel J, Chemical bath deposition of zinc sulphide thin films, in *Proceedings of the 11th European Photovoltaic Solar Energy Conference*, Montreux, Switzerland, 1992, p. 862.

273. Kushiya K, Nii T, Sugiyama I et al., Application of Zn-compound buffer layer for polycrystalline CuInSe$_2$-based thin-film solar cells. *Japanese Journal of Applied Physics* **35** (1996) 4383.

274. Naghavi N, On a better understanding of post-treatment on Ci(G)S/Zn(S,O,OH)/ZnMgO based solar cells, in *Proceedings of the 23rd European Photovoltaic Solar Energy Conference*, Valencia, Spain, 2008, p. 2160.

275. Saez-Araoz R, Ennaoui A, Niesen TP et al., Scaling up of efficient Cd-free thin film Cu(In,Ga)(S,Se)$_2$ and CuInS$_2$ PV-devices, in *Proceedings of the 22nd European Photovoltaic Solar Energy Conference*, Milan, Italy, 2007, p. 2360.

276. Chaparro AM, Gutiérrez MT, Herrero J et al., Characterisation of CuInS$_2$/Zn(Se,O)/ZnO solar cells as a function of Zn(Se,O) buffer deposition kinetics in a chemical bath. *Progress in Photovoltaics: Research and Applications* **10** (2002) 465–480.

277. Subbaiah YPV, Prathap P, Reddy KTR et al., Studies on ZnS$_{0.5}$Se$_{0.5}$ buffer based thin film solar cells. *Thin Solid Films* **516** (2008) 7060–7064.

278. Armstrong S, Datta PK, and Miles RW, Properties of zinc sulfur selenide deposited using a close-spaced sublimation method. *Thin Solid Films Proceedings of Symposium P on Thin Film Materials for Photovoltaics* **403–404** (2002) 126–129.

279. Ennaoui A, Weber M, Scheer R et al., Chemical-bath ZnO buffer layer for CuInS$_2$ thin-film solar cells. *Solar Energy Materials and Solar Cells* **54** (1998) 277–286.

280. Boxman R, Review of TCO films, in *Proceedings of Thin Films 2008*, Singapore, 2008.

281. Morales-Acevedo A, Thin film CdS/CdTe solar cells: Research perspectives. *Solar Energy* **80** (2006) 675–681.

282. Minami T, Transparent conducting oxide semiconductors for transparent electrodes. *Semiconductor Science and Technology* **20** (2005) S35.

283. Minami T, Sato H, Nanto H et al., Highly conductive and transparent silicon doped zinc oxide thin films prepared by RF magnetron sputtering. *Japanese Journal of Applied Physics* **25** (1986) 776–779.

284. Bae JW, Lee SW, and Yeom GY, Doped-fluorine on electrical and optical properties of tin oxide films grown by ozone-assisted thermal CVD. *Journal of the Electrochemical Society* **154** (2007) D34–D37.

285. Jin M, Feng J, De-heng Z et al., Optical and electronic properties of transparent conducting ZnO and ZnO:Al films prepared by evaporating method. *Thin Solid Films* **357** (1999) 98–101.

286. Lamb DA and Irvine SJC, Growth properties of thin film ZnO deposited by MOCVD with n-butyl alcohol as the oxygen precursor. *Journal of Crystal Growth* **273** (2004) 111–117.

287. Natsume Y and Sakata H, Zinc oxide films prepared by sol-gel spin-coating. *Thin Solid Films* **372** (2000) 30–36.

288. Lin S-S, Huang J-L, and Sajgalik P, Effects of substrate temperature on the properties of heavily Al-doped ZnO films by simultaneous r.f. and d.c. magnetron sputtering. *Surface and Coatings Technology* **190** (2005) 39–47.

289. Khandelwal R, Singh AP, Kapoor A et al., Effects of deposition temperature on the structural and morphological properties of thin ZnO films fabricated by pulsed laser deposition. *Optics & Laser Technology* **40** (2008) 247–251.

290. Li X-Y, Li H-J, Wang Z-J et al., Effect of substrate temperature on the structural and optical properties of ZnO and Al-doped ZnO thin films prepared by dc magnetron sputtering. *Optics Communications* **282** (2009) 247–252.

291. Agura H, Suzuki A, Matsushita T et al., Low resistivity transparent conducting Al-doped ZnO films prepared by pulsed laser deposition. *Thin Solid Films* **445** (2003) 263–267.

292. Gupta RK, Ghosh K, Patel R et al., Preparation and characterization of highly conducting and transparent Al doped CdO thin films by pulsed laser deposition. *Current Applied Physics* **9** (2009) 673–677.

293. Sravani C, Reddy KTR, and Jayarama Reddy P, Preparation and properties of CdO/CdTe thin film solar cells. *Journal of Alloys and Compounds* **215** (1994) 239–243.

294. Chu T and Chu S, Degenerate cadmium oxide films for electronic devices. *Journal of Electronic Materials* **19** (1990) 1003–1005.

295. Kawazoe H, Yasukawa M, Hyodo H et al., P-type electrical conduction in transparent thin films of CuAlO$_2$. *Nature* **389** (1997) 939–942.

296. Banerjee AN and Chattopadhyay KK, Recent developments in the emerging field of crystalline p-type transparent conducting oxide thin films. *Progress in Crystal Growth and Characterization of Materials* **50** (2005) 52–105.

297. Yamamoto T and Yoshida HK, Solution using a codoping method to unipolarity for the fabrication of p-type ZnO. *Japanese Journal of Applied Physics* **38** (1999) 166–169.

298. Joseph M, Tabata H, and Kawai T, p-Type electrical conduction in ZnO thin films by Ga and N codoping. *Japanese Journal of Applied Physics* **38** (1999) 1205–1207.

299. David T, Goldsmith S, and Boxman R, P-type Sb-doped ZnO thin films prepared with filtered vacuum arc deposition, in *Proceedings of the 47th Annual Technical Conference*, Society of Vacuum Coaters, Albuquerque, NM, 2008, pp. 27–31.

2 Anodized Titania Nanotube Array and Its Application in Dye-Sensitized Solar Cells

Lidong Sun, Sam Zhang, Xiao Wei Sun, and Xiaodong He

CONTENTS

2.1 ANODIZED TITANIA NANOTUBE ARRAY

Titanium dioxide (TiO_2) has been widely employed in dye-sensitized solar cells (DSSCs) [1,2], biosensors [3], hydrogen sensors [4], and water photolysis [5] due to its various intriguing properties. Among diverse one-dimensional nanostructures, highly oriented TiO_2 nanotube arrays prepared by electrochemical anodization have been vigorously investigated since it was first reported in 1999 [6]. The highly ordered structure obtained is much more favored comparing to the randomly packed configuration presented by the sol-gel method [7], e.g., for application in DSSCs. And the nanotube geometry can be tailored by changing experimental parameters [8–13], superior to template-assisted synthesis where the geometry is limited and controlled by the original template [14].

In 1999, Zwilling et al. [6] reported the formation of porous film with regular columnar structure in fluorinated electrolyte during the electrochemical anodization of pure titanium and TA6V (Ti-6% Al-4%V) alloy. In 2001, Gong et al. [15] reported titanium dioxide nanotube arrays by the electrochemical anodization of pure titanium foil in HF solution. Thereafter, this kind of TiO$_2$ nanotube array was successfully prepared in various fluoride-containing electrolytes [16–21] and also fluoride-free electrolytes [22–24]. Geometries of the nanotube array could be modulated by adopting different anodic parameters.

In addition, other metals (so-called valve metals) with highly ordered nanoporous or nanotube oxide structures, such as aluminum [25–28], hafnium [29], niobium [30], tantalum [31–33], tungsten [34,35], zirconium [36–38], and also silicon [39,40], InP [41,42], and titanium alloys [43–46], were produced by electrochemical anodization.

In this section, the concept of electrochemical anodization related to the formation of compact oxide layer on aluminum substrate is introduced. To better understand the mechanism of growth of TiO$_2$ nanotubes, porous alumina films formed by electrochemical anodization is first presented to establish a foundation because of its process similarity with TiO$_2$ nanotubes and its well-established growth mechanism. Anodizing parameters that influence the geometry of the anodized titania nanotube arrays are discussed in the section to follow.

2.1.1　Electrochemical Anodization

Electrochemical anodization is an electrochemical process to form passivated oxide layer on metal surface originally used for the protection of metal components against corrosion, especially for aluminum and aluminum alloys. In the case of pure aluminum, a native dense amorphous aluminum oxide with a thickness of 2–3 nm [47] is formed at room temperature on the surface once it is exposed to air, or any environment containing oxygen. This layer prevents aluminum from further oxidation or corrosion. But for aluminum alloy, the corrosion resistance is greatly decreased because of the alloying elements or impurities, such as copper, iron, and silicon [47]. Therefore, electrochemical anodization is employed to increase the thickness of passivated oxide layer for better protection.

Figure 2.1 is a schematic of the formation of an anodic alumina film on an aluminum substrate, where a piece of aluminum foil is employed as the working electrode (anodic electrode) and a piece of

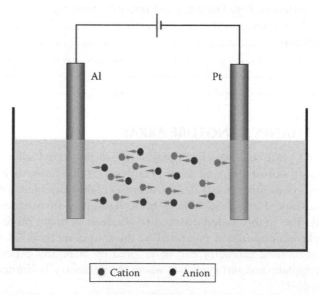

FIGURE 2.1　Schematic of formation of compact alumina films on aluminum substrate by electrochemical anodization; only ions between the electrodes are illustrated.

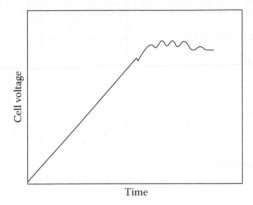

FIGURE 2.2 Typical voltage–time behavior during anodizing of aluminum in neutral solution by galvanostatic method. (Reprinted from Thompson, G.E., *Thin Solid Films*, 297, 194, 1997. With permission.)

Pt foil is used as the counter electrode (cathodic electrode). To form compact barrier-type (compared to porous-type) films, near-neutral aqueous electrolytes are generally employed [48]. The anodization process can be performed by galvanostatic method (control the current passing the working electrode while measuring the potential applied to it) or potentiostatic method (control the potential applied to the working electrode while measuring the current passed). In both cases, the anions in the electrolyte migrate to the working electrode (anodic electrode) and the cations to the counter electrode (cathodic electrode) under electric field. The main anodic reaction can be presented as [49]

$$2Al + 3H_2O \rightarrow Al_2O_3 + 6H^+ + 6e \tag{2.1}$$

At the same time, hydrogen gas emits from the Pt counter electrode because

$$2H^+ + 2e \rightarrow H_2 \tag{2.2}$$

The driving force for growth of the Al_2O_3 film is the electric field strength (the potential drop across the alumina film to the film thickness [48]) which drives the ions (metal ions from the film and anions in the electrolyte) to pass through it [50]. Figure 2.2 shows a typical voltage–time curve for the anodizing of aluminum. The initial voltage surge reveals the presence of air-formed oxide film on aluminum surface. The curve displays a linear increase during the primary anodizing process, indicating uniform thickening of compact alumina film at this stage. After that, vibration of the curve arises, and then the anodization process terminates. This is due to dielectric breakdown which results in the termination of film thickening, with frequently observed sparking over the anodic film surface [48].

The above discussion is about the whole growth process of anodic compact alumina films (also called barrier-type films).

To understand the formation of porous-type anodic films, the understanding of the growth mechanism via ion transport in the compact anodic films is needed, which is presented below.

The ionic transport processes in the anodic films were investigated by inert marker (e.g., Xe) that could define the location of film growth [48]. According to the location of the marker in anodic alumina films, three types of film growth are identified, as shown in Figure 2.3a through c. Under high current efficiency (the ratio of the substance mass liberated by a current to the theoretical mass predicted by Faraday's law), the marker locating at metal–film interface (Figure 2.3a) reveals ion transport by cation egress, and the marker locating at film–electrolyte interface (Figure 2.3b) indicates ion transport by anion ingress. The combined transport of cation egress and anion ingress results in the marker locating between the two interfaces (Figure 2.3c). The investigation results

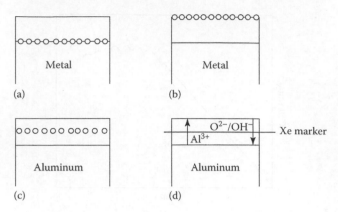

FIGURE 2.3 Schematic diagram of sections of films, revealing locations of marker for film growth by (a) cation transport; (b) anion transport; (c) cation and anion transport; and (d) mobile ions for aluminum anodic oxidation. (Reprinted from Thompson, G.E., *Thin Solid Films*, 297, 194, 1997. With permission.)

[51,52] show that the ion transport in anodic alumina films during the anodizing of aluminum is via Al^{3+} egress and O^{2-}/OH^- ingress, as illustrated in Figure 2.3d, and the transport number is about 0.4 for Al^{3+}. The distribution of the marker in the anodic film is uniform and limited across the film thickness, suggesting that the anodic film growth takes place simultaneously only at metal–film and film–electrolyte interfaces by O^{2-}/OH^- ingress and Al^{3+} egress under electric field [48]. In other words, there is no film growth within the film section; otherwise the marker will be distributed nonuniformly.

As discussed above, the driving force for anodic film growth is the high electric field under which Al^{3+} and O^{2-}/OH^- migrate through the film. On the other hand, other electrolyte species can also be incorporated into the films under this electric filed, thus influencing the film compositions and structures. Figure 2.4 is the distributions of B, P, and W species in the film section when borate, phosphate, and tungstate electrolytes are employed, respectively, during electrochemical anodization. It reveals that B species spread over 0.4 of the film thickness from the film–electrolyte interface (Figure 2.4a), P species spread over two-thirds of the film thickness (Figure 2.4b), and W species locates just at the outermost film region (Figure 2.4c). Considering the marker locations in Figure 2.4d through f and ionic transport processes in Figure 2.3, it indicates that B species are present above the marker due to Al^{3+} egress, P species are present both above and below the marker due to O^{2-}/OH^- ingress, and W species are present above the marker and located at the outermost film region owing to Al^{3+} egress [48]. The distributions of B, P, and W species in the film sections can also be identified from Figure 2.4g through i. This is because the film regions with free electrolyte species crystallize more rapidly than those regions doped with electrolyte species under electron irradiation [48].

The distributions of different electrolyte species in the film sections suggest that B species are immobile in anodic film, P species are mobile inward, and W species are mobile outward under the electric field [48,53]. In other words, the B elements exist in neutral species, the P elements exist in negatively charged ions, and the W elements present in positively charged ions. Possible reactions between Al^+ ions and electrolyte anions adsorbed on the film surfaces are as follows [48]:

$$2Al^{3+} + HB_4O_7^- + 2H_2O \rightarrow Al_2O_3 + 2B_2O_3 + 5H^+ \left(\text{borate solution}\right)$$

$$Al^{3+} + PO_4^{3-} \rightarrow AlPO_4 \left(\text{phosphate solution}\right)$$

$$2Al^{3+} + WO_4^{2-} + 2H_2O \rightarrow Al_2O_3 + WO_3 + 4H^+ \left(\text{tungstate solution}\right)$$

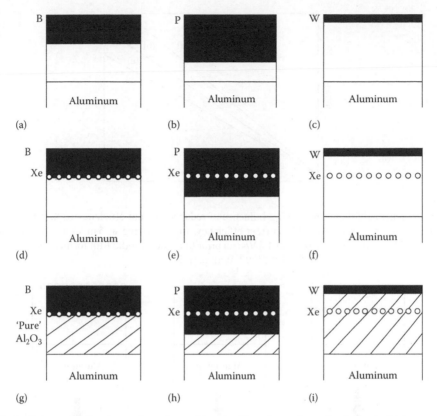

FIGURE 2.4 Schematic diagrams of sections of barrier films formed on aluminum in (a) borate; (b) phosphate; and (c) tungstate electrolytes. The location of xenon markers are displayed in (d)–(e), and the consequence of differential electron-beam-induced crystallization (hatched region) in (g)–(i). (Reprinted from Thompson, G.E., *Thin Solid Films*, 297, 195, 1997. With permission.)

In fact, the main anodic reactions are based on Equation 2.1, since the surface covered by adsorbed ions is relatively small compared to that covered by water molecules.

Consequently, the electrolyte species are incorporated into the anodic films by means of Al^{3+} egress. It is suggested that chemical bonds are polarized and weakened under electric field. If the field is sufficient to break the bonds, for example, Al–O bonds of Al_2O_3 and W–O bonds of WO_3, the resultant ions will migrate oppositely under the electric field. If the field is insufficient to break the bonds, for example, B–O bonds of B_2O_3, the incorporated species will be immobile, though polarized, under the field. As for the PO_4^{3-} species, the P–O bonds are also unaffected by the field; however, these species are mobile inward as a result of the negative charge. In view of these, the distributions are understandable of different species in anodic film sections in Figure 2.4.

The voltage–time behavior during anodizing of aluminum is shown in Figure 2.5. At high current efficiency, i.e., 100%, the slope of the curve is 2.3 V/s (curve (a) in Figure 2.5) for anodizing at $5 \, mA/cm^2$ [48]. As the current density decreases, the slope reduces proportionally from curve (a) to curve (c), which indicates a critical value prior to the formation of anodic porous-type film. With reduction of the slope, more charge amount is required to achieve a selected voltage or film thickness, suggesting decreased current efficiency from 100% [48]. Similar behavior is also present when reducing the electrolyte pH from near-neutral levels.

Figure 2.6a is the marker location when anodizing aluminum at a constant current density in a typical near-neutral electrolyte. Figure 2.6b and c display the respective marker locations in electrolytes with gradually reduced pH. It is apparent that the marker location moves outwardly with reduced pH value of the electrolyte toward the film–electrolyte interface, revealing progressively

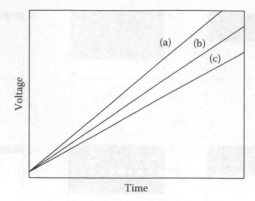

FIGURE 2.5 Schematic diagram showing influence of reduced current efficiency of film formation (a)–(c) on the voltage–time response. At 100% current efficiency, for anodizing at $5\,mA/cm^2$, the slop of the V–t curve of (a) is 2.3 V/s; this reduces to about 1.4 V/s; (c) prior to porous anodic film formation. (Reprinted from Thompson, G.E., *Thin Solid Films*, 297, 196, 1997. With permission.)

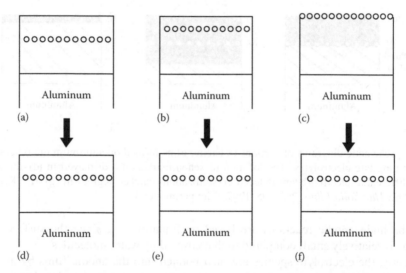

FIGURE 2.6 Schematic diagram of marker locations during anodizing of aluminum at constant current density in a typical near-neutral solution (a), and solution of progressively reduced pH (b) and (c). And marker positions after transforming respective aluminum ion concentration in electrolyte into an effective film thickness (d)–(f). (Reprinted from Thompson, G.E., *Thin Solid Films*, 297, 196, 1997. With permission.)

decreased transport number of Al^{3+} ions. In particular, a critical current density or pH value is achieved when the marker is just located at film–electrolyte interface, suggesting zero cation transport number. Parallel studies [48] of the electrolyte composition demonstrate the absence of aluminum in the solution for the case of Figure 2.6a, and gradually increased aluminum concentration in the solution for Figure 2.6b and c. Converting the corresponding aluminum in the solution into effective film thicknesses, the respective marker locations are shown in Figure 2.6d through f, implying the transport numbers of Al^{3+} ions for all the three cases are comparable of ~0.4, which is identical with the value discussed before.

The above results indicate that the growth of barrier-type anodic film takes place only at metal–film and film–electrolyte interfaces by O^{2-}/OH^- ingress and Al^{3+} ingress under the electric field. However, formation of anodic films at film–electrolyte interface depends on the interfacial conditions. At high current efficiency, all the outwardly mobile Al^{3+} ions transform into alumina film at film–electrolyte interface. At critical current density or electrolyte pH value, all the outwardly mobile Al^{3+} ions are ejected into the electrolyte directly without film formation [54]. In the case of intermediate conditions, partial Al^{3+} ions form film at the interface while the others are ejected into the electrolyte.

In the anodic film growth process, the chemical dissolution of the alumina at film–electrolyte interface is relatively insignificant, and the presence of aluminum in the solution is from direct ejection [48]. This has been identified by reanodizing the films formed under the condition of Figure 2.6c [48], showing non-disappeared marker at film–electrolyte interface.

The fact that no outwardly mobile Al^{3+} ions can form films at film–electrolyte interface under critical anodizing conditions is essential for the formation of porous anodic alumina films. This is because the presence of preferential penetration paths on the film surface can no longer be restored under the critical conditions, resulting in pore initiation and further propagation.

For the formation of porous anodic alumina films, the voltage–time (under galvanostatic mode) and current–time (under potentiostatic mode) curves during electrochemical anodization are shown in Figure 2.7. The whole growth process of porous anodic films is normally divided into four main stages, as displayed in the curves and illustrated in the schematic diagrams below the curves. To interpret the process more clearly, only potentiostatic method is considered in the following description. As for galvanostatic method, the same growth mechanism governs but monitored and explained using voltage–time rather than current–time curve.

As discussed before, there is no film formation at film–electrolyte interface induced by outwardly mobile Al^{3+} ions under the critical conditions. In the initial anodizing stage (Stage I), compact anodic alumina films form on aluminum by Al^{3+} egress and O^{2-}/OH^- ingress, resulting in progressively decreased anodic current, as shown in Figure 2.7b. Thereafter, preferential penetration paths of ion transport appear on the film surface (Stage II), owing to chemical dissolution that is assisted by the electric field. This gives rise to relatively small slope of the decreased current (Figure 2.7b). Since no outwardly mobile Al^{3+} ions contribute to the films at film–electrolyte interface, the penetration pathways can no longer be healed, which act as the precursors of the regular pores [48]. The distribution of potential lines within the compact films keeps uniform between the penetration pathways, while concentrates immediately beneath [48], revealing increased field strengths at these regions. As a consequence, both processes of Al^{3+} egress and O^{2-}/OH^- ingress are accelerated

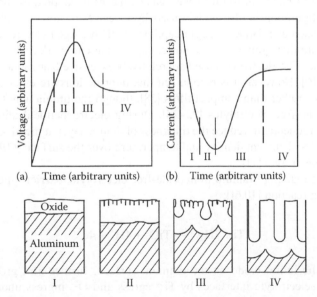

FIGURE 2.7   Schematic diagram of the kinetics of porous oxide growth on aluminum by (a) galvanostatic and (b) potentiostatic methods. The stages of porous structure development are also shown. (Reprinted from Parkhutik, V.P. and Shershulsky, V.I., *J. Phys. D: Appl. Phys.*, 25, 1258, 1992. With permission.)

beneath the penetration paths, thus causing more rapid inward movement of metal–film interface and fast ejection of aluminum ions into the electrolyte compared to that in the regions between the pathways. On the other hand, the increased local field strengths also polarize the Al–O bonds effectively, hence facilitates the chemical dissolution of the films at the interface. These result in the development of the penetration paths in both vertical and lateral directions and thus the formation of embryo pores at these sites (Stage III). At the same time, the anodic current increases because of film dissolution with the assistance of electric field (Figure 2.7b). The lateral expansions of the pathways are induced by lateral components of the field strength [48].

As the pore develops, the electric field concentrates beneath the main pores, thus facilitating both anodic film growth at metal–film interface and field-assisted dissolution at pore base–electrolyte interface in contrast to the regions between pores. This eventually gives rise to a localized scalloped-shape metal–film interface. In the end, steady-state growth of porous anodic films reaches when the scalloped interfaces merge [48].

With respect to the film growth under galvanostatic mode, the same mechanism dominates the process, with the kinetics during electrochemical anodization reflected by voltage–time behavior.

2.1.2 Titania Nanotube Growth Mechanism

In Section 2.1.1, the growth mechanism of porous alumina films by electrochemical anodization is introduced. This section presents the growth mechanism of TiO$_2$ nanotube array based on the discussion of the formation of anodic alumina films. Titanium and aluminum have many common features in electrochemical behaviors [56]. For example, both titanium [57] and aluminum [53] can form barrier-type films with similar cation and anion transport number of ~0.4 and ~0.6 [51,52,57], respectively. Foreign species from alloy substrates (cations) or electrolytes (anions) can be incorporated into the anodic films under the electric field. Moreover, porous-type anodic films can also form on both titanium and aluminum. The distinct difference is that nanotube rather than nanoporous structures are obtained in the anodizing of titanium.

A simple two-electrode configuration is illustrated in Figure 2.1, where a piece of titanium foil (commercially obtainable with common thickness of 100 or 250 μm) is employed as the working electrode and a piece of platinum foil (or sheet, gauze, mesh) is adopted as the counter electrode. Three-electrode configuration is also widely used by researchers, for example [16–22], where a reference electrode, such as a Haber–Luggin capillary with Ag/AgCl electrode, is added into the cell. The electrodes are then connected to a power supply that can work under either galvanostatic or potentiostatic mode. The electrolytes used are relatively extensive from aqueous solution [17–20] to organic solution [16]. However, it is recognized that fluorine ions are essential in the formation of nanotube structures rather than compact films [58], though the titania nanotube arrays can also be obtained in fluorine-free electrolytes [22–24]. During electrochemical anodization, magnetic stirring is employed frequently to reduce the thickness of double layer at metal–electrolyte interface [59], ensure uniform local current density and temperature over the surface [60], and also facilitate ion diffusion in the electrolyte [11].

In the process of anodizing titanium in fluoride-free electrolyte, barrier-type anodic films form through the following reaction [10,61]:

$$Ti + 2H_2O \rightarrow TiO_2 + 4H^+ + 4e \qquad (2.3)$$

After formation of the initial oxide on the titanium surface, further film growth takes place at metal–film and film–electrolyte interfaces by Ti^{4+} egress and O^{2-} ingress under the electric field [57,58], respectively, as illustrated in Figure 2.8a. This process is similar to that during anodizing aluminum. The outwardly mobile Ti^{4+} ions can also form hydroxide on the film surface, which is revealed by XPS measurement [19]. The anodic current density decreases with prolonged duration,

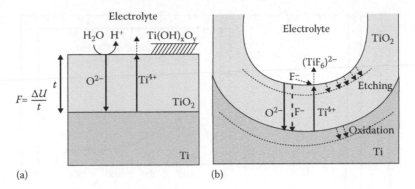

FIGURE 2.8 Schematic presentation of anodizing titanium in (a) absence of fluorides (results in flat layers), and (b) presence of fluorides (results in the tube growth). (Reprinted from Macak, J.M. et al., *Curr. Opin. Solid State Mater. Sci.*, 11, 7, 2007. With permission.)

as a result of progressively increased film thickness. The film growth terminates eventually by dielectric breakdown.

The anodic film growth process in fluoride-containing electrolyte is totally different, which gives rise to formation of titania nanotubes rather than barrier-type films (Figure 2.8b). In this scenario, the fashioned anodic films at film–electrolyte interface experience chemical dissolution as follows [10,61,62]:

$$TiO_2 + 4H^+ + 6F^- \rightarrow TiF_6^{2-} + 2H_2O \tag{2.4}$$

In addition, direct ejection (or direct complexation) of the outwardly mobile Ti^{4+} ions into the electrolyte also occurs under the electric field, which can be presented as [58,62]

$$Ti^{4+} + 6F^- \rightarrow TiF_6^{2-} \tag{2.5}$$

A typical current–time curve during titania nanotube growth (potentiostatic mode) is displayed in Figure 2.9. The trend of the curve is similar to that of the formation of porous anodic alumina films (Figure 2.7b), implying similar electrochemical anodizing behaviors. To date, the mechanism of titania nanotube growth [11,13,19,56,59,61] is not well established as the porous anodic alumina films. Possible growth mechanism is to be presented based on porous anodic alumina films subsequently. Mor et al. [59] studied the surface morphology of titania nanotube array prepared at 20 V for different anodizing durations and revealed the evolution of the nanotubes. Accordingly, a possible mechanism of nanotube growth [59] is proposed, as shown in Figure 2.10.

Upon the commencement of the anodization, the current density declines progressively (Figure 2.9), corresponding to the formation of the initial oxide layer on metal surface (Figure 2.10a) by ion transport under the electric field. As discussed in Section 2.1.1, the preferential penetration paths of ion transport can no longer be healed on alumina film surface under critical anodizing conditions, thus resulting in the initiation of embryo pores for porous anodic film formation. As for the case of titania, it is relatively simple to generate the embryo pores, since titania can be dissolved by HF (originates from the fluoride-containing electrolyte), especially under the electric field that can polarize and weaken the Ti–O bonds [59]. In view of these, small pits are produced by subsequent chemical dissolution and field-assisted dissolution on the oxide surface (Figure 2.10b), which act as the embryo pores. This can be reflected from the mildly decreased curve slope, as displayed in Figure 2.9 (the region between 0 and P1), similar to the Stage II in Figure 2.7.

The field strength beneath the embryo pores is relatively large compared to the region between the pores [48], hence giving rise to rapid ion transport at the pore base. This effectively accelerates

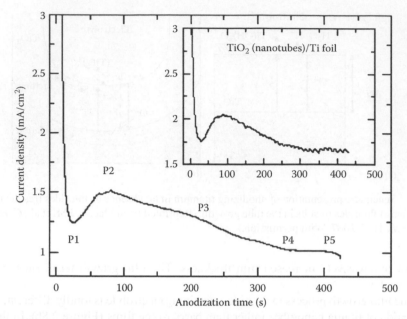

FIGURE 2.9 A typical current density–time behavior during anodizing titanium. (Reprinted from Mor, G.K. et al., *Sol. Energy Mater. Sol. Cells*, 90, 2024, 2006. With permission.)

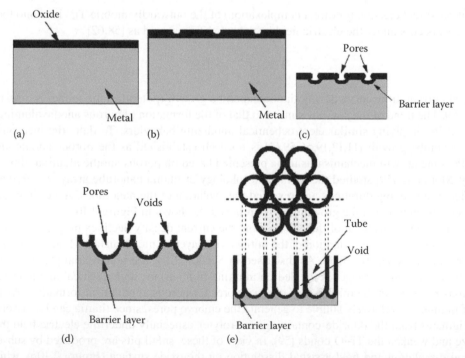

FIGURE 2.10 Schematic diagram of the evolution of straight nanotubes at a constant anodizing voltage, as follows: (a) oxide layer formation; (b) pit formation on the oxide layer; (c) growth of the pit into scallop-shape pores; (d) the metallic part between the pores undergoes oxidation and field-assisted dissolution; and (e) fully developed nanotubes with corresponding top views. (Reprinted from Mor, G.K. et al., *Sol. Energy Mater. Sol. Cells*, 90, 2028, 2006. With permission.)

the development of the embryo pores both in vertical and lateral directions, as illustrated in Figure 2.10c. In this process, the anodic current density increases (the range from P1 to P2 in Figure 2.9).

All the above processes are similar to the formation of porous anodic alumina films. However, it is obvious that the titania nanotube structure is unique, since the individual tubes are separated by interstices. The formation mechanism of intertube spacing is still unclear. Mor et al. [59] assumed that only pores with thin walls could form owing to relatively low ion mobility and high chemical solubility of the oxides. Consequently, unanodized metallic portions can initially present between the pores. With pores deepened, the electric field in these metallic regions is enhanced, thus resulting in the formation of intertube spacing, as exhibited in Figure 2.10d. Thereafter, the pores and voids (or intertube spacing) develop equally and yield an exclusive nanotube structure (Figure 2.10e). On the other hand, the anodic current gradually decreases with prolonged duration (Figure 2.9 in the range of P2–P4), as a consequence of increased nanotube length.

In addition, chemical dissolution plays an important role during nanotube growth, since the fashioned nanotubes can be dissolved by fluoride-containing solution, especially by aqueous solution. In general, there are three essential processes during nanotube growth, i.e., field-assisted oxidation at metal–film interface, field-assisted dissolution at film–electrolyte interface, and chemical dissolution at tube mouth–electrolyte interface. On the other hand, these three processes contribute to two important rates. One is electrochemical etching rate at tube base, determined by field-assisted oxidation and dissolution. At table anodization stage, this rate is constant since the filed-assisted oxidation and dissolution are equal to each other. The other rate is the chemical dissolution rate at tube mouth, which progressively consumes the formed nanotubes. If these two rates are comparable during the anodizing of titanium, the nanotube length will not change with extended duration, as illustrated in Figure 2.9 (from P4 to P5). The effect of the three process and the two rates will be discussed in detail in Section 2.1.3.

2.1.3 CONTROL OF THE GEOMETRY OF THE ANODIZED TITANIA NANOTUBE ARRAYS

In application, the cell performance is seriously affected by the geometry of the anodized titania nanotube arrays (see Section 2.2.5 for detail). Thus the tailoring of the nanotube geometry is important in efficiency enhancement. It was reported that the optimal length of nanotube array was close to 30 μm (see Section 2.2.5.2). By considering the different situations under open-circuit and short-circuit conditions, electron diffusion length of around 100 μm in nanotube array–based DSSCs was also estimated recently [64]. Therefore, it is crucial to produce high-aspect-ratio nanotubes in view of application in DSSCs for more dye loading and thus higher conversion efficiency. To date, the length of anodized titania nanotube array achieved progressively increased from the initial 250 nm to hundreds of micrometers [15,65–67], and to recent 720 μm [68]. Essentially, to achieve such a long nanotube array, it was to minimize the water content in the electrolyte by using organic electrolytes, as a result of reduced chemical dissolution of the oxide [65,67]. High-aspect-ratio nanotube arrays were also produced by employing buffer species to adjust the dissolution rate [69]. All these results suggest that chemical dissolution plays an important role in determining the length of nanotube arrays. This process, in turn, is influenced by a variety of factors, such as applied potential, working distance, electrolyte composition and pH value, magnetic stirring, anodizing temperature, and duration. On the other hand, other configurations of the nanotubes (e.g., pore diameter, wall thickness, intertube spacing) are also affected by these parameters. To better control the nanotube geometry, the effect of these anodizing parameters on the features (particularly the pore diameter and length) of the titania nanotube array is introduced in the following sections.

2.1.3.1 Effect of Applied Potential and Working Distance

Section 2.1.2 has discussed that the initiation and evolution of pore diameter are due to the lateral components of the electric field strength. Two direct methods can modulate the electric field strength: changing potentials that applied to the working electrode and the counter electrode, or varying the distance between the two electrodes, which are discussed in detail below.

In [70], nanotubes with different geometries were produced in ethylene glycol + 0.3 wt% ammonium fluoride + 2 vol% H_2O at different applied potentials and working distances. The current transients show that the current density at steady-state anodization increases with the increasing applied potential, owing to larger driving force of ion transport induced by higher potential. In addition, the initial current density at high potential is also larger than that at low potential. The relation between anodic current density (i) and field strength (E) can be presented as [48,50,71]

$$i = A \exp(BE) \tag{2.6}$$

where

A and B are constants related to temperature

E is the field strength, which is the ratio of the potential drop across the oxide layer to barrier layer thickness

Accordingly, electric field strength can be reflected from the current density level for a given temperature. Therefore, on the basis of the results of current transients, it is obvious that the relevant field strength increases with applied potential.

For nanotubes prepared at different working distances, the current transients reveal that the current density at steady state decreases gradually as the distance increases, due to increased IR drop in the electrolyte. According to the current density level and the above discussion, the electric field strength reduces at increase of working distance.

Equation 2.6 can be rewritten in a natural logarithm form, i.e.,

$$\ln i = \ln A + BE \tag{2.7}$$

Therefore, the field strength is proportional to $\ln i$ at a given temperature. In [70], $\ln i$ vs. time for different applied potentials and working distances is illustrated. It clearly elucidates the field strength increases with applied potential whereas decreases with increase in working distance. In addition, the variation of field strength when changing working distance in the range of 13–40 mm at 60 V corresponds to that when changing applied potential in the range of 60–50 V (keeping distance constant at about 13 mm).

The changes in field strength at different applied potentials and working distances directly affect the nanotube morphologies. Relevant field emission scanning electron microscopy (FESEM) results reveal that the pore diameter of the nanotubes increases with applied potential whereas decreases with increase in working distance, which can be attributed to the variation of effective field strength at the Ti electrode.

Bauer et al. [8] tailored the geometry of the anodic nanotubes in H_3PO_4/HF electrolytes by changing applied potentials, as illustrated in Figure 2.11. In Figure 2.12, it is clearly shown that both pore diameter and length (layer thickness) of these nanotubes increase linearly with applied potential. Similar results were also reported by other researchers [10,11,15,21,58,72].

The length of as-prepared TiO_2 nanotube arrays as a function of applied potential and working distance is summarized in [70]. It reveals that the length increases with applied potential and decreases with increased working distance. The variation of length for all situations is also coincident with the relevant changes in field strength.

As mentioned in Section 2.1.2, there are three key processes during the growth of nanotube arrays, i.e., field-assisted oxidation at metal–film interface, field-assisted dissolution at film–electrolyte interface, and chemical dissolution at pore mouth [15,59–66]. These processes are responsible for two essential rates that in turn determine the final length of nanotube arrays. One is the electrochemical etching rate, determined by field-assisted oxidation and dissolution, being constant when

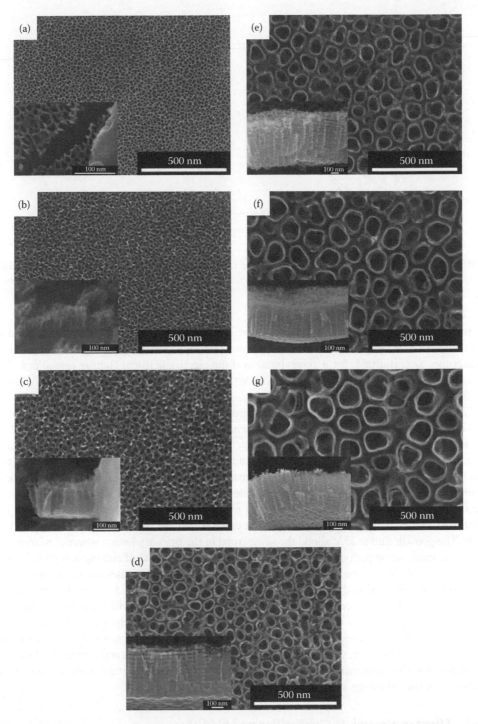

FIGURE 2.11 SEM images of TiO$_2$ nanotubes (top view and cross section) formed in 1 M H$_3$PO$_4$ + 0.3 wt% HF at 1 V (a), 2.5 V (b), 5 V (c), 10 V (d), 15 V (e), 20 V (f), and 25 V (g) for 1 h. (Reprinted from Bauer, S. et al., *Electrochem. Commun.*, 8, 1323, 2006. With permission.)

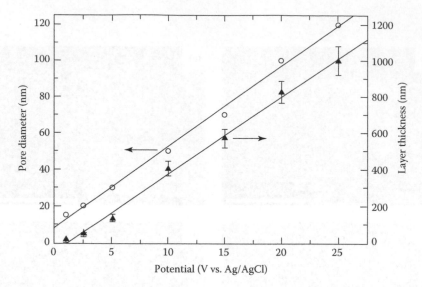

FIGURE 2.12 Diameters and thickness of nanotube layers formed in 1 M H_3PO_4 + 0.3 wt% HF at different potentials (values extracted from Figure 2.11). (Reprinted from Bauer, S. et al., *Electrochem. Commun.*, 8, 1324, 2006. With permission.)

these two processes reach equilibrium. The other is the chemical dissolution rate, the speed under which the fashioned TiO_2 nanotube array is dissolved through reaction (2.4). As a consequence, longer nanotubes are attainable for a given duration if the electrochemical etching rate at pore bottom is faster than the chemical dissolution rate at pore mouth, otherwise, shorter or even no nanotubes are obtained [15].

In [70], the length of the nanotube array as a function of time under different anodization potentials is also investigated. The growth rate of nanotube array can be attained from the slope of the curve, which is the combined result of electrochemical etching rate and chemical dissolution rate. The slope is positive if the former is faster than the latter, and it is negative if the former is slower than the latter, and equals to zero if these two rates are comparable. Moreover, the larger is the slope, the more apparent is the discrepancy between these two rates. The figure reveals that the growth rate at 60 V in the initial two hours is ~7.81 μm/h, more than nine times higher than that at 25 V (which is only ~0.80 μm/h), and it increases slightly as duration extends and decreases finally for long anodization periods. The growth rate at 25 V shows the same trend in the primary hours, whereas becomes negative for long anodization periods. The results suggest the electrochemical etching at high field strength is prominent, while the chemical dissolution at low field strength dominates the growing process.

The above results reveal that the pore diameter of the anodic titania nanotubes increases with applied potential whereas decreases with increased working distance, which can be attributed to the enhanced lateral components of the field strength. The growth rate of the nanotubes also increases at increased potential or decreased working distance, therefore longer nanotubes are obtainable at high field strength. However, the final length of the nanotubes is significantly influenced by electrolyte composition that determines the chemical dissolution process during nanotube growth, which is discussed in the following section.

2.1.3.2 Effect of Electrolyte Composition and pH Value

In general, the highly ordered titania nanotube arrays can only be produced in fluoride-containing electrolyte by electrochemical anodization, as illustrated in Figure 2.8. In consideration of the essential role of the fluorides, geometry evolution with variation of fluoride composition is discussed here. The pH value also plays an important role during nanotube growth. In addition, water content

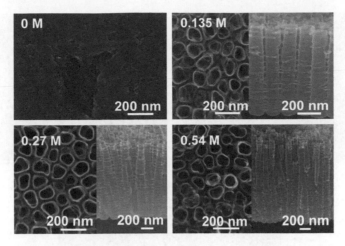

FIGURE 2.13 SEM top views (and cross sections in insets) of samples anodized in water/glycerol electrolyte (50:50 vol%) at 20 V during 3 h with different NH₄F concentrations and with a sweep rate of 250 mV/s. (Reprinted from Macak, J.M. et al., *J. Electroanal. Chem.*, 621, 257, 2008. With permission.)

in organic electrolyte is another key factor in determining the nanotube configuration; therefore, related topics are introduced.

Macak et al. [11] investigated the influence of fluoride concentration in the electrolyte on the configuration of the anodic titania nanotubes. The results reveal that only a compact oxide layer is formed in fluoride-free electrolyte, as shown Figure 2.13, similar to that discussed in Section 2.1.2. In all other cases displayed in the figure, nanotube structures are produced in the presence of fluoride in the electrolyte. It also indicates the length of the nanotubes increases with the fluoride concentration, as illustrated in the insets of Figure 2.13, suggesting promoted growth rate at increased concentration. However, the formation of the nanotubes was not homogeneous over the entire Ti surface at very high F⁻ concentration (e.g., 0.54 M), in agreement with Bauer et al. [8]. As shown in Figure 2.14, the anodization with concentrations of fluoride ions <0.1 wt% resulted in compact oxide layer (see Figure 2.14a), while with a concentration of 0.5 wt% HF or more etching of the surface was observed (see Figure 2.14c). In view of these, it seems that there exists an optimal concentration of fluorine ions for a given electrolyte to generate nanotube structure while with a desired growth rate. Below a lower critical concentration, no nanotubes can be obtained, whereas the etching away of nanotubes takes place above certain fluoride concentration.

Figure 2.13 suggests that the pore diameter of the nanotubes is not strongly influenced by concentration of the fluorine ions (0.135; 0.27; 0.54 M NH₄F), echoing the results from Macak et al. [18], in which anodic titania nanotubes were grown in Na₂SO₄ (1 M) electrolyte with different NaF concentration (0.5; 0.7; 1 wt% NaF). The pore diameter is not significantly depending on the NaF concentration. However, these results are obtained within one individual electrolyte regime, i.e., in a particular solution while varying the fluoride concentration. Nevertheless, the pore diameters of the nanotubes differ in different electrolytes. For example, pore diameters of approx. 20 nm are obtained at 10 V in acetic acid electrolyte containing 0.5 wt% NH₄F [20], whereas approx. 50 nm at 10 V and 100 nm at 20 V in 1 M phosphoric acid electrolyte containing 0.3 wt% HF [8], approx. 140 nm at 20 V in 1 M sulfuric acid electrolyte containing 0.15 wt% HF [17]. These examples indicate that the pH value of the electrolyte plays an important role in determination of pore diameter.

The geometry of the nanotubes is also affected seriously by the electrolyte pH value. Ghicov et al. [73] studied titania nanotubes prepared in phosphate electrolytes. The results show that nanotube pore diameter increases with HF concentration from 30 nm in 0.3 wt% HF to 50 nm in 0.5 wt% HF, and to 70 nm in 0.7 wt% HF. Feng et al. [74] prepared titania nanotubes over a wide pH range. Their results indicate the nanotube length increases significantly with the electrolyte pH value, and

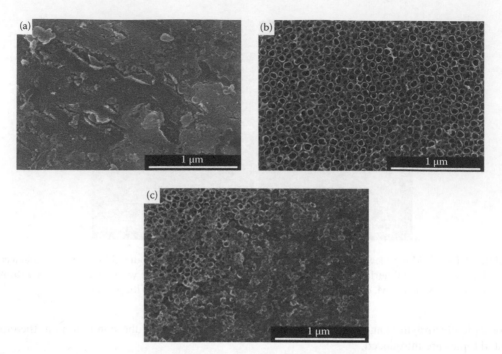

FIGURE 2.14 SEM images of TiO_2 layers formed at 10 V for 2 h in 1 M H_3PO_4 with different additions of HF: (a) 0.025 wt%, (b) 0.3 wt%, and (c) 0.5 wt%. (Reprinted from Bauer, S. et al., *Electrochem. Commun.*, 8, 1322, 2006. With permission.)

the corresponding growth rate is larger in basic conditions than in acidic conditions. The authors attributed it to the different chemical dissolution rates at pore mouth under different pH values in consideration of the three processes during tube growth mentioned in Sections 2.1.2 and 2.1.3.1.

Since the electrochemical formation of titania nanotubes was first reported in 1999 [6], many researchers have devoted themselves to promoting the nanotube length. It was revealed that the tube length of only hundreds of nanometers (so-called first generation) could be obtained in aqueous or acidic HF solutions [8,15,17,20,59], as a result of the high chemical dissolution rate of the fashioned nanotubes. To solve this problem, researchers proposed to use pH buffer species and adopt relatively mild solutions in preparation of anodic nanotubes [18,19,69,73]. In this case, nanotubes of a few micrometers (so-called second generation) were achieved. However, this length was still far from the ideal values for the sake of applications. To further suppress the chemical dissolution process, organic electrolyte was employed [65–68], which boomed the tube length to hundreds of micrometers (so-called third generation).

In the growth of anodic titania nanotubes in organic electrolyte, water content becomes a critical factor in producing and tailoring the tube structure. Raja et al. [75] investigated the effect of water content on formation of the nanotubes. Their results demonstrate that a minimum of about 0.18 wt% water is required to produce well-ordered nanotubes in the ethylene glycol+0.2 wt% NH_4F solution. With water content more than 0.18 wt%, no significant changes in the morphology of the nanotubes are observed. However, in very high water content (>0.5 wt%) ripples are present on the nanotube walls and the number of the ripples increases with water content. In general, the ripples are commonly observed in the nanotubes prepared in aqueous solutions, owing to the current oscillation induced by pH burst during nanotube growth [76], whereas smooth wall is normally obtained in organic electrolyte (see Figure 2.15). The same results were also reported by Macak et al. [11], as shown in Figure 2.15, where ripples on the nanotube walls become more obvious as water content increases. Additionally, these authors claimed that the pore diameter of the nanotubes increases with water content, i.e., 45±6 nm for 0 vol% H_2O, 60±7 nm for 0.67 vol% H_2O, 75±8 nm for 6.7 vol% H_2O, and 105±10 nm for 16.7 vol%

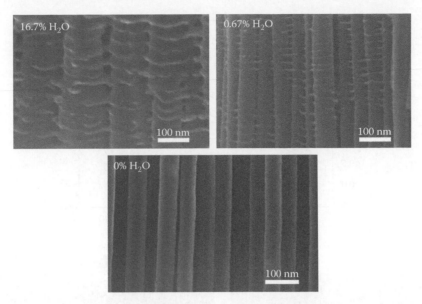

FIGURE 2.15 SEM high-magnification images of the tube walls showing different features depending on the water content. (Reprinted from Macak, J.M. et al., *J. Electroanal. Chem.*, 621, 265, 2008. With permission.)

H_2O. This may be attributed to the decreasing IR drop with increased water content, thus producing higher effective potential on the working electrode (Ti foil) at higher water addition.

In addition, it was reported that the nanotube wall thickness decreases with increasing NH_4F concentration during anodization of Ti-Zr alloy at 20 V in 1 M $(NH_4)_2SO_4$ solution [13].

2.1.3.3 Effect of Temperature and Stirring

Yasuda and Schmuki [13] demonstrated that the length-limiting factor of the nanotube growth is controlled by the diffusion of ionic species in the electrolyte. The ion diffusion process, in turn, is affected by anodizing temperature and magnetic stirring, therefore these two factors are introduced in this section.

Macak and Schmuki [16] investigated the influence of electrolyte viscosity and temperature on morphology of anodic titania nanotubes. Clearly, their results reveal that both pore diameter and tube length increase with elevated temperature, as illustrated in Figure 2.16. In the case of anodization at a temperature higher than 40°C, only unstable bundles of the tubes are formed, whereas no regular and mechanically stable nanotube architectures are obtained. The strong temperature dependence of the nanotube geometry reflects the important role of ion diffusion in the growth of anodic nanotubes.

To further clarify the effect of diffusion process, magnetic stirring is employed during nanotube growth and studied in comparison to that without stirring [11,16,77]. The results indicate the current density is higher in stirred electrolyte than that in static. Moreover, the length of the nanotubes produced in the stirred electrolyte is much longer. For example [13], the nanotubes anodized at 20 V in 1 M $(NH_4)_2SO_4 + 0.5$ wt% NH_4F electrolyte has a length of 16.9 μm with magnetic stirring. In contrast, only 10.7 μm is obtained in static electrolyte. This again suggests ion diffusion controls the nanotube growth process.

2.2 APPLICATION IN DYE-SENSITIZED SOLAR CELLS

2.2.1 INTRODUCTION

To date, the world's total energy consumption is 4.1×10^{20} J/year, equivalent to a continuous power consumption of 13×10^{12} W, or 13 TW [78]. World demand for energy is estimated to be more

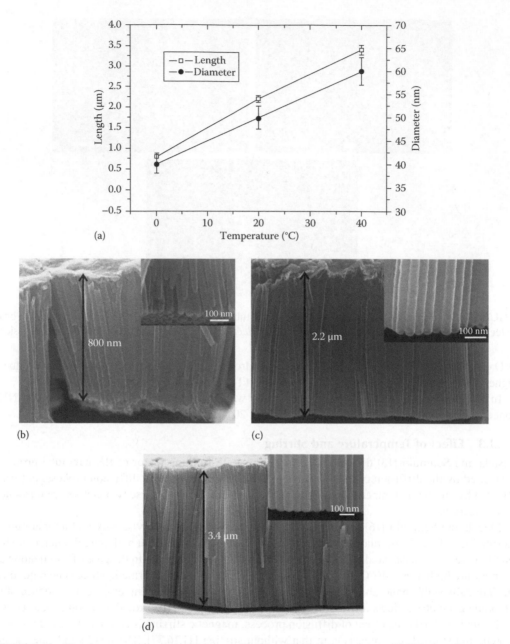

FIGURE 2.16 Titania nanotubes prepared in glycerol +0.5 wt% NH_4F electrolytes at 20 V during 6 h at different temperatures. (a) An evaluation of the tube diameter and length as a function of the temperature for glycerol electrolytes. The average tube diameters are 40 ± 5, 50 ± 6, and 60 ± 7 nm and lengths are 800 ± 50 nm, 2.2 ± 0.1 μm, and 3.4 ± 0.1 μm for 0°C, 20°C, and 40°C. SEM cross-sectional images anodized at 0°C (b), 20°C (c), and 40°C (d). (Reprinted from Macak, J.M. and Schmuki, P., *Electrochim. Acta*, 52, 1261, 2006. With permission.)

than double by 2050 and more than triple by the end of this century. Incremental improvements in existing energy networks are obviously inadequate in meeting this demand. Therefore, looking for sufficient supplies of clean energy for the future becomes one of the most challenging endeavors of the present mankind.

Fortunately, the sun supplies 120,000 TW of radiation on the surface of the earth, far exceeding human needs even in the most aggressive energy demand scenarios [78]. In other words, 20 TW of power is obtainable provided that 0.16% of the land on earth is covered with 10% efficient solar conversion systems, nearly twice the world's consumption of fossil energy or equivalent to 20,000 nuclear fission plants of 1 GW. It is obvious that the solar energy can be exploited on the needed scale to meet global energy needs without significantly influencing the solar resource.

Conventional photovoltaic devices are based on the concept of charge separation at an interface of two materials with different conduction mechanisms, which are capable of converting solar energy directly into electricity. Currently, the photovoltaic market is dominated by solid-state p–n junction devices (usually silicon based), and profits from the experience and material availability from semiconductor industry. For the sake of reducing production cost, thin-film technologies are attractive and effective. To date, there are a variety of thin-film solar cells, such as [79] copper indium gallium diselenide (CIGS, maximum cell efficiency of 19.4% ± 0.6%), cadmium telluride (CdTe, maximum cell efficiency of 16.7% ± 0.5%), gallium arsenide (GaAs, maximum efficiency of 26.1% ± 0.8%), amorphous (maximum efficiency of 9.5% ± 0.3%), and nanocrystalline (maximum efficiency of 10.1% ± 0.2%) silicon thin-film solar cells. Yet all of them have their own problems. The use of indium and gallium in CIGS increases the manufacturing cost due to limited supply. The cadmium in CdTe solar cells is toxic and environment unfriendly. Although GaAs achieves the highest cell efficiency, it is the most expensive [80]. Low conversion efficiency and low stability as well as questionable durability are the main limiting factors of amorphous silicon thin film solar cells [81]. As for micro-/nanocrystalline silicon thin-film solar cells, the low growth rate with high hydrogen dilution may be problematic in manufacturing [80].

Alternatively, DSSC is one of the most promising photovoltaic devices for cost reduction, due to a large number of merits compared to inorganic p–n junction solar cells. These advantages include ease of fabrication, low cost of scale-up without high-vacuum, high-temperature and ultraclean environment, compatibility with flexible substrates, etc. [2].

The current version of DSSCs was first reported in 1991 by O'Regan and Grätzel with an overall conversion efficiency of 7.1%–7.9% [1]. In the last decades, the DSSCs were widely investigated and further improved. The overall conversion efficiency of the cell reached 10.4% in 2001 [82] by Grätzel's research group. An efficiency of 10.2% was also achieved by another group in 2005 through optimization of the design of each cell components [83]. Certified efficiencies exceeding 11% were demonstrated for laboratory cells (Sharp) and 6%–7% of efficiency for modules [2,84–87]. Tandem versions of the cell reached 15% conversion efficiencies [88]. However, an assessment from U.S. Department of Energy concluded that by developing the scientific underpinning to exploit the unique properties of sensitized systems, an efficiency of 20% was attainable, and that such studies may further lay the scientific foundation for developing nanostructural systems with efficiencies beyond the Shockley–Queisser limit of 32% [78]. Apparently, the current version is far below the limiting ceiling, and there is still a large room for improvement. Although the challenge of realizing this vision is considerable and requires substantial advances in understanding each component as well as physical process affecting the performance and stability of the devices, it is yet the most intriguing and promising way to offer cheap, renewable, and clean energy for mankind on earth.

Though the conversion efficiency is seemingly still low, DSSCs have already generated considerable scientific, technological, and industrial interests, as a result of their intriguing merits over other technologies [89]:

1. Tolerance to impurities. It is well known that conventional p–n junction–based solar cells are produced under ultraclean environment, since the cell performance is seriously affected by the impurities. In contrast, the DSSCs are tolerant to impurities. This is because light absorption, charge separation, and electron and hole transport take place at different parts of the system. Furthermore, the large surface area of photoanode (e.g., nanoparticle or nanotube photoanode) enables more dye loading and facilitates light absorption.
2. Simple and cost-effective production process. Since the system is tolerant to impurities, the solar cells can be produced in absence of high vacuum, high temperature, and ultraclean

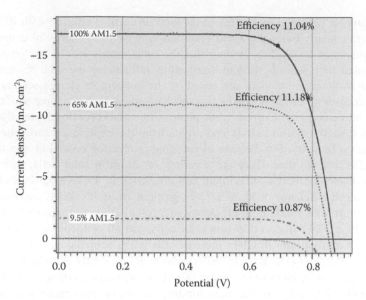

FIGURE 2.17 Photocurrent–voltage curve of a dye-sensitized solar cell at different light intensities. (Reprinted from Grätzel, M., *J. Photochem. Photobiol. A: Chem.*, 164, 11, 2004. With permission.)

environment, thus reducing the cost. This allows the cells to be manufactured through continuous processes, such as screen printing, pressing, and roll-to-roll production.

3. Abundant and environment-friendly materials. Titania is also widely used in daily life in toothpaste and white paint. There is no shortage of supply of titania on earth.
4. Operation over a wide range of temperature, approximately from ambient temperature to 60°C.
5. Efficiency insensitive to the angle or intensity of the incident light, as illustrated in Figure 2.17.
6. Extensive applications. The sensitizers have a variety of presentations, appearances, and colors with a full range of transparencies from ultraviolet to infrared. This allows for building-integrated windows, walls, and roofs of varying color and transparency that simultaneously generate electricity even in diffuse light or at relatively low light levels in addition to whatever other function they serve [89].
7. Compatible with lightweight and flexible (G24I, Konarka) or rigid substrates (e.g., plastic, fabric, metal, glass, and ceramic). An example is shown in Figure 2.18.

These merits of DSSCs have generated worldwide interests of research and development, especially the great challenge for industrialization. In 2007, the first manufacturing plant (G24I) to produce sensitized solar cells on a commercial scale began operations. By the year 2007, there were eight industrial corporations that had gained the patent right, including Konarka in the United States, Aisin Serki in Japan, RWE in Germany, Solaronix in Switzerland, Sustainable Technologies International in Australia, etc. On the other hand, the protective period of the patent owned by Grätzel, one of the inventors of DSSCs, matured in 2008; therefore, more and more companies will enter this area in the future.

2.2.2 Typical Structure and Working Principle of Dye-Sensitized Solar Cells

A typical nanocrystalline-based DSSC consists of transparent conducting oxide (TCO), dye molecules anchored to the surface of TiO_2 nanoparticle network employed as the photoanode, redox electrolyte penetrating through the photoanode, and platinized counter electrode, as depicted in Figure 2.19.

FIGURE 2.18   Lightweight flexible solid-state dye-sensitized solar cells: a stainless steel supported flexible cell with a conversion efficiency of 4.2%. (Reprinted from Kang, M.G. et al., *Sol. Energy Mater. Sol. Cells*, 90, 576, 2006. With permission.)

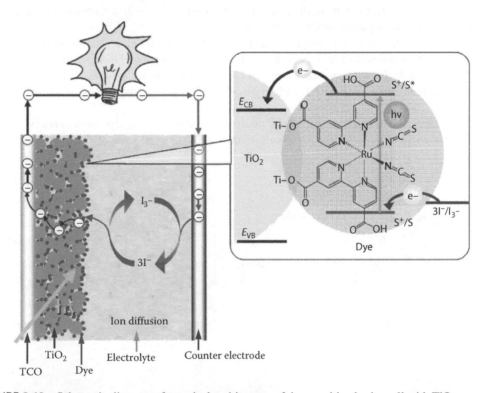

FIGURE 2.19   Schematic diagram of a typical architecture of dye-sensitized solar cell with TiO_2 nanocrystalline photoanode. Inset is local magnification as well as corresponding energy level of the system. (Adapted from Grätzel, M., *Inorg. Chem.*, 44, 6841, 2005. With permission.)

As the material of TCO, indium-tin oxide (ITO), fluorine-doped tin oxide (FTO), and recently developed aluminum-doped zinc oxide (AZO) are commonly used as the front contact of DSSC for light trapping and electron collection.

The core component of DSSC system is the mesoporous semiconductor photoanode, which is generally placed in contact with a redox electrolyte or an organic hole conductor. The material used has been anatase TiO_2, though alternative ZnO [92–94], SnO_2 [95,96], and Nb_2O_5 [97,98] have also been investigated. The mesoporous TiO_2 photoanode is sintered at 450°C for a short time after doctor blade or screen printing on the TCO substrate to ensure that the particles are electronically interconnected. The TiO_2 particle size is typically 15–20 nm, and the real surface areas of the films with a thickness of 10 μm is 1000 times larger than that of the projected area [2,85,99]. The much higher surface area renders larger dye-loading ability for sufficient absorption of the incident photon energy.

Attached to the surface of the mesoporous TiO_2 film is a monolayer of sensitizer or dye. The sensitizer is usually ruthenium compounds, which are known for their outstanding stability. *cis*-Di(thiocyanato)bis(2,2′-bipyridyl)-4,4′-dicarboxylate) ruthenium-(II), coded as N3 or N-719 dye depending on whether it contains four or two protons, is found to be an excellent solar light absorber and charge-transfer sensitizer [2,100]. Figure 2.20 shows the structures of three different ruthenium sensitizers and their corresponding appearances loaded by mesoporous TiO_2 film, and the center one is the so-called N3 sensitizer (also shown in inset of Figure 2.19). The sensitizers are adsorbed onto the mesoporous film surface by dip coating in the dye solutions for certain time (e.g., 48 h).

The redox electrolyte usually containing I^-/I_3^- ions is filled in the pores of the mesoporous film after sensitization by dye molecules. The oxidized dye molecules are regenerated by electron donation from the redox couple $(3I^- \rightarrow I_3^- + 2e)$. On the other hand, regeneration of the redox couple occurs via collecting electrons from the counter electrode $(I_3^- + 2e \rightarrow 3I^-)$. Certain hole-conducting polymers can also be employed as the redox couple [101–103].

The platinized TCO is widely employed as the standard counter electrode to reduce the redox couple or collect the holes from hole-conducting materials in solid-state DSSCs, where the platinum acts as catalyst [104]. In addition, other kinds of counter electrode are also studied, such as single-wall carbon nanotube on glass [105], carbon black on FTO [106], and carbon black on stainless steel [107].

Once a DSSC is soaked under the sun, photons are absorbed by dye molecules (sensitizer), resulting in the excitation of electrons from ground state to excited state of the dye molecules, as

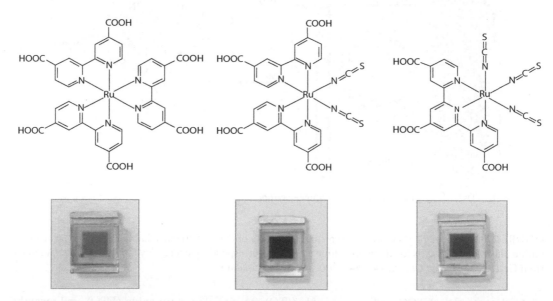

FIGURE 2.20 Structures of three different ruthenium sensitizers RuL_3 (left), *cis*-$RuL_2(NCS)_2$ (middle), and $RuL'(NCS)_3$ (right), where L = 2,2′-bipyridyl-4,4′-dicarboxylic acid, and L′ = 2,2′,2″-terpyridyl-4,4′,4″-tricarboxylic acid. The lower part of the picture shows nanocrystalline TiO_2 films loaded with a monolayer of the respective sensitizer. The film thickness is 5 μm. (Reprinted from Grätzel, M., *Inorg. Chem.*, 44, 6841, 2005. With permission.)

illustrated in the inset of Figure 2.19. Subsequently, electron injection occurs from dye molecules into the conduction band of TiO_2 photoanode, due to its relatively lower energy level. Thereafter, the injected electrons percolate to the collecting electrode through the nanoparticle network via trapping/detrapping mechanism [108–111]. At the same time, the original state of the dye molecules is restored by electron donation from the I^-/I_3^- redox couple in the electrolyte, while the iodide is regenerated in turn by the reduction of triiodide at the platinized counter electrode. The circuit is eventually completed via electron migration through the external load. The theoretical maximum open-circuit photovoltage produced under illumination corresponds to the difference between the quasi-Fermi level of TiO_2 photoanode and the chemical potential of I^-/I_3^- in the electrolyte. The conduction band level of the TiO_2 photoanode and the redox potential of I^-/I_3^- are estimated to be −0.5 V versus normal hydrogen electrode (NHE) and 0.4 V versus NHE [112], respectively. Therefore, in the case of a DSSC using TiO_2 photoanode and I^-/I_3^- redox mediator, the maximum photovoltage is expected to be approximately 0.9 V, depending on the electrolyte components because the Fermi level of the TiO_2 electrode depends on the electrolyte components and their concentrations [113].

2.2.3 EFFICIENCY LIMITING FACTORS OF DYE-SENSITIZED SOLAR CELLS

The overall conversion efficiency (η) of DSSCs can be described as follows [84,85]:

$$\eta = \frac{J_{sc} \times V_{oc} \times FF}{P_{in}} \qquad (2.8)$$

where

J_{sc} is short-circuit photocurrent density
V_{oc} is open-circuit photovoltage
FF is fill factor
P_{in} is intensity of incident light

The equation suggests that it is necessary to maximize these three parameters (numerators) so as to promote global efficiency of the cell. Figure 2.21 is a typical $I–V$ curve (solid line) for DSSC, from which the J_{sc}, V_{oc}, and FF can be obtained, as illustrated in the figure where the dotted line represents dark current.

The FF is mainly related to the series resistance of DSSCs, which is determined by the following relation [113]:

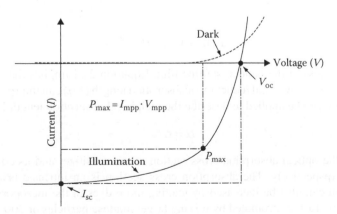

FIGURE 2.21   A typical $I–V$ curve for efficiency calculation of a DSSC.

$$FF = \frac{J_{mpp} \cdot V_{mpp}}{J_{sc} \cdot V_{oc}} \qquad (2.9)$$

Herein, the J_{mpp} and V_{mpp} are photocurrent density and photovoltage at maximum power point, respectively, both of which can be obtained from the I–V curve. Han et al. [114] proposed an equivalent circuit for modeling DSSCs based on the electrochemical impedance spectroscopy measurement. Their further study showed that the series resistance of DSSCs consisted of three resistance elements, namely, the sheet resistance of TCO, the resistance of ionic diffusion in electrolyte, and the resistance at the interface of counter electrode and electrolyte [83]. The FF increases with decrease in the internal resistance elements, and a conversion efficiency of 10.2% was obtained by the optimization design of each DSSC component [83]. Consequently, reduction of series resistance can result in high FF.

The J_{sc} can be expressed by integrating the product of incident photon flux density, $F(\lambda)$, and incident photon to current efficiency (IPCE) spectra, presented as [84]

$$J_{sc} = \int qF(\lambda)(1 - r(\lambda))\text{IPCE}(\lambda)d\lambda \qquad (2.10)$$

where
  q is the electron elementary charge
  $r(\lambda)$ is the energy loss of incident light, such as absorption and reflection at the TCO

As for the IPCE(λ), it can be given by [2]

$$\text{IPCE}(\lambda) = \text{LHE}(\lambda)\phi_{inj}\eta_{cc} \qquad (2.11)$$

where
  LHE(λ) is the light-harvesting efficiency of photons with wavelength λ
  ϕ_{inj} is the quantum yield of electron injection from the excited sensitizer to the conduction band of the semiconductor oxide
  η_{cc} is the charge collection efficiency

Equation 2.10 shows that the photocurrent density, for a given photon flux density, is determined only by $r(\lambda)$ and IPCE(λ).

$r(\lambda)$ is a property of the TCO. To reduce $r(\lambda)$, optimal doping concentration and incremental light trapping by surface texturing are robust and effective [115]. Consequently, J_{sc} can be increased by TCO improvement.

IPCE is dominated by three terms. The LHE is derived from the reciprocal absorption length (α) via [2]

$$\text{LHE}(\lambda) = 1 - 10^{-\alpha \cdot d} \qquad (2.12)$$

where d is the thickness of the nanocrystalline film. Equation 2.12 implies that increment of film thickness (d) to load more dyes and reduction of absorption length ($1/\alpha$) can improve LHE. Moreover, Beer–Lambert's law can be applied to describe the reciprocal absorption length [2]

$$\alpha = \sigma \cdot c \qquad (2.13)$$

Here σ and c are the optical absorption cross section of the sensitizer and its concentration in the mesoporous film, respectively. The absorption cross section is an intrinsic property of the dye, while concentration can also be increased by loading more dyes in the mesoporous film. In addition, the LHE can be further promoted by mixing larger anatase particles of 200–400 nm radius to enhance light scattering in the mesoporous film [2]. Chiba et al. [84] reported an overall conversion

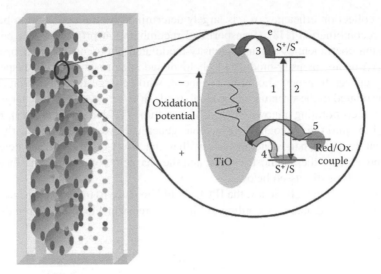

FIGURE 2.22 Photoinduced processes occurring during photovoltaic energy conversion at the surface of the nanocrystalline titania films: (1) sensitizer (S) excitation by light; (2) radiative and nonradiative deactivation of the sensitizer; (3) electron injection in the conduction band, followed by electron trapping and diffusion to the particle surface; (4) recapture of the conduction band electron by the oxidized sensitizer (S$^+$); (5) recapture of the conduction band electrons by the oxidized form of the redox couple regenerating the sensitizer and transporting the positive charge to the counter electrode. Big spheres: Titania nanoparticles. Elliptical dots: sensitizer. Light and dark small dots: oxidized and reduced forms of the redox couple. (Reprinted from Grätzel, M., *Inorg. Chem.*, 44, 6841, 2005. With permission.)

efficiency of 11.1% and J_{sc} of 21 mA/cm^2 by using TiO$_2$ electrode with high haze controlled via addition of submicron particles (400 nm diameter), which exhibited the best performance currently. Chou et al. [116] published the same effect of light scattering on efficiency enhancement of DSSCs. Furthermore, to achieve light harvesting, investigation of new sensitizer with effective light absorption is necessary. In view of these, the promotion of light harvesting is attainable by increasing mesoporous film thickness to load enough dye, incorporating submicron (200–400 nm) particles into the film to enhance light scattering especially in the red and near-infrared region, and investigating new sensitizers with larger light absorption coefficient.

Another impact factor of IPCE is quantum yield of charge injection, ϕ_{inj}, which denotes the fraction of the photons absorbed by the dye that are converted into conduction band electrons. Charge injection from the electronically excited sensitizer into the conduction band of the semiconductor (Process 3 shown in Figure 2.22) is in competition with other radiative or nonradiative deactivation channels (Process 2 shown in Figure 2.22). The ϕ_{inj} can be expressed as [2]

$$\phi_{inj} = \frac{k_{inj}}{k_{deact} + k_{inj}} \tag{2.14}$$

where
   k_{inj} is the fraction of excited electrons injected into the conduction band of the photoanode
   k_{deact} is the fraction of the deactivated electrons

To achieve a good quantum yield, ϕ_{inj}, the rate constant for charge injection should be at least 100 times higher than k_{deact} [2]. This is achievable by employing the sensitizers with special functional group such as, for instance, carboxylate, hydroxamate, or phosphonate moieties that anchor the sensitizer to the oxide surface [117].

The charge collection efficiency, η_{cc}, is largely determined by the competition between charge transport and recombination [118]. Transport and recombination properties can be measured by electron diffusion coefficient (D_n) using intensity-modulated photocurrent spectroscopy and electron lifetime (τ_n) using intensity-modulated photovoltage spectroscopy [119], respectively. So as to increase η_{cc}, charge transport should be enhanced, while charge recombination should be suppressed. This is related to the structure of the photoanode. It is believed that the charge transport can be improved by incorporating nanowire [120,121], nanotube [122], nanosheet [123], nanorod [124], and nanobelt [125] into the photoanode, which has generated great interest recently. Investigation results pertinent to these nanostructures show that the short-circuit current density has been improved by introducing these nanostructures, and the charge recombination has been remarkably hindered, which will be discussed below.

To attain high conversion efficiency, the IPCE should be close to unity over the near-UV, visible, and near-IR wavelength domain. According to Einstein's relationship, the electron diffusion length is [2,124]

$$L_n = \sqrt{D_n \tau_n} \tag{2.15}$$

where, D_n and τ_n are the electron diffusion coefficient and lifetime, respectively. The quantitative collection of charge carriers can be achieved only if the electron diffusion length is greater than the film thickness (d) [2]

$$L_n > d \tag{2.16}$$

Otherwise, the electrons cannot be collected. On the other hand, the mesoporous film should be thick enough to ensure light harvesting in the spectral absorption range of the sensitizer, which has been discussed above, i.e., [2]

$$d > 1/\alpha \tag{2.17}$$

where $1/\alpha$ is the light absorption length. This has restricted the thickness of the conventional nano-crystalline DSSCs to an order of a few microns, which depends on the optical cross section of the sensitizer or quantum dot and their concentration in the film.

The open-circuit voltage is determined, to a large extent, by the charge recombination. It can be calculated from the diode equation [99]:

$$V_{oc} = \left(\frac{nRT}{F} \right) \ln \left[\frac{i_{sc}}{i_o} - 1 \right] \tag{2.18}$$

where

  n is the ideality factor whose value is between 1 and 2 for the DSSCs
  i_o is the reverse saturation current induced by recombination
  R is the gas constant
  T is the Kelvin temperature
  F is the Faraday constant

Thus, for each order of magnitude decrease in the dark current, the gain in V_{oc} would be 59 mV at room temperature [90]. The injected electrons in the conduction band of photoanode can recombine with oxidized sensitizer (Process 4 shown in Figure 2.22) and oxidized form of the redox couple regenerating the sensitizer (Process 5 shown in Figure 2.22). The corresponding reactions are displayed as follows [126]:

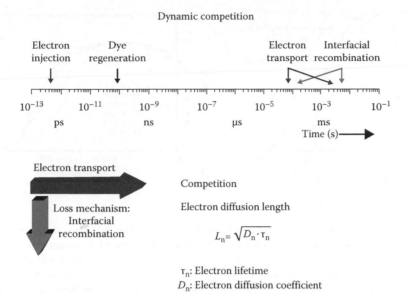

FIGURE 2.23 Dynamics of redox processes involved in the conversion of light to electric power by dye-sensitized solar cells. (Reprinted from Grätzel, M., *J. Photochem. Photobiol. A: Chem.*, 164, 8, 2004. With permission.)

$$TiO_2 \mid S^+ + e_{cb}^- \ \rightarrow TiO_2 \mid S \tag{2.19}$$

$$I_3^- + 2e_{cb}^- \ \rightarrow 3I^- \tag{2.20}$$

Herein, Equations 2.19 and 2.20 are associated with Process 4 and 5 in Figure 2.22, respectively. Compared with the former, the recombination occurring between injected electrons and the electrolyte is dominant, since the oxidized dye molecules can be promptly reduced by the surrounding electrolyte that takes place at an order of nanosecond, as shown in Figure 2.23. Accordingly, a blocking layer is employed to hinder the electron back reaction with the electrolyte [127,128]. The results of incorporating one-dimension nanostructures, such as nanowire, nanotube, and nanorod, indicate that the recombination can be highly suppressed, which will be discussed in the following sections.

In summary, the following aspects should be taken into account to achieve high overall conversion efficiency. The FF can be increased by decreasing the series resistance, i.e., the sheet resistance of TCO, the resistance of ionic diffusion in the electrolyte, and the resistance at the interface of the counter electrode and the electrolyte. The short-circuit current density can be promoted by optimal design of the TCO to reduce light loss via absorption and reflection; by employing novel photoanode with high dye-loading ability, large electron diffusion coefficient, and strong light scattering; and by investigating new sensitizer with short absorption length and high quantum yield for electron injection. The open-circuit voltage can be enhanced by photoanode modification to suppress charge recombination.

2.2.4 From Nanoparticle Network to Nanotube Array–Based Photoanode

It is widely recognized that two of the major factors have limited the overall conversion efficiency of conventional nanocrystalline DSSCs. One is the slow percolation of electrons through the random polycrystalline networks. The injected electrons repeatedly interact with a distribution of traps as they undertake a random walk through the mesoporous film [110]. The structural disorder at the contact between two crystalline nanoparticles leads to the aggravated scattering of free electrons

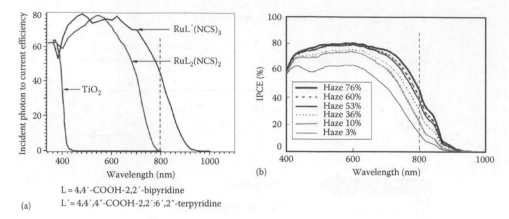

(a)   L = 4,4'-COOH-2,2'-bipyridine
L' = 4,4',4''-COOH-2,2':6',2''-terpyridine

FIGURE 2.24 (a) Photocurrent action spectra obtained with the N3 (ligand L) and the black dye (ligand L') as sensitizers. The photocurrent response of bare TiO$_2$ films is also shown for comparison. (Reprinted from Grätzel, M., *J. Photochem. Photobiol. A: Chem.*, 164, 3, 2004. With permission.) (b) Dependence of IPCE spectra on haze of TiO$_2$ electrodes. Haze in the figure was measured at 800 nm. (Reprinted from Chiba, Y. et al., *Jpn. J. Appl. Phys.*, 45, L638, 2006. With permission.) The black dye is employed as the sensitizer. The vertical dashed line is added for convenience of view.

[129], thus reducing electron mobility. The electron diffusion coefficient, $D_n \leq 10^{-4}$ cm^2/s, is several orders of magnitude smaller than that in TiO$_2$ and ZnO single crystal [130,131]. The other factor is the poor absorption of low-energy photons by available dyes, especially in red and near-IR region. The IPCE as a function of wavelength for widely used sensitizers in DSSCs, N3 (ligand L), and black dye (ligand L'), is shown in Figure 2.24a. The absorption spectra of the sensitizers cover a large area, even extending to 800 nm for N3 and 900 nm for black dye. For comparison, the IPCE spectra of DSSCs with commonly nanocrystalline TiO$_2$ photoanode employing black dye as the sensitizer is also displayed in Figure 2.24b. The solid line marked with haze 3% presents the performance of the traditional nanocrystalline DSSCs. Haze, defined as the ratio of diffuse transmittance to total transmittance, is controlled by the addition of submicron particles (400 nm diameter) to the TiO$_2$ photoanode in the experiment [84]. Although the IPCE of black dye can reach 80% in the visible light region, the normal nanocrystalline DSSCs only have a value of 60%. Especially for the wavelength in the region of 800–900 nm, the IPCE of conventional nanocrystalline DSSCs is relatively lower than that achieved by the black dye, as illustrated in Figure 2.24.

In addition, high charge recombination rate in conventional DSSCs, due to considerable electron trap sites (interparticle contacts, grain boundaries, defects, surface states, self-trapping, etc.), is also believed to limit the power conversion efficiency.

In view of these, high-aspect-ratio nanostructures (such as nanotube, nanowire, nanorod, and nanosheet) have generated great interest, owing to their intriguing properties (e.g., accelerated electron transport and suppressed electron recombination) that induced by stretched grown structure with specified directionality and highly decreased interparticle contacts, which in turn enables a much thicker photoanode layer to maximize photon absorption.

Law et al. [110] reported a new DSSC architecture employing ZnO nanowire array as the photoanode prepared by dip coating and seeded growth, as shown in Figure 2.25. In this array, the nanowire length and diameter varied from 16–17 μm and 130–200 nm, respectively. The authors estimated an electron diffusion coefficient of $D_n = 0.05–0.5$ cm^2/s for single dry nanowire, which was several hundred times larger than that for the operating cells ($D_n \leq 10^{-4}$ cm^2/s, as mentioned above), suggesting fast electron transport in the nanowire structure. Moreover, the conductivity of the nanowire arrays increased by 5%–20% when bathed in the standard DSSC electrolyte. Besides, the electron injection in nanowire (~5 ps) was much faster than that in nanoparticle (~100 ps). All these characteristics would enable high IPCE.

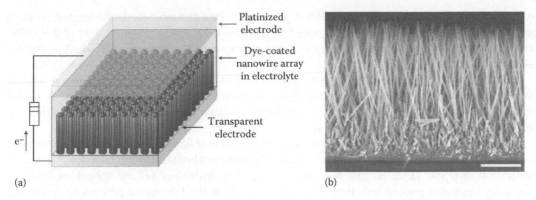

(a)   (b)

FIGURE 2.25 The nanowire dye-sensitized cell, based on a ZnO wire array. (a) Schematic diagram of the cell. Light is incident through the bottom electrode. (b) Typical scanning electron microscopy cross section of a cleaved nanowire array on FTO. The wires are in direct contact with the substrate, with no intervening particle layer. Scale bar, 5 μm. (Reprinted from Law, M. et al., *Nat. Mater.*, 4, 455, 2005. With permission.)

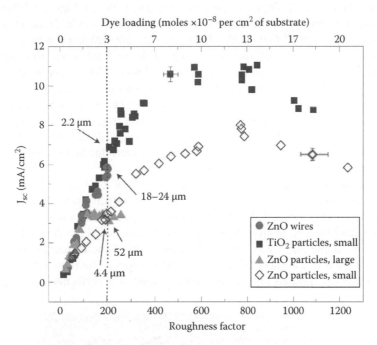

FIGURE 2.26 Comparative performance of nanowire and nanoparticle cells. (Reprinted from Law, M. et al., *Nat. Mater.*, 4, 455, 2005. With permission.)

However, the ZnO nanowire DSSCs exhibited a performance of $J_{sc} = 5.3$–$5.85\,\mathrm{mA/cm^2}$, $V_{oc} = 0.61$–$0.71\,\mathrm{V}$, FF $= 0.36$–0.38, and overall conversion efficiency $\eta = 1.2\%$–1.5% under AM 1.5 illumination. The relatively small J_{sc} resulted from low dye loading of the nanowire film, as shown in Figure 2.26. It indicated that, for the same J_{sc}, e.g., $6\,\mathrm{mA/cm^2}$, the dye loading of ZnO nanowire film with a thickness of 18–24 μm was comparable with that of conventional TiO_2 nanoparticle film with an order of magnitude smaller in thickness, i.e., 2.2 μm, which could be attributed to the 130–200 nm diameter of ZnO nanowire, in comparison to the conventional TiO_2 nanoparticle size of 15–20 nm, leading to much lower surface area as well as the dye-loading ability, though the discrimination in crystal structure as well as other properties of ZnO and TiO_2 may also be the reasons. On the other hand, the comparable J_{sc} also implied the desirable electron transport property in nanowire structure. The poor FF suggested high potential for further improvement of this DSSC architecture, and it was robust with respect to the alteration of the nanowire electrical properties, electrolyte concentration, and substrate.

Considering the superior electron injection and transport properties but relatively low dye-loading ability of the nanowire structure discussed above, the nanotube architecture is proposed as a result of similar architecture to the nanowire, much higher surface area induced by additional internal surfaces, and diverse preparation methods.

Conventional methods for TiO_2 nanotube preparation include hydrothermal, sol-gel, and template synthesis. However, the TiO_2 nanotubes fabricated by these methods are not suitable for photoanode utilization in DSSC [132]. Titania nanotubes prepared by hydrothermal and sol-gel methods are generally randomly packed with limited length. In addition, the fabrication process for hydrothermal or sol-gel method needs to be prosecuted in rigorous conditions, and usually takes a long time. The template method requires porous alumina or certain organic polymer as template; moreover, the dimension of the titania nanotubes fabricated is restricted and determined by the original template features, thus resulting in limited output and bad patterns.

Figure 2.27 shows the transmission electron microscopy (TEM) images of TiO_2 nanotubes prepared by the sol-gel method [7]. The disordered nanotubes with closed (Figure 2.27b) or open (Figure 2.27c) tips are viewed. Employing these titania nanotubes as photoanode, Flores et al. demonstrated an obtained efficiency of $\eta = 5.64\%$ (under light intensity $100\,mW/cm^2$) with $J_{sc} = 12.47\,mA/cm^2$, $V_{oc} = 0.829\,V$, and FF $= 0.55$. Though the morphology of the nanotubes was disordered, the J_{sc} and V_{oc} were still highly promoted compared to nanoparticle-based DSSCs. The superior performances was attributed to the high surface area of the photoanode that allowed more ruthenium dyes to be anchored to the oxide surface and created ample ion path for fast ionic motion from the bulk to the inner parts of the film. The outstanding property of electron transport of TiO_2 nanotube photoanode also renders an intriguing alternative for more efficient solid-state DSSCs. However, randomly packed structure was not preferred because of the electron scattering at the connections of the nanotubes as well as the uncontrollable tip features.

Figure 2.28 shows scanning electron microscopy images of ZnO nanotubes fabricated by template-assisted synthesis, based on commercially available $60\,\mu m$ thick anodic aluminum oxide (AAO) membrane as the template [133]. The nominally $60\,\mu m$ thick membrane with $200\,nm$ pores was coated with ZnO by atomic layer deposition (ALD) and then $1\,\mu m$ thick electrode composed of transparent conducting AZO was deposited on one side of the membrane by ALD. Then the membrane was applied to assemble DSSCs as the photoanode. The most efficient cell with $7\,nm$ thick ZnO nanotubes, deposited on the pore wall, exhibited a performance of $J_{sc} = 3.3\,mA/cm^2$,

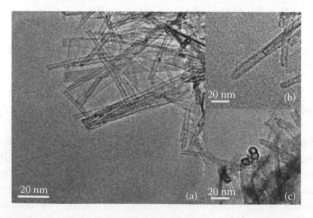

FIGURE 2.27 TEM images of the TiO_2 nanotubes prepared by sol-gel method at different magnifications. (Reprinted from Flores, I.C. et al., *J. Photochem. Photobiol. A*, 189, 156, 2007. With permission.)

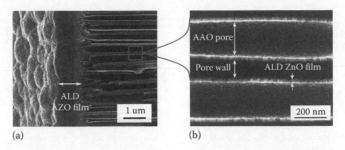

FIGURE 2.28 SEM images of ZnO nanotube photoanode prepared by template synthesis. (a) Cross-sectional SEM image of commercial anodic aluminum oxide membrane face coated with transparent conductive oxide AZO. (b) Cross-sectional SEM image of commercial AAO membrane pores coated with 20 nm of ZnO by atomic layer deposition. (Adapted from Martinson, A.B.F. et al., *Nano Lett.*, 7, 2183, 2007. With permission.)

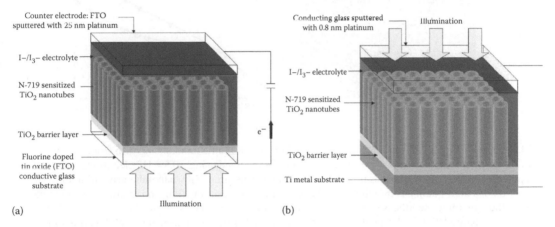

FIGURE 2.29 Schematic diagram of anodized TiO_2 nanotube–based DSSCs. (a) Front-side illuminated type; (b) back-side illuminated type. (Reprinted from Mor, G.K. et al., *Sol. Energy Mater. Sol. Cells*, 90, 2059, 2006. With permission.)

$V_{oc} = 0.739$ V, FF $= 0.64$, and $\eta = 1.6\%$ under light intensity 100 mW/cm². The lower J_{sc}, resulted from low dye-loading ability due to large pore diameter and low charge collection efficiency due to the thick photoanode, had limited the performance of the cell. Whereas, the two limiting factors largely depend on the initial template properties.

In contrast, the electrochemical anodization method for TiO_2 nanotube preparation exhibits unique properties when applied in DSSC, making it of considerable scientific interest as well as practical importance. As discussed before, the electrochemical anodization method is a simple, rapid, and inexpensive approach for TiO_2 nanotube preparation. The TiO_2 nanotube arrays formed are highly oriented with controllable pore diameter [10], wall thickness [5], intertube spacing, and tube length [5,65,66,68], which are desirable virtues for application in DSSCs.

There are two types of anodized TiO_2 nanotube array–based DSSCs, front-side and back-side illuminated [63], as shown in Figure 2.29. In the front-side illuminated DSSC, the TCO is used as the substrate on which a layer of Ti thin film is sputtered, which is subsequently undertaken anodization in fluoride-containing electrolyte to form TiO_2 nanotube array. After post-treatment, the nanotube array is employed as photoanode and assembled in DSSC. Since light can enter directly from the TCO substrate, this kind of DSSC is termed the front-side illuminated DSSC. In the back-side illuminated DSSC, Ti foil is used as the substrate, in which TiO_2 nanotube array is grown. The obtained array is exploited as photoanode of DSSC. As the opaque nature of the substrate (usually

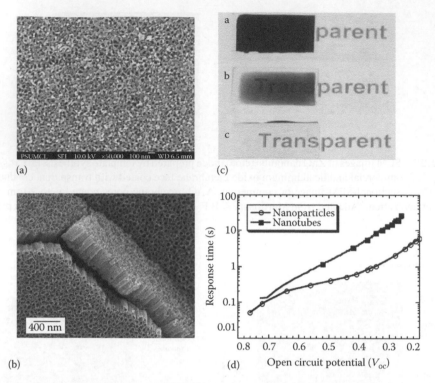

FIGURE 2.30 Top (a) and lateral (b) view FESEM images of titania nanotubes grown from a 500 nm thick Ti thin film (sputtered onto SnO$_2$:F coated glass at 500°C) anodized using a 0.5% HF electrolyte concentration at a potential of 12 V. (c) Key stages in the fabrication of a transparent TiO$_2$ nanotube-array film: (top) sputter deposition of a high-quality Ti thin film; (middle) anodization of resulting film; and (bottom) heat treatment to oxidize the remaining metallic islands. (d) Comparison of response time vs. V_{oc} for TiO$_2$ nanotube–based and nanoparticle-based DSSCs. (Adapted from Mor, G.K. et al., *Nano Lett.*, 6, 215, 2006. With permission.)

with a thickness of 100 or 250 μm), illumination needs to come from the platinized counter electrode, on top of the nanotube array. As the electrons generated are collected on the substrate side (the "front" side), while in this case the illumination is on the opposite side of the DSSC, this type of DSSC is termed "back side" illuminated DSSC.

Mor et al. [134] prepared highly ordered TiO$_2$ nanotube arrays with particulars of 46 nm in pore diameter, 17 nm in wall thickness, and 360 nm in length, which were perpendicular to the FTO substrate, by means of the anodic oxidation of Ti thin film sputtered onto the substrate beforehand. After that, front-side illuminated DSSCs were obtained using these nanotube arrays as photoanodes. Figure 2.30a and b display the top and lateral view images of titania nanotube arrays attained by FESEM, where the highly oriented structure can be clearly observed. The key stages in the fabrication of a transparent TiO$_2$ nanotube array are shown in Figure 2.30c, i.e., sputter deposition of high-quality Ti thin film (top), the anodization of the resulting film (middle), and heat-treatment to oxidize the remaining metallic islands (bottom). These front-side illuminated DSSCs presented J_{sc} of 7.87 mA/cm^2, V_{oc} of 0.75 V, and FF of 0.49, with an overall conversion efficiency of η = 2.9% under light intensity 100 mW/cm^2.

It is necessary to highlight that the length of the titania nanotubes was only 0.36 μm, whereas the performance of photovoltaic device was superior to the 60 μm thick ZnO nanotube DSSCs prepared by template synthesis as mentioned above. The low J_{sc} was attributed to the short tube length with less dye loadings, while the low FF was due to the inevitable barrier layer and poor contact between barrier layer and FTO substrate. In comparison to conventional TiO$_2$ nanoparticle–based DSSCs, the transparent nanotubes exhibited superior recombination characteristics, with the longer lifetimes indicating fewer recombination centers in the nanotubular films, illustrated in Figure 2.30d. In summary, the highly ordered architecture allows for improved charge separation and transport properties compared with typical TiO_2 nanoparticle DSSCs, in addition, comparable dye loadings are achievable if the nanotubes are long enough.

Zhu et al. [118] investigated the back-side illuminated DSSCs. The nanotube arrays prepared from electrochemically anodized Ti foils were found to consist of closely packed nanotubes, several micrometers in length (~5.7 μm), with typical wall thicknesses as well as intertube spaces of 8–10 nm and pore diameters of 30 nm. Dye molecules were shown to cover both the interior and exterior walls of the nanotubes. Besides favorable electron transport and recombination properties reported by others, the author demonstrated stronger internal light-scattering effects were discovered in nanotube-based DSSCs, which improved the light-harvesting efficiencies.

The thickness of conventional nanoparticle photoanodes is typically 10 μm [90], which is determined by Equations 2.15 through 2.17. As mentioned above, the highly oriented TiO_2 nanotube photoanode exhibits desirable electron transport and recombination properties. In view of this, it is believed that the highly ordered TiO_2 nanotube array photoanodes can be made much thicker. Grimes' group [66] studied the performance of back-side illuminated DSSCs with a 20 μm long anodized TiO_2 nanotube array (Figure 2.31). The cells presented J_{sc} of 12.72 mA/cm^2, V_{oc} of 0.817 V, and FF of 0.663, with an overall conversion efficiency of $\eta = 6.89\%$ under light intensity 100 mW/cm^2. The authors found that the use of nanotube arrays both shorter and longer than 20 μm resulted in lower photoconversion efficiencies. It seems that 20 μm is an ideal length. But their

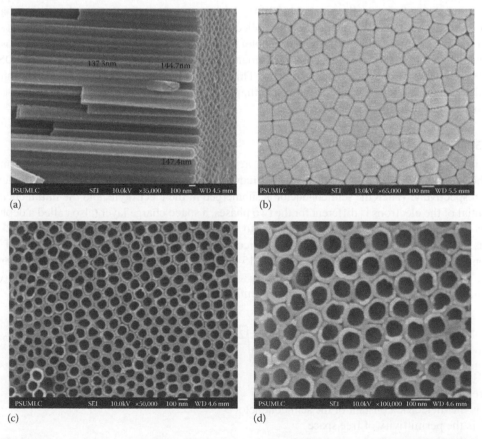

FIGURE 2.31   Lateral (a), bottom (b), and top (c) and (d) view FESEM images of the anodized TiO_2 nanotube arrays grown from Ti foil. (Reprinted from Shankar, K. et al., *Nanotechnology*, 18, 065707, 2007. With permission.)

previous work [63], for the same DSSC structure, showed a cell performance of $J_{sc} = 15\,mA/cm^2$, $V_{oc} = 0.842\,V$, FF = 0.43, and $\eta = 5.44\%$ under light intensity $100\,mW/cm^2$, where the length of the array was only $6\,\mu m$. Though the overall efficiency of this cell is lower than that of the former, both values of the J_{sc} and V_{oc} are higher, especially for the open-circuit voltage which approaches the theoretical value $0.9\,V$ [113] (some references giving a value of $\sim 1\,V$ [135]). The high V_{oc} value indicates the electron recombination in this photovoltaic device is effectively suppressed due to the existence of inevitable barrier layer incorporated during fabrication insulating the charge-collecting electrode from the electrolyte, and the negative shift of conduction band with respect to that of the nanocrystalline electrodes [136,137]. Moreover, the low FF reveals a large potential for further improvement of this cell. Consequently, there must be optimal values for nanotube geometries, such as length, wall thickness, pore diameter, and intertube spacing, to maximize LHE and overall conversion efficiency of DSSCs, which will be discussed later.

2.2.5 EFFECT OF TiO₂ NANOTUBE GEOMETRIES ON THE PERFORMANCE OF DYE-SENSITIZED SOLAR CELLS

The nanotube arrays as the photoanodes eliminate the particle-to-particle contacts necessary in the randomly packed configuration for electron transport. This largely reduces electron scattering at the joining points [7,136]. In nanotubes, usually the grains are stretched in the tube growth direction. This further decreases possible electron losses at grain boundaries [124,138–140]. Other obvious benefit in nanotube photoanode includes vectorial transport of electrons to minimize trapping/detrapping events before collection [124,138,141]. In the case of anodized nanotube array, a barrier layer exists between the photoanode and the collecting electrode. Electrons collected are thus separated from the redox couple (usually I^-/I_3^- couple) in the electrolyte [136,137], making it impossible for further recombination. However, the performance of the anodized nanotube array–based DSSCs is greatly affected by the nanotube geometry. This section dissects this geometry to look into the details of the influence of wall thickness, length, pore diameter, and intertube spacing to cast light into better design for high efficiency.

2.2.5.1 Effect of Wall Thickness

The initial chemical potential of electrons in a semiconductor and an electrolyte is determined by the Fermi level of the semiconductor and redox potential of the redox couples in the electrolyte, respectively. When the semiconductor is submerged in the electrolyte, and the initial chemical potential of the electrons is different for the two phases, a space charge layer (also called a depletion layer) develops in the semiconductor adjacent to the electrolyte, and a Helmholtz layer [142–144] forms in the electrolyte in contact with the semiconductor. As illustrated below, in the case of TiO₂ nanoparticles, the size (usually in the range of 15–20 nm) of the particle is too small to support the space charge layer [90,110,145,146].

The width of the space charge layer in the semiconductor can be determined by [142]

$$w = \sqrt{\frac{2\varepsilon\varepsilon_0 V_B}{qN}} \qquad (2.21)$$

where
  ε is the dielectric constant of the semiconductor
  ε_0 is the permittivity of free space
  q is the electronic charge
  N is the charge carrier concentration in the semiconductor
  V_B is the amount of band bending in the depletion layer

The value of V_B is given by [142]

$$V_B = U - U_{fb} \qquad (2.22)$$

where

U is the electrode potential (Fermi level) of the semiconductor

U_{fb} is the flat-band potential of the semiconductor

For titanium dioxide with anatase structure that is widely employed in the conventional DSSCs, the width of space charge layer is about 15–20 nm, using $\varepsilon = 31$ [147,148], $N = 10^{18}–10^{19}$ cm^{-3} [147–149], $U = -4.2$ eV (quasi-Fermi level of anatase TiO$_2$ under illumination) [150,151], and U_{fb} (usually depends on pH value of the electrolyte) [149,152–155]. Consequently, the size of a TiO$_2$ particle must be at least 30–40 nm to support space charge layer in the particle.

Now consider the photoanode of highly oriented TiO$_2$ nanotube array. If the wall thickness of the nanotube exceeds the critical value discussed above, the space charge layer forms. As a result, the conduction and valence band edges are upwardly bent [146] so that potential barriers are created near both inner and outer surfaces of the nanotubes against further electron transfer into the redox couples in the electrolyte [142,144], as shown in Figure 2.32a. Once the electrons are injected into

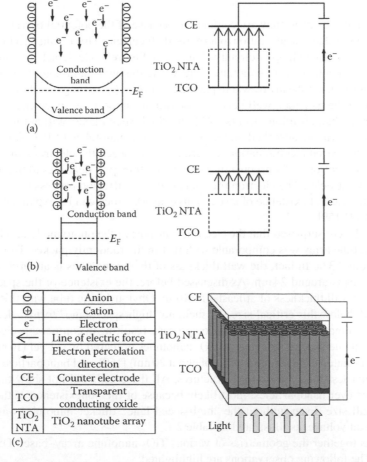

FIGURE 2.32 Schematic diagram of electron percolation in a thick-walled (a) and thin-walled (b) nanotube, and its corresponding band structure as well as line of electric force in the cell. (c) Instruction of the symbols and presentation of the 3D structure of the cell as well as the circuit under illumination. (Reprinted from Sun, L. et al., *J. Nanosci. Nanotechnol.*, 10, 1, 2010. With permission.)

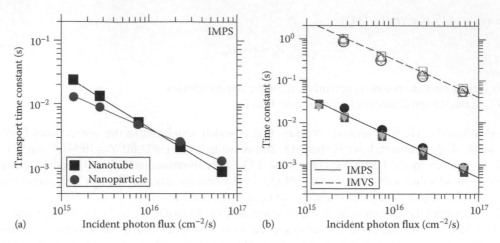

FIGURE 2.33   (a) Comparison of transport time constants for nanotube- and nanoparticle-based DSSCs as a function of the incident photon flux; (b) transport (IMPS, solid symbols) and recombination (IMVS, open symbols) time constants for dye-sensitized nanotube films of various thicknesses as a function of the incident photon flux. (Adapted from Zhu, K. et al., *Nano Lett.*, 7, 69, 2007. With permission.)

this "high-speed lane" by dyes, they are confined in because of the potential barrier and move away from the source of the electron generation to form the drift current. This results in a high short-circuit current density (J_{sc}) and a high open-circuit voltage (V_{oc}). Figure 2.32c depicts the illumination process at DSSC with a photoanode of nanotube array, where the light illuminates through a TCO layer to be absorbed in the nanotube array, where photon absorption/electron generation takes place.

In the event that the nanotube wall becomes too thin to support the space charge layer, negligibly small or even no band bending occurs [146], the "high-speed lane" will not form. In this case, cations in the electrolyte are adsorbed on the surfaces of the nanotubes [2,157,158] (i.e., Helmholtz layer) leading to increased recombination (Figure 2.32b) between the electrons in the channel and the triiodide in the electrolyte. Consequently, the electron transport degrades to that in the photoanode based on nanoparticles. The electron transport is thus a diffusion process (as in contrast to the drift process in the case of existence of space charge layer), which can be explained by the random walk model [111,141,159].

Zhu et al. [118] were surprised that the electron transport in the photoanode of a highly oriented anodic TiO_2 nanotube array was comparable with that in the randomly packed TiO_2 nanoparticles, as shown in Figure 2.33a. In fact, the wall thickness of their nanotubes is about 8 nm and the size of the nanoparticles is around 24 nm. As discussed before, the existence of the space charge layer needs a particle or wall thickness of at least 30–40 nm. Obviously, the tube wall thickness and particle size are both below this critical size. As such, no "high-speed lane" forms; electron transport in both cases is dominated by the same diffusion process, resulting in comparable electron transport time. Similar phenomena were reported in TiO_2 nanotubes prepared by sol-gel (10 nm in diameter) [138], randomly oriented nanowire (diameter of about 20 nm) [160], and nanoparticle-fused nanorod (effective diameter less than 5 nm) [124] structures. All these cells did not show the advantage of electron transport over nanoparticles, most likely because of the nonexistence of the space charge layer due to small size. Even without the "high-speed lane" thicker nanotube walls give rise to higher open-circuit voltage, as illustrated in Table 2.1.

Table 2.1 puts together the geometries of various TiO_2 nanotube array–based DSSCs and their performances. The following observations are highlighted:

1. There seems to be no direct relationship between the open-circuit voltage and the length of the nanotubes (V_{oc} vs. L).

TABLE 2.1
Performance of DSSCs with TiO$_2$ Nanotube Array Photoanodes

Cell No.	L (μm)	T (nm)	D (nm)	J_{sc} (mA/cm^2)	V_{oc} (V)	FF	η (%)	Ref.
			Back-side illumination					
1	0.36	17 ± 2	46 ± 8	2.40	0.786	0.69	1.3	[136]
2	1.9	8 ± 1	30 ± 4	4.4	0.64	0.60	1.7	[118]
3	3.2	8 ± 1	30 ± 4	6.6	0.63	0.58	2.4	[118]
4	4.3	8 ± 1	30 ± 4	7.8	0.61	0.57	2.7	[118]
5	5.7	8 ± 1	30 ± 4	9.0	0.61	0.55	3.0	[118]
6	6 ± 0.3	20	110	8.79	0.84	0.57	4.24	[137]
7	6.2 ± 0.3	20 ± 2	110 ± 10	10.6	0.82	0.51	4.4	[136]
8	20	24	132	12.72	0.817	0.663	6.89	[66]
			Front-side illumination					
9	0.36	17 ± 2	46 ± 8	7.87	0.75	0.49	2.9	[134]
10	3.6	17	46	10.3	0.84	0.54	4.7	[136]
11	5	20	100	16	0.43	0.70	4.8	[14]

Source: Reprinted from Sun, L. et al., *J. Nanosci. Nanotechnol.*, 10, 1, 2010. With permission.

Notes: Cell #1–10 by electrochemical anodization, cell # 11 by template assistant method. L, length of nanotube array; T, wall thickness; D, pore diameter; J_{sc}, short-circuit current density; V_{oc}, open-circuit voltage; FF, fill factor; η, overall conversion efficiency.

2. The pore diameters varied from about 30 to 130 nm while the open-circuit voltage only varied from 0.6 to 0.8 V. Thus the pore diameter does not affect V_{oc} noticeably.

3. The influence of the wall thickness can be extracted from separating the cells into two groups: group I with the same wall thickness of 8 nm (cell #2, 3, 4, and 5), and Group II with thickness more than its double, i.e., 17–24 nm (cell #1, 6, 7, and 8). Though wall thicknesses of both groups are below the critical thickness (30 nm) for "high-speed lane" formation, thicker-walled nanotubes still result in higher open-circuit voltage (0.8 vs. 0.6 V), as a result of lower chances of electrons interfacing with redox electrolyte thus lower chances of recombination.

4. Cell #11 achieved very high current density (16 mA/cm^2). However, it only achieved a conversion efficiency of 4.8%, likely due to its thin wall thickness (20 nm) which restricted the open-circuit voltage (only 0.43 V).

To allow fast and uninterrupted transportation of electrons through the nanotube, it is crucial that the tube wall is thick enough for the space charge layer to form.

2.2.5.2 Effect of Length

Table 2.1 also shows that the short-circuit current density (from 2.4 to about 13 mA/cm^2) and the overall conversion efficiency increase (from 1.3% to about 7%) as the length of the nanotube array increases from about 0.4 to 20 μm. This is a direct result of increased dye-loading capacity.

To obtain high conversion efficiency, the electrons injected into the conduction band of the TiO$_2$ photoanode should be collected as much as possible. A key parameter determining this process is the electron diffusion length as presented by Equation 2.15. Effective collection of electrons can be realized only if the electron diffusion length is larger than the film thickness (or nanotube length, L) [2,85], i.e., $L_n > L$. That means shorter nanotube length would result in more electrons being collected. On the other hand, however, the film must be thick enough to allow more photon/sensitizer reaction to generate more electrons. In a typical TiO$_2$ nanoparticle photoanode, the electron diffusion length is about 10 μm [108], i.e.,

$$L_{NP} = 10 \ \mu m \tag{2.23}$$

Figure 2.33b displays virtually no change in electron transport time as the length of the nanotube array changes from 1.9 to 5.7 μm. As discussed before, under thin-walled condition, the electron transport time (τ_c) is about the same as in nanoparticle-based DSSCs (cf. Figure 2.33a), which can be expressed as [141]

$$\tau_c \approx \frac{L^2}{2.35 D_n} \tag{2.24}$$

Therefore, the electron diffusion coefficient in nanotubes is, to some extent, similar to that in nanoparticles. Based on the results of Zhu et al. [118], the recombination time in nanotubes is an order of magnitude (10 times) greater than that in nanoparticles (similar result was also obtained in nanorod structure [124]). If so, Equation 2.15 thus gives rise to the electron diffusion length in nanotubes being $\sqrt{10}$ times that in a typical nanoparticle photoanode. As such, using Equations 2.15 and 2.23, the optimal length of TiO_2 nanotubes (L_{NT}) should be around

$$L_{NT} = \sqrt{10} L_{NP} = \sqrt{10} \times 10 \ \mu m \approx 30 \ \mu m \tag{2.25}$$

Shankar et al. [66] suggested the optimal nanotube length is nominally 20 μm (cell #8 in Table 2.1). Kim et al. [161] assembled nanotube array–based DSSCs with different lengths (Table 2.2). Maximum conversion efficiency seems to be with nanotube length of close to 30 μm. An evaluation of the photocurrent for dye-sensitized TiO_2 nanotube arrays displayed in Figure 2.34 [58,162] also indicates that the maximum photocurrent occurs at close to a nanotube length of 30 μm.

Depending on the orientation of the illumination with respect to the electron collecting electrode, a DSSC can be illuminated from the front side or the back side [63,136,163]. Electrons start their journey toward the load from the collecting electrode, thus this side is considered "front." Consequently, the counter electrode (CE) side is the "back side," where the electrons have to travel all the way through the whole photoanode pathway to be collected at the "front." Usually, TCO is used as collecting electrode in the front-side illuminated DSSC because of its light transmission and electrical conduction properties, as shown in Figure 2.32c. When Ti foil is employed as collecting electrode, the "front" side is no longer transparent (the foil is usually 100 or 250 μm), illumination then comes from the transparent platinized counter electrode (the "back-side"), thus the term "back-side illuminated."

It is noteworthy that the optimal nanotube length of about 30 μm discussed before is based on back-side illuminated DSSCs. Usually back-side illuminated DSSCs have much lower efficiency

---

TABLE 2.2
Performance of DSSCs with Different Nanotube Lengths

	L (μm)	J_{sc} (mA/cm^2)	V_{oc} (V)	FF	η (%)
NT	10	7.77	0.65	0.458	2.33
NT	20	9.62	0.66	0.450	2.88
NT	30	10.40	0.64	0.430	2.87
NGF-NT	10	7.52	0.71	0.539	2.87
NGF-NT	20	8.56	0.69	0.516	3.05
NGF-NT	30	9.39	0.67	0.468	2.94

Source: Reprinted from Kim, D. et al., *Electrochem. Commun.*, 10, 1837, 2008. With permission.

Note: L, length of nanotube array; J_{sc}, short-circuit current density; V_{oc}, open-circuit voltage; FF, fill factor; η: overall conversion efficiency.

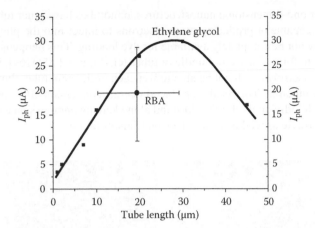

FIGURE 2.34 Photocurrent evaluation of the dye-sensitized nanotube layers grown to different tube lengths based on the photocurrent transients at 650 nm. (Reprinted from Macak, J.M. et al., *Curr. Opin. Solid State Mater. Sci.*, 11, 13, 2007. With permission.)

than the front-side illuminated counterpart. For example, cell #1 and 9 in Table 2.1 have the same geometry, but #1 is back-side illuminated and #9 is front-side illuminated. Cell #9 has much higher short-circuit current density (7.87 vs. 2.4 mA/cm^2) and much higher conversion efficiency (2.9% vs. 1.3%), which is more than twofold increase in efficiency.

The lower efficiency of the back-illumination type comes because of the following drawbacks: (1) energy loss due to reflection by the platinized counter electrode; (2) absorption of the photon by the iodine in the electrolyte [136,137] before reaching the nanotube array; and (3) after the electrons are generated, they have to travel all the way through the photoanode to reach the collecting electrode. The increased distance of travel drastically increased the chances of recombination. In contrast, front-side illumination has the advantage of not having the above problems. In addition, since there is no such loss of photon energy, the electron density in the front-side illuminated photoanode is much higher under light soaking. As such, some of the electrons fall into the trapping states in the bandgap of the oxide, paving way for other electrons to move more freely. Since the traps that participate in the electron motion affect the value of the diffusion coefficient [2], the reduction of quantity of traps results in increase in the electron diffusion coefficient, D_n. As illustrated in Equation 2.15, the electron diffusion length, L_n, is also related to electron lifetime, τ_n. According to Paulose et al. [136], the electron lifetime for both illumination types is comparable. Consequently, the electron diffusion length in the front-side illuminated photoanode should be larger than that in the back-side illuminated architecture, which in turn gives rise to longer optimal tube length. One recent estimation [64] of the electron diffusion length suggests that the optimal tube length is in the order of 100 μm in TiO$_2$ nanotube array–based DSSCs, using a method that takes into consideration the fact that the occupancies of electron traps are substantially different under open-circuit and short-circuit conditions.

2.2.5.3 Effect of Pore Diameter and Intertube Spacing

The pore diameter (or inner tube diameter) and intertube spacing have significant influence on surface roughness factor and porosity of nanotube arrays [66,118], thus affecting the performance of DSSCs. Nanotube arrays prepared by anodization have a unique close-end structure in which air may be trapped in during the infiltration of the dye solution and the redox electrolyte. This is more serious in the case of nanotubes with small pore diameters. Conventional titanium dioxide nanoparticles used in DSSCs are of 15–20 nm in diameter, which gives rise to very large surface area for possible generation of electrons. In nanotube arrays, however, the surface area is drastically reduced due to the tube structure and it becomes worse when large pore diameter tubes are used.

Compared to other one-dimensional nanostructures, nanotubes have inner tube surface available for dye loading, thus capable of producing more electrons to inject into the photoanode. The outer surface, however, may not be completely available for dye loading. This happens when the tubes are touching one another to form a strand or bundle of tubes (cf., Figure 2.35) [164]. In this case, the dye molecules and redox electrolyte will not be able to wet each of the nanotube, thus drastically reduce the effective dye-loading area. The result is reduced overall conversion efficiency, as illustrated in Table 2.3. Ideally, each individual tube should stand alone leaving enough space to facilitate infiltration of dyes and electrolyte as well as faster electron transportation.

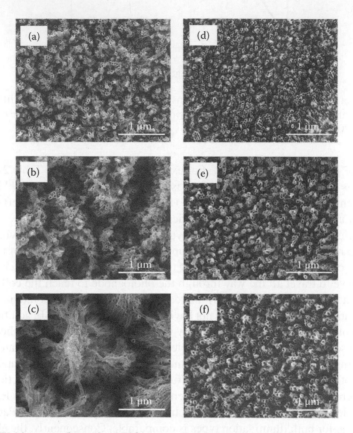

FIGURE 2.35   SEM images of anodized TiO_2 nanotube array with bundle structure (a) 1.1 μm, (b) 2.8 μm, (c) 6.1 μm, and bundle-free structure (d) 1.1 μm, (e) 2.8 μm, and (f) 6.1 μm. (Adapted from Zhu, K. et al., *Nano Lett.*, 7, 3739, 2007. With permission.)

TABLE 2.3
Performance of Dye-Sensitized Solar Cells with Bundle and Bundle-Free Nanotube Structures

	L (μm)	J_{sc} (mA/cm²)	V_{oc} (V)	FF	η (%)
Bundle structure	6.1	4.9	0.60	0.53	1.6
Bundle-free structure	6.1	5.7	0.58	0.56	1.9

Source: Zhu, K. et al., *Nano Lett.*, 7, 3739, 2007. With permission.

L, length of nanotube array; J_{sc}, short-circuit current density; V_{oc}, open-circuit voltage; FF, fill factor; η, overall conversion efficiency.

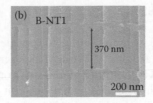

FIGURE 2.36 SEM images of anodized TiO_2 nanotube array with smooth wall (a), and bamboo-type wall (b) and (c). (Adapted from Kim, D. et al., *J. Am. Chem. Soc.*, 130, 16454, 2008. With permission.)

TABLE 2.4
Performance of Dye-Sensitized Solar Cells with Bamboo-Type Nanotubes

	L (μm)	J_{sc} (mA/cm²)	V_{oc} (V)	FF	η (%)	L_{dye} (au)
NT	8	5.93	0.59	0.51	1.90	52
B-NT1	8	7.85	0.60	0.51	2.48	69
B-NT2	8	8.76	0.62	0.52	2.96	78

Source: Adapted from Kim, D. et al., *J. Am. Chem. Soc.*, 130, 16454, 2008. With permission.

Note: L, length of nanotube array; J_{sc}, short-circuit current density; V_{oc}, open-circuit voltage; FF, fill factor; η, overall conversion efficiency; L_{dye}, dye loading per unit volume.

The roughness of the outer surface of the nanotubes also affects the efficiency in DSSCs. The bamboo-type nanotubes prepared by modulating electrochemical anodization process [165], as illustrated in Figure 2.36, result in increased short-circuit current density (8.76 vs. 5.93 mA/cm²) and overall conversion efficiency (2.96% vs. 1.90%) as compared to smooth nanotube outer surface, as shown in Table 2.4. This is because of the increased dye loading per unit volume. The additional area provided by the bamboo rings may allow both inner and outer surfaces to be covered by dye molecules as well [165].

2.2.5.4 Summary

This section reviewed the effect of geometry of TiO_2 nanotube arrays synthesized by electrochemical anodization on the performance of the DSSCs using these nanotube arrays as the photoanodes. The following points are worth highlighting:

1. There seems to exist a critical wall thickness for fast electron transportation. The critical thickness comes from the formation of the space charge layer. If the wall is too thin to support this layer, the electron transport will be degraded to that of the nanoparticle case. The estimated critical wall thickness is about 30–40 nm.
2. There also exists an optimal nanotube length for optimal electron collection in consideration of dye-loading capacity and electron diffusion length. In back-side illuminated DSSCs, the optimal length is close to 30 μm. Under front-side illumination, the value could even be a lot greater.
3. The pore diameter should be as small as possible to increase surface area and the nanotubes should ideally be free of bundling to facilitate dye and electrolyte infiltration.

2.2.6 THE CHALLENGE

Although the anodized TiO_2 nanotube arrays are widely investigated for applications in DSSCs, to date, there are still many challenges as listed below.

For the front-side illuminated DSSCs:

1. The relatively short nanotube array gives rise to considerably less photoabsorption than the conventional DSSCs with an optimal 10 μm thick layer of TiO_2 nanoparticles, since the nanotube array is grown from the Ti thin film sputtered on the TCO, whereas it is difficult to deposit high-quality Ti films suitable for anodization of greater thickness. Currently, only 3.6 μm long arrays have been achieved [136], which is far from the ideal length that is believed is larger than 10 μm because of the prominent electron transport property and electron lifetime.
2. Nonuniform dye adsorption within the pores of nanotube array caused by trapped air [63] that is also the case for back-side illuminated DSSCs. The surface of the connected nanoparticles can be totally covered by dye molecules for conventional DSSCs, even at the bottom of the nanoparticulate electrode, since the air can be driven out easily from the mesoporous structure. Whereas the nanotube geometry has only one opening while the bottom of nanotubes is closed. The air could be trapped in the tubes, such that it prevents the dye molecules from entering the bottom of the nanotubes, resulting in reduced dye loadings.
3. Low fill factor as a result of unavoidable barrier layer and poor contact between the barrier layer and the fluorine-doped tin oxide substrate [63]. The appearance of the barrier layer increases the internal resistance of the DSSC system. The adhesion between barrier layer and FTO substrate is a function of initial Ti film quality.
4. The resistance of the FTO substrate increases at least one order of magnitude during the oxygen annealing step to crystallize the nanotube array at 450°C [136].

For the back-side illuminated DSSCs:

1. The platinized counter electrode partially reflects incident light. Since the back-side-type DSSCs need illuminating from the counter electrode as discussed above, the TCO is usually adopted as substrate of the counter electrode, and a thin layer of Pt is deposited on the substrate to catalyze the reduction of iodide/triiodide redox couple in the electrolyte, thus leading to the loss of incident light via reflection by the thin platinum layer.
2. The iodine in the electrolyte absorbs photons at lower wavelength (400–600 nm) [132,137].
3. Low fill factor due to relatively thick barrier layer. The annealing step used to crystallize the nanotube arrays can significantly increase the barrier layer thickness up to, approximately, 1 μm [166]. The presence of a thick barrier layer causes increase of the series resistance in photovoltaic device, thus reducing the levels of the fill factor.
4. Electron injection occurs predominantly at the "back" of the nanotube arrays (opposite the barrier layer) far from the collecting electrode, thereby the injected electrons undergo more trapping/detrapping events, compared to that in the front-side illuminated, before reaching the collecting electrode.

Accordingly, the front-side illuminated architecture is much superior and preferable, as the undesirable energy loss is inevitable for the back-side illuminated structure.

To date, the highest efficiency achieved by titania nanoparticle-based DSSCs is ~11% [79,82–84]. Whereas the most efficient DSSCs based on anodized titania nanotube array is ~7%, as shown in Table 2.1. It is expected that the performance of the latter should be better as a result of superior electron transport and recombination behaviors. However, the use of nanotube array is not as effective as that of conventional nanoparticles. The overall conversion efficiency of the cells can be influenced by various factors, such as the dye (e.g., N3 or black dye), the electrolyte (e.g., liquid or solid), the series resistance of the system, the phase of titania (e.g., anatase or rutile), the preferred orientation of anatase structure (e.g., whether in the desired (101) direction), the grain size, the particle size, and so on. Apparently, the performance of DSSCs based on the titania nanotube array

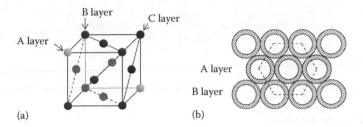

FIGURE 2.37 (a) Face-centered cubic structure of close-packed nanoparticles; (b) hexagonal close-packed structure of ideal nanotube arrays.

is restricted by the low short-circuit current density (which is about maximum $13\,mA/cm^2$ in Table 2.1 compared to that over $20\,mA/cm^2$ in high-efficiency nanoparticle DSSCs [82–84]). Although adopting different dyes and illumination types are two important aspects, difference in effective surface areas between nanoparticle and nanotube photoanodes may be another key factor because the surface area directly results in different dye loading.

To illustrate this point, surface areas of nanoparticle and nanotube photoanodes are calculated based on the following assumptions:

1. The nanoparticles are spherical and the nanotubes are regular cylindrical tubes.
2. The wall thickness and bottom area of the nanotubes are negligible.
3. The contact areas between nanoparticles or nanotubes are ignored (though this may not be negligible).

The face-centered cubic (FCC) structure is the most closely packed form of spheres in the space with *ABCABC...* sequence, as illustrated in Figure 2.37a. In this architecture, the occupying percentage of nanoparticles (η_{NP}) can be calculated:

$$n = \frac{1}{8} \times 8 + \frac{1}{2} \times 6 = 4 \tag{2.26}$$

$$4r = \sqrt{2}a \tag{2.27}$$

$$\eta_{NP} = \frac{V_{NP}}{V} = \frac{(4/3)\pi r^3 \cdot n}{a^3} = 74.05\% \tag{2.28}$$

where
 n is number of nanoparticles in the unit cell
 r is the radius of the nanoparticle
 a is the lattice constant of the unit cell

With nanotubes, the space arrangement can only be *ABABAB...* sequence, as shown in Figure 2.37b. To facilitate further calculation, the space filling factor of nanotubes (η_{NT}) is calculated assuming that the inner tube is solid instead of hollow (i.e., nanorod structure), or

$$n = 1 + \frac{1}{3} \times 6 = 3 \tag{2.29}$$

$$2R = a \tag{2.30}$$

$$\eta_{NT} = \frac{V_{NT}}{V} = \frac{\pi R^2 h \cdot n}{\left(\sqrt{3}/4\right) a^2 \cdot 6 \cdot h} = 90.69\% \tag{2.31}$$

where

n is number of solid nanotubes (i.e., nanorods) in a hexagonal column (see dashed line in Figure 2.37b)

R is the radius of the nanotube

a is the lattice constant of the hexagonal unit cell

h is the height of the column

The key point that is concerned herein is how to obtain comparable surface area of nanoparticle and nanotube photoanodes under the same volume. Assuming there are x nanoparticles and y nanotubes in respective photoanode under the same volume V,

$$\frac{4}{3}\pi r^3 \cdot x = \eta_{NP}V = 74.05\%V \left(\text{for nanoparticles}\right) \tag{2.32}$$

$$\pi R^2 h \cdot y = \eta_{NT}V = 90.69\%V \left(\text{for nanotubes}\right) \tag{2.33}$$

$$4\pi r^2 \cdot x = 2\pi R \cdot h \cdot 2 \cdot y \left(\text{for the same surface area}\right) \tag{2.34}$$

Equation 2.34 presents x nanoparticles and y nanotubes having the same surface area. Note that the second number 2 on the right of the equation originates from the two surfaces (both inner and outer surface) of each nanotube. According to Equations 2.32 through 2.34, the following relation can be obtained:

$$\frac{r}{R} \approx \frac{3}{5} \tag{2.35}$$

Equation 2.35 indicates that the surface areas of nanoparticle and nanotube photoanodes are comparable under the same photoanode volume, provided that the radius ratio of nanoparticle to nanotube is about 3:5. In conventional titania nanoparticle–based DSSCs, the particle size is normally in the range of 15–20 nm. Taking the particle size of 18 nm for example, it requires the nanotube outer diameter to go down to 30 nm to have comparable surface areas. As presented and discussed in Section 2.1.3, both the outer and the inner diameter of the nanotubes prepared by electrochemical anodization are affected by electric field strength. The small outer diameter (e.g., 30 nm) can only be achieved at low field strength. However, the growth rate is relatively small at such low field strength. In other words, it is difficult to produce long nanotube arrays (e.g., optimized ~30 μm discussed in Section 2.2.5.2) with 30 nm in outer diameter, which can be reflected from Table 2.1 (up to date, most anodic nanotubes employed in DSSCs are larger than 30 nm in tube outer diameter). Even though the required 30 nm is achieved, the utilization of both inner and outer nanotube surfaces remains another challenging issue, as discussed in Section 2.2.5.3. In view of these factors, in general, the effective surface area per volume of the nanotube array photoanode is relatively small compared to that of the conventional nanoparticle photoanode, thus resulting in less dye loading which, in turn, results in the low short-circuit current density and overall low conversion efficiency.

The above discussion is based on the calculation without taking into account the nanotube wall thickness and bottom area (assumption #2). If these factors are taken into consideration, in the case of funnel shaped [13,67,167,168] nanotube inner walls, Equation 2.35 becomes

$$\frac{r}{R} > \frac{3}{5} \qquad (2.36)$$

This is because the second number 2 on the right of Equation 2.34 should be substituted with a relatively smaller number due to reduced inner tube surface area. In this scenario, it requires even smaller nanotube outer diameter to produce comparable surface area with reference to the conventional nanoparticle photoanode, which is even more difficult to achieve.

Therefore, as it stands, it is difficult for anodized titania nanotube array–based solar cell to "beat" the nanoparticle-based counterpart unless drastic improvements in surface area is realized to enable drastic increase in the generation of electrons. Only then can the advantage be capitalized of the superior electron transport and recombination properties for high-efficiency DSSCs.

ABBREVIATIONS

AAO	Anodic aluminum oxide
ALD	Atomic layer deposition
AZO	Aluminum-doped zinc oxide
CIGS	Copper indium gallium diselenide
DSSC	Dye-sensitized solar cell
FCC	Face-centered cubic
FESEM	Field emission scanning electron microscope
FF	Fill factor
FTO	Fluorine-doped tin oxide
IPCE	Incident photon to current efficiency
ITO	Indium-tin oxide
LHE	Light-harvesting efficiency
NHE	Normal hydrogen electrode
SEM	Scanning electron microscope
TCO	Transparent conducting oxide
TEM	Transmission electron microscope
TW	Terawatts
XPS	X-ray photoelectron spectroscopy
XRD	X-ray diffraction

SYMBOLS

a	Lattice constant
c	Concentration
d	Thickness
D_n	Electron diffusion coefficient
E	Electric field strength
F	Faraday constant
$F(\lambda)$	Incident photon flux density
i	Current density
i_0	Reverse saturation current
$IPCE(\lambda)$	Incident photon to current efficiency
J_{mpp}	Photocurrent density at maximum power point
J_{sc}	Short-circuit photocurrent density
k_{deact}	Fraction of deactivated electrons
k_{inj}	Fraction of injected electrons
L	Nanotube length

L_n	Electron diffusion length
L_{NT}	Optimal nanotube length
$LHE(\lambda)$	Light-harvesting efficiency
N	Charge carrier concentration
P_{in}	Intensity of incident light
q	Electron elementary charge
r	Radius of nanoparticle
R	Gas constant, radius of nanotube
$r(\lambda)$	Energy loss of incident light
T	Kelvin temperature
U	Electrode potential
U_{fb}	Flat-band potential
V	Volume
V_B	Amount of band bending
V_{mpp}	Photovoltage at maximum power point
V_{oc}	Open-circuit photovoltage
α	Reciprocal absorption length
ε	Dielectric constant
ε_0	Permittivity of free space
ϕ_{inj}	Quantum yield
η	Overall-conversion efficiency
η_{cc}	Charge-collection efficiency
η_{NP}	Occupying percentage of nanoparticles
η_{NT}	Occupying percentage of nanotubes
σ	Optical absorption cross section
τ_n	Electron lifetime
τ_c	Electron transport time

REFERENCES

1. B. O'Regan and M. Grätzel. 1991. A low-cost, high-efficiency solar cell based on dye-sensitized colloidal TiO_2 films. *Nature* 353: 737–739.
2. M. Grätzel. 2005. Solar energy conversion by dye-sensitized photovoltaic cells. *Inorg. Chem.* 44: 6841–6851.
3. P. Xiao, B. B. Garcia, Q. Guo, D. Liu, and G. Cao. 2007. TiO_2 nanotube arrays fabricated by anodization in different electrolytes for biosensing. *Electrochem. Commun.* 9: 2441–2447.
4. M. Paulose, O. K. Varghese, G. K. Mor, C. A. Grimes, and K. G. Ong. 2006. Unprecedented ultra-high hydrogen gas sensitivity in undoped titania nanotubes. *Nanotechnology* 17: 398–402.
5. G. K. Mor, K. Shankar, M. Paulose, O. K. Varghese, and C. A. Grimes. 2005. Enhanced photocleavage of water using titania nanotube arrays. *Nano Lett.* 5: 191–195.
6. V. Zwilling, M. Aucouturier, and E. Darque-Ceretti. 1999. Anodic oxidation of titanium and TA6V alloy in chromic media. An electrochemical approach. *Electrochim. Acta* 45: 921–929.
7. I. C. Flores, J. Nei de Freitas, C. Longo, M. A. De Paoli, H. Winnischofer, and A. F. Nogueira. 2007. Dye-sensitized solar cells based on TiO_2 nanotubes and a solid-state electrolyte. *J. Photochem. Photobiol. A* 189: 153–160.
8. S. Bauer, S. Kleber, and P. Schmuki. 2006. TiO_2 nanotubes: Tailoring the geometry in H_3PO_4/HF electrolytes. *Electrochem. Commun.* 8: 1321–1325.
9. W. Chanmanee, A. Watcharenwong, C. R. Chenthamarakshan, P. Kajitvichyanukul, N. R. de Tacconi, and K. Rajeshwar. 2007. Titania nanotubes from pulse anodization of titanium foils. *Electrochem. Commun.* 9: 2145–2149.
10. Q. Cai, M. Paulose, O. K. Varghese, and C. A. Grimes. 2005. The effect of electrolyte composition on the fabrication of self-organized titanium oxide nanotube arrays by anodic oxidation. *J. Mater. Res.* 20: 230–236.

11. J. M. Macak, H. Hildebrand, U. Marten-Jahns, and P. Schmuki. 2008. Mechanistic aspects and growth of large diameter self-organized TiO_2 nanotubes. *J. Electroanal. Chem.* 621: 254–266.

12. H. Tsuchiya, J. M. Macak, A. Ghicov, L. Taveira, and P. Schmuki. 2005. Self-organized porous TiO_2 and ZrO_2 produced by anodization. *Corros. Sci.* 47: 3324–3335.

13. K. Yasuda and P. Schmuki. 2007. Control of morphology and composition of self-organized zirconium titanate nanotubes formed in $(NH_4)_2SO_4/NH_4F$ electrolytes. *Electrochim. Acta* 52: 4053–4061.

14. N. N. Bwana. 2009. Improved short-circuit photocurrent densities in dye-sensitized solar cells based on ordered arrays of titania nanotubule electrodes. *Curr. Appl. Phys.* 9: 104–107.

15. D. Gong, C. A. Grimes, O. K. Varghese et al. 2001. Titanium oxide nanotube arrays prepared by anodic oxidation. *J. Mater. Res.* 16: 3331–3334.

16. J. M. Macak and P. Schmuki. 2006. Anodic growth of self-organized anodic TiO_2 nanotubes in viscous electrolytes. *Electrochim. Acta* 52: 1258–1264.

17. R. Beranek, H. Hildebrand, and P. Schmuki. 2003. Self-organized porous titanium oxide prepared in H_2SO_4/HF electrolytes. *Electrochem. Solid-State Lett.* 6: B12–B14.

18. J. M. Macak, K. Sirotna, and P. Schmuki. 2005. Self-organized porous titanium oxide prepared in Na_2SO_4/NaF electrolytes. *Electrochim. Acta* 50: 3679–3684.

19. L. V. Taveira, J. M. Macák, H. Tsuchiya, L. F. P. Dick, and P. Schmuki. 2005. Initiation and growth of self-organized TiO_2 nanotubes anodically formed in $NH_4F/(NH_4)_2SO_4$ electrolytes. *J. Electrochem. Soc.* 152: B405–B410.

20. H. Tsuchiya, J. M. Macak, L. Taveira et al. 2005. Self-organized TiO_2 nanotubes prepared in ammonium fluoride containing acetic acid electrolytes. *Electrochem. Commun.* 7: 576–580.

21. I. Paramasivam, J. M. Macak, T. Selvam, and P. Schmuki. 2008. Electrochemical synthesis of self-organized TiO_2 nanotubular structures using an ionic liquid (BMIM-BF4). *Electrochim. Acta* 54: 643–648.

22. R. Hahn, J. M. Macak, and P. Schmuki. 2007. Rapid anodic growth of TiO_2 and WO_3 nanotubes in fluoride free electrolytes. *Electrochem. Commun.* 9: 947–952.

23. C. Richter, Z. Wu, E. Panaitescu, R. Willey, and L. Menon. 2007. Ultrahigh-aspect-ratio titania nanotubes. *Adv. Mater.* 19: 946–948.

24. C. Richter, E. Panaitescu, R. Willey, and L. Menon. 2007. Titania nanotubes prepared by anodization in fluorine-free acids. *J. Mater. Res.* 22: 1624–1631.

25. G. E. Thompson, R. C. Furneaux, G. C. Wood, J. A. Richardson, and J. S. Goode. 1978. Nucleation and growth of porous anodic films on aluminium. *Nature* 272: 433–435.

26. G. E. Thompson, K. Shimizu, and G. C. Wood. 1980. Observation of flaws in anodic films on aluminium. *Nature* 286: 471–472.

27. G. E. Thompson and G. C. Wood. 1981. Porous anodic film formation on aluminium. *Nature* 290: 230–232.

28. H. Masuda and K. Fukuda. 1995. Ordered metal nanohole arrays made by a two-step replication of honeycomb structures of anodic alumina. *Science* 268: 1466–1468.

29. H. Tsuchiya and P. Schmuki. 2005. Self-organized high aspect ratio porous hafnium oxide prepared by electrochemical anodization. *Electrochem. Commun.* 7: 49–52.

30. I. Sieber, H. Hildebrand, A. Friedrich, and P. Schmuki. 2005. Formation of self-organized niobium porous oxide on niobium. *Electrochem. Commun.* 7: 97–100.

31. I. V. Sieber and P. Schmuki. 2005. Porous tantalum oxide prepared by electrochemical anodic oxidation. *J. Electrochem. Soc.* 152: B405–B410.

32. I. Sieber, B. Kannan, and P. Schmuki. 2005. Self-assembled porous tantalum oxide prepared in H_2SO_4/HF electrolytes. *Electrochem. Solid-State Lett.* 8: J10–J12.

33. W. Wei, J. M. Macak, and P. Schmuki. 2008. High aspect ratio ordered nanoporous TaO_5 films by anodization of Ta. *Electrochem. Commun.* 10: 428–432.

34. H. Tsuchiya, J. M. Macak, I. Sieber et al. 2005. Self-organized porous WO_3 formed in NaF electrolytes. *Electrochem. Commun.* 7: 295–298.

35. Y.-C. Nah, A. Ghicov, D. Kim, and P. Schmuki. 2008. Enhanced electrochromic properties of self-organized nanoporous WO_3. *Electrochem. Commun.* 10: 1777–1780.

36. H. Tsuchiya and P. Schmuki. 2004. Thick self-organized porous zirconium oxide formed in H_2SO_4/NH_4F electrolytes. *Electrochem. Commun.* 6: 1131–1134.

37. H. Tsuchiya, J. M. Macak, L. Taveira, and P. Schmuki. 2005. Fabrication and characterization of smooth high aspect ratio zirconia nanotubes. *Chem. Phys. Lett.* 410: 188–191.

38. S. Berger, F. Jakubka, and P. Schmuki. 2008. Formation of hexagonally ordered nanoporous anodic zirconia. *Electrochem. Commun.* 10: 1916–1919.

39. V. Lehmann and H. Föll. 1990. Formation mechanism and properties of electrochemically etched trenches in n-type silicon. *J. Electrochem. Soc.* 137: 653–659.

40. L. T. Canham. 1990. Silicon quantum wire array fabrication by electrochemical and chemical dissolution of wafers. *Appl. Phys. Lett.* 57: 1046–1048.

41. H. Tsuchiya. M. Hueppe, T. Djenizian, and P. Schmuki. 2003. Electrochemical formation of porous superlattices on n-type (100) InP. *Surf. Sci.* 547: 268–274.

42. H. Tsuchiya, M. Hueppe, T. Djenizian, P. Schmuki, and S. Fujimoto. 2004. Morphological characterization of porous InP superlattices. *Sci. Technol. Adv. Mater.* 5: 119–123.

43. K. Yasuda and P. Schmuki. 2007. Electrochemical formation of self-organized zirconium titanate nanotube multilayers. *Electrochem. Commun.* 9: 615–619.

44. H. Tsuchiya, J. M. Macak, A. Ghicov, et al. 2006. Nanotube oxide coating on Ti-29Nb-13Ta-4.6Zr alloy prepared by self-organizing anidization. *Electrochim. Acta* 52: 94–101.

45. X. Feng, J. M. Macak, and P. Schmuki. 2007. Flexible self-organization of two size-scales oxide nanotube on Ti45Nb alloy. *Electrochem. Commun.* 9: 2403–2407.

46. H. Tsuchiya, S. Berger, J. M. Macak, A. Ghicov, and P. Schmuki. 2007. Self-organized porous and tubular oxide layers on TiAl alloys. *Electrochem. Commun.* 9: 2397–2402.

47. P. G. Sheasby and R. Pinner. 2001. *The Surface Treatment and Finishing of Aluminium and Its Alloys*, Vol. 1 (6th edn.). ASM International & Finishing Publications, Middlesex, U.K.

48. G. E. Thompson. 1997. Porous anodic alumina: Fabrication, characterization and applications. *Thin Solid Films* 297: 192–201.

49. G. C. Wood, P. Skeldon, G. E. Thompson, and K. Shimizu. 1996. A model for the incorporation of electrolyte species into anodic alumina. *J. Electrochem. Soc.* 143: 74–83.

50. N. Cabrera and N. F. Mott. 1948–1949. Theory of the oxidation of metals. *Rep. Prog. Phys.* 12: 163–184.

51. F. Brown and W. D. Mackintosh. 1973. The use of Rutherford backscattering to study the behavior of ion-implanted atoms during anodic oxidation of aluminium: Ar, Kr, Xe, K, Rb, Cs, Cl, Br, and I. *J. Electrochem. Soc.* 120: 1096–1102.

52. K. Shimizu, K. Kobayashi, G. E. Thompson, and G. C. Wood. 1991. A novel marker for the determination of transport numbers during anodic barrier oxide growth on aluminium. *Philos. Mag. B* 64: 345–353.

53. P. Skeldon, K. Shimizu, G. E. Thompson, and G. C. Wood. 1985. Fundamental studies elucidating anodic barrier-type film growth on aluminium. *Thin Solid Films* 123: 127–133.

54. J. Siejka and C. Ortega. 1977. An O^{18} Study of field–assisted pore formation in compact anodic oxide films on aluminium. *J. Electrochem. Soc.* 124: 883–891.

55. V. P. Parkhutik and V. I. Shershulsky. 1992. Theoretical modelling of porous oxide growth on aluminium. *J. Phys. D: Appl. Phys.* 25: 1258–1263.

56. D. J. LeClere, A. Velota, P. Skeldon et al. 2008. Tracer investigation of pore formation in anodic titania. *J. Electrochem. Soc.* 155: C487–C494.

57. H. Habazaki, K. Shimizu, S. Nagata, P. Skeldon, G. E. Thompson, and G. C. Wood. 2002. Ionic transport in amorphous anodic titania stabilised by incorporation of silicon species. *Corros. Sci.* 44: 1047–1055.

58. J. M. Macak, H. Tsuchiya, A. Ghicov et al. 2007. TiO_2 nanotubes: Self-organized electrochemical formation, properties and applications. *Curr. Opin. Solid State Mater. Sci.* 11: 3–18.

59. G. K. Mor, O. K. Varghese, M. Paulose, N. Mukherjee, and C. A. Grimes. 2003. Fabrication of tapered, conical-shaped titania nanotubes. *J. Mater. Res.* 18: 2588–2593.

60. G. Patermarakis and K. Moussoutzanis. 1995. Mathematical models for the anodization conditions and structural features of porous anodic Al_2O_3 films on aluminium. *J. Electrochem. Soc.* 142: 737–743.

61. K. Yasuda, J. M. Macak, S. Berger, A. Ghicov, and P. Schmuki. 2007. Mechanistic aspects of the self-organization process for oxide nanotube formation on valve metals. *J. Electrochem. Soc.* 154: C472–C478.

62. J. Kunze, A. Seyeux, and P. Schmuki. 2008. Anodic TiO_2 layer conversion: Fluoride-induced rutile formation at room temperature. *Electrochem. Solid-State Lett.* 11: K11–K13.

63. G. K. Mor, O. K. Varghese, M. Paulose, K. Shankar, and C. A. Grimes. 2006. A review on highly ordered, vertically oriented TiO_2 nanotube arrays: Fabrication, material properties, and solar energy applications. *Sol. Energy Mater. Sol. Cells* 90: 2011–2075.

64. J. R. Jennings, A. Ghicov, L. M. Peter, P. Schmuki, and A. B. Walker. 2008. Dye-sensitized solar cells based on ordered TiO_2 nanotube arrays: Transport, trapping, and transfer of electrons. *J. Am. Chem. Soc.* 130: 13364–13372.

65. M. Paulose, K. Shankar, S. Yoriya et al. 2006. Anodic growth of highly ordered TiO_2 nanotube arrays to 134 µm in length. *J. Phys. Chem. B* 110: 16179–16184.

66. K. Shankar, G. K. Mor, H. E. Prakasam et al. 2007. Highly-ordered TiO_2 nanotube arrays up to 220 μm in length: Use in water photoelectrolysis and dye-sensitized solar cells. *Nanotechnology* 18: 065707.

67. S. P. Albu, A. Ghicov, J. M. Macak, and P. Schmuki. 2007. 250 μm long anodic TiO_2 nanotubes with hexagonal self-ordering. *Phys. Stat. Sol. (RRL)* 1: R65–R67.

68. H. E. Prakasam, K. Shankar, M. Paulose, O. K. Varghese, and C. A. Grimes. 2007. A new benchmark for TiO_2 nanotube array growth by anodization. *J. Phys. Chem. C* 111: 7235–7241.

69. J. M. Macák, H. Tsuchiya, and P. Schmuki. 2005. High-aspect-ratio TiO_2 nanotubes by anodization of titanium. *Angew. Chem. Int. Ed.* 44: 2100–2102.

70. L. Sun, S. Zhang, and X. W. Sun, X. He. 2009. Effect of electric field strength on the length of anodized titania nanotube arrays. *J. Electroanal. Chem.* 637: 6–12.

71. E. J. W. Verwey. 1935. Electrolytic conduction of a solid insulator at high fields. *Physica* 2: 1059–1063.

72. Y. Yang, X. Wang, and L. Li. 2008. Synthesis and photovoltaic application of high aspect-ratio TiO_2 nanotube arrays by anodization. *J. Am. Ceram. Soc.* 91: 3086–3089.

73. A. Ghicov, H. Tsuchiya, J. M. Macak, and P. Schmuki. 2005. Titanium oxide nanotubes prepared in phosphate electrolytes. *Electrochem. Commun.* 7: 505–509.

74. X. Feng, J. M. Macak, and P. Schmuki. 2007. Robust self-organization of oxide nanotubes over a wide pH range. *Chem. Mater.* 19: 1534–1536.

75. K. S. Raja, T. Gandhi, and M. Misra. 2007. Effect of water content of ethylene glycol as electrolyte for synthesis of ordered titania nanotubes. *Electrochem. Commun.* 9: 1069–1076.

76. J. M. Macak, H. Tsuchiya, L. Taveira, S. Aldabergerova, and P. Schmuki. 2005. Smooth anodic TiO_2 nanotubes. *Angew. Chem. Int. Ed.* 44: 7463–7465.

77. D. Kim, F. Schmidt-Stein, R. Hahn, and P. Schmuki. 2008. Gravity assisted growth of self-organized anodic oxide nanotubes on titanium. *Electrochem. Commun.* 10: 1082–1086.

78. U.S. Department of Energy. 2005. Basic research needs for solar energy utilization. Report on the basic energy sciences workshop on solar energy utilization. April 18–21, 2005.

79. M. A. Green, K. Emery, Y. Hishikawa, and W. Warta. 2009. Solar cell efficiency tables. *Prog. Photovolt. Res. Appl.* 17: 85–94.

80. K. L. Chopra, P. D. Paulson, and V. Dutta. 2004. Thin-film solar cells: An overview. *Prog. Photovolt. Res. Appl.* 12: 69–92.

81. M. A. Green. 2007. Thin-film solar cells: Review of materials, technologies and commercial status. *J. Mater. Sci.: Mater. Electron.* 18: S15–S19.

82. M. K. Nazeeruddin, P. Péchy, T. Renouard et al. 2001. Engineering of efficient panchromatic sensitizers for nanocrystalline TiO_2-based solar cells. *J. Am. Chem. Soc.* 123: 1613–1624.

83. L. Han, N. Koide, Y. Chiba et al. 2005. Improvement of efficiency of dye-sensitized solar cells by reduction of internal resistance. *Appl. Phys. Lett.* 86: 213501.

84. Y. Chiba, A. Islam, Y. Watanabe, R. Komiya, N. Koide, and L. Han. 2006. Dye-sensitized solar cells with conversion efficiency of 11.1%. *Jpn. J. Appl. Phys.* 45: L638–L640.

85. M. Grätzel. 2005. Mesoscopic solar cells for electricity and hydrogen production from sunlight. *Chem. Lett.* 34: 8–13.

86. M. Späth, P. M. Sommeling, J. A. M. van Roosmalen et al. 2003. Reproducible manufacturing of dye-sensitized solar cells on a semi-automated baseline. *Prog. Photovolt. Res. Appl.* 11: 207–220.

87. L. Han, A. Fukui, N. Fuke, N. Koide, and R. Yamanaka. 2006. High efficiency of dye-sensitized solar cell and module. *Conference Record of the 2006 IEEE 4th World Conference on Photovoltaic Energy Conversion,* Waikoloa, HI, pp. 179–182.

88. P. Liska, K. R. Thampi, M. Grätzel et al. 2006. Nanocrystalline dye-sensitized solar cell/copper indium gallium selenide thin-film tandem showing greater than 15% conversion efficiency. *Appl. Phys. Lett.* 88: 203103.

89. U.S. Department of Energy. 2007. National solar technology roadmap: Sensitized solar cells. Management report NREL/MP-520-41739, June 2007.

90. M. Grätzel. 2004. Conversion of sunlight to electric power by nanocrystalline dye-sensitized solar cells. *J. Photochem. Photobiol. A: Chem.* 164: 3–14.

91. M. G. Kang, N. G. Park, K. S. Ryu, S. H. Chang, and K. J. Kim. 2006. A 4.2% efficient flexible dye-sensitized TiO_2 solar cells using stainless steel substrate. *Sol. Energy Mater. Sol. Cells* 90: 574–581.

92. Z. Chen, Y. Tang, L. Zhang, and L. Luo. 2006. Electrodeposited nanoporous ZnO films exhibiting enhanced performance in dye-sensitized solar cells. *Electrochim. Acta* 51: 5870–5875.

93. M. Law, L. E. Greene, A. Radenovic, T. Kuykendall, J. Liphardt, and P. Yang. 2006. ZnO-Al_2O_3 and ZnO-TiO_2 core-shell nanowire dye-sensitized solar cells. *J. Phys. Chem. B* 110: 22652–22663.

94. J. J. Wu, G. R. Chen, H. H. Yang, C. H. Ku, and J. Y. Lai. 2007. Effects of dye adsorption on the electron transport properties in ZnO-nanowire dye-sensitized solar cells. *Appl. Phys. Lett.* 90: 213109.

95. Y. Fukai, Y. Kondo, S. Mori, and E. Suzuki. 2007. Highly efficient dye-sensitized SnO_2 solar cells having sufficient electron diffusion length. *Electrochem. Commun.* 9: 1439–1443.

96. N. G. Park, M. G. Kang, K. S. Ryu, K. M. Kim, and S. H. Chang. 2004. Photovoltaic characteristics of dye-sensitized surface-modified nanocrystalline SnO_2 solar cells. *J. Photochem. Photobiol. A: Chem.* 161: 105–110.

97. P. Guo and M. A. Aegerter. 1999. Ru(II) sensitized Nb_2O_5 solar cells made by the sol-gel process. *Thin Solid Films* 351: 290–294.

98. M. L. Cantu and F. C. Krebs. 2006. Hybrid solar cells based on MEH-PPV and thin film semiconductor oxides (TiO_2, Nb_2O_5, ZnO, CeO_2 and CeO_2-TiO_2): Performance improvement during long-time irradiation. *Sol. Energy Mater. Sol. Cells* 90: 2076–2086.

99. M. Grätzel. 2003. Dye-sensitized solar cells. *J. Photochem. Photobiol. C: Photochem. Rev.* 4: 145–153.

100. S. Nakade, W. Kubo, Y. Saito et al. 2003. Influence of measurement conditions on electron diffusion in nanoporous TiO_2 films: Effects of bias light and dye adsorption. *J. Phys. Chem. B* 107: 14244–14248.

101. Y. Saito, N. Fukuri, R. Senadeera, T. Kitamura, Y. Wada, and S. Yanagada. 2004. Solid state dye-sensitized solar cells using in situ polymerized PEDOTs as hole conductor. *Electrochem. Commun.* 6: 71–74.

102. F. F. Santiago, J. Bisquert, E. Palomares, S. A. Haque, and J. R. Durrant. 2006. Impedance spectroscopy study of dye-sensitized solar cells with undoped spiro-OMeTAD as hole conductor. *J. Appl. Phys.* 100: 034510.

103. Y. Saito, T. Azechi, T. Kitamura, Y. Hasegawa, and S. Yanagada. 2004. Photo-sensitizing ruthenium complexes for solid state dye solar cells in combination with conducting polymers as hole conductors. *Coord. Chem. Rev.* 248: 1469–1478.

104. T. N. Murakami and M. Grätzel. 2008. Counter electrodes for DSC: Application of functional materials as catalysts. *Inorg. Chim. Acta* 361: 572–580.

105. K. Suzuki, M. Yamaguchi, M. Kumagai, and S. Yanagida. 2003. Application of carbon nanotubes to counter electrodes of dye-sensitized solar cells. *Chem. Lett.* 32:28–29.

106. T. N. Murakami, S. Ito, Q. Wang et al. 2006. Highly efficient dye-sensitized solar cells based on carbon black counter electrodes. *J. Electrochem. Soc.* 153: A2255–A2261.

107. Y. Jun, J. Kim, and M. G. Kang. 2007. A study of stainless steel-based dye-sensitized solar cells and modules. *Sol. Energy Mater. Sol. Cells* 91: 779–784.

108. L. Peter. 2007. Transport, trapping and interfacial transfer of electrons in dye-sensitized nanocrystalline solar cells. *J. Electroanal. Chem.* 599: 233–240.

109. W. R. Duncan and O. V. Prezhdo. 2007. Theoretical studies of photoinduced electron transfer in dye-sensitized TiO_2. *Annu. Rev. Phys. Chem.* 58:143–184.

110. M. Law, L. Greene, J. Johnson, R. Saykally, and P. Yang. 2005. Nanowire dye-sensitized solar cells. *Nat. Mater.* 4: 455–459.

111. N. Kopidakis, E. A. Schiff, N. G. Park, J. van de Lagemaat, and A. J. Frank. 2000. Ambipolar diffusion of photocarriers in electrolyte-filled, nanoporous TiO_2. *J. Phys. Chem. B.* 104: 3930–3936.

112. A. Hagfeldt and M. Grätzel. 2000. Molecular photovoltaics. *Acc. Chem. Res.* 33: 269–277.

113. L. Antonio and H. Steven. 2003. *Handbook of Photovoltaic Science and Engineering*. John Wiley & Sons, New York.

114. L. Han, N. Koide, Y. Chiba, and T. Mitate. 2004. Modeling of an equivalent circuit for dye-sensitized solar cells. *Appl. Phys. Lett.* 84: 2433–2435.

115. M. Berginski, J. Hüpkes, M. Schulte, G. Schöpe, H. Stiebig, and B. Rech. 2007. The effect of front ZnO: Al surface texture and optical transparency on efficient light trapping in silicon thin-film solar cells. *J. Appl. Phys.* 101: 074903.

116. T. P. Chou, Q. Zhang, G. E. Fryxell, and G. Cao. 2007. Hierarchically structured ZnO films for dye-sensitized solar cells with enhances energy conversion efficiency. *Adv. Mater.* 19: 2588–2592.

117. H. G. Agrell, J. Lindgren, and A. Hagfeldt. 2004. Coordinative interactions in a dye-sensitized solar cell. *J. Photochem. Photobiol. A* 164: 23–27.

118. K. Zhu, N. R. Neale, A. Miedaner, and A. J. Frank. 2007. Enhanced charge-collection efficiencies and light scattering in dye-sensitized solar cells using oriented TiO_2 nanotubes arrays. *Nano Lett.* 7: 69–74.

119. J. van de Lagemaat, N.-G. Park, and A. J. Frank. 2000. Influence of electrical potential distribution, charge transport, and recombination on the photopotential and photocurrent conversion efficiency of dye-sensitized nanocrystalline TiO_2 Solar cells: A study by electrical impedance and optical modulation techniques. *J. Phys. Chem. B.* 104: 2044–2052.

120. B. Tan and Y. Wu. 2006. Dye-sensitized solar cells based on TiO_2 nanoparticle/nanowire composite. *J. Phys. Chem. B*. 110: 15932–15938.

121. D. I. Suh, S. Y. Lee, T. H. Kim, J. M. Chun, E. K. Suh, and S. K. Lee. 2007. The fabrication and characterization of dye-sensitized solar cells with a branched structure of ZnO nanowire. *Chem. Phys. Lett.* 442: 348–353.

122. S. H. Kang, J. Y. Kim, Y. Kim, H. S. Kim, and Y. E. Sung. 2007. Surface modification of stretched TiO_2 nanotubes for solid state dye-sensitized solar cells. *J. Phys. Chem. C.* 111: 9614–9623.

123. F. Wang, R. Liu, A. Pan et al. 2007. The optical properties of ZnO sheets electrodeposited on ITO glass. *Mater. Lett.* 61: 2000–2003.

124. S. H. Kang, S. H. Choi, M. S. Kang et al. 2008. Nanorod-based dye-sensitized solar cells with improved charge collection efficiency. *Adv. Mater.* 20: 54–58.

125. C. Lin, H. Lin, J. Li, and X. Li. 2008. Electrodeposition preparation of ZnO nanobelt array films and application to dye-sensitized solar cells. *J. Alloys Compd.* 462: 175–180.

126. A. F. Nogueira, C. Longo, and M. A. De Paoli. 2004. Polymers in dye sensitized solar cells: overview and perspectives. *Coord. Chem. Rev.* 248: 1455–1468.

127. E. Palomares, J. N. Clifford, S. A. Haque, T. Lutz, and J. R. Durrant. 2003. Control of charge recombination dynamics in dye-sensitized solar cells by the use of conformally deposited metal oxide blocking layers. *J. Am. Chem. Soc.* 125: 475–482.

128. P. J. Cameron and L. M. Peter. 2003. Characterization of titanium dioxide blocking layers in dye-sensitized nanocrystalline solar cells. *J. Phys. Chem. B* 107: 14394–14400.

129. T. Y. Peng, A. Hasegawa, J. Qiu, and K. Hirao. 2003. Fabrication of titania tubules with high surface area and well-developed mesostructural walls by surfactant-mediated templating method. *Chem. Mater.* 15: 2011–2016.

130. L. Kavan, M. Grätzel, S. E. Gilbert, C. Klemenz, and H. J. Scheel. 1996. Electrochemical and photoelectrochemical investigation of single crystal anatase. *J. Am. Chem. Soc.* 118: 6716–6723.

131. P. Wagner and R. Helbig. 1974. The Hall effect and the anisotropy of the mobility of the electrons in ZnO. *J. Phys. Chem. Solid* 35: 327–335.

132. B. Liu, B. Zhou, B. Xiong, J. Bai, and L. Li. 2007. TiO_2 nanotube arrays and TiO_2-nanotube-array based dye-sensitized solar cell. *Chin. Sci. Bull.* 52: 1585–1589.

133. A. B. F. Martinson, J. W. Elam, J. T. Hupp, and M. J. Pellin. 2007. ZnO nanotube based dye-sensitized solar cells. *Nano Lett.* 7: 2183–2187.

134. G. K. Mor, K. Shankar, M. Paulose, O. K. Varghese, and C. A. Grimes. 2006. Use of highly-ordered TiO_2 nanotube arrays in dye-sensitized solar cells. *Nano Lett.* 6: 215–218.

135. G. R. R. A. Kumara, K. Tennakone, V. P. S. Perera, A. Konno, S. Kaneko, and M. Okuya. 2001. Suppression of recombinations in a dye-sensitized photoelectrochemical cell made from a film of tin IV oxide crystallites coated with a thin layer of aluminium oxide. *J. Phys. D: Appl. Phys.* 34: 868–873.

136. M. Paulose, K. Shankar, O. K. Varghese, G. K. Mor, and C. A. Grimes. 2006. Application of highly-ordered TiO_2 nanotube-arrays in heterojunction dye-sensitized solar cells. *J. Phys. D: Appl. Phys.* 39: 2498–2503.

137. M. Paulose, K. Shankar, O. K. Varghese, G. K. Mor, B. Hardin, and C. A. Grimes. 2006. Backside illuminated dye-sensitized solar cells based on titania nanotube array electrodes. *Nanotechnology* 17: 1446–1448.

138. Y. Ohsaki, N. Masaki, T. Kitamura et al. 2005. Dye-sensitized TiO_2 nanotube solar cells: Fabrication and electronic characterization. *Phys. Chem. Chem. Phys.* 7: 4157–4163.

139. M. Adachi, Y. Murata, I. Okada, and S. Yoshikawa. 2003. Formation of titania nanotubes and applications for dye-sensitized solar cells. *J. Electrochem. Soc.* 150: G488–G493.

140. J. Jiu, S. Isoda, F. Wang, and M. Adachi. 2006. Dye-sensitized solar cells on a single-crystalline TiO_2 nanorod film. *J. Phys. Chem. B* 110: 2087–2092.

141. J. van de Lagemaat and A. J. Frank. 2001. Nonthermalized electron transport in dye-sensitized nanocrystalline TiO_2 films: Transient photocurrent and random-walk modeling studies. *J. Phys. Chem. B* 105: 11194–11205.

142. A. J. Nozik. 1978. Photoelectrochemistry: Applications to solar energy conversion. *Ann. Rev. Phys. Chem.* 29: 189–222.

143. C. M. A. Brett and A. M. O. Brett. 1994. *Electrochemistry Principles, Methods, and Applications.* Oxford University Press, New York.

144. M. Grätzel. 2001. Photoelectrochemical cells. *Nature* 414: 338–344.

145. B. O'Regan, J. Moser, M. Anderson, and M. Grätzel. 1990. Vectorial electron injection into transparent semiconductor membranes and electric field effects on the dynamics of light-induced charge separation. *J. Phys. Chem.* 94: 8720–8726.

146. A. Hagfeldt and M. Grätzel. 1995. Light-induced redox reaction in nanocrystalline systems. *Chem. Rev.* 95: 49–68.

147. H. Tang, K. Prasad, R. Sanjinès, P. E. Schmid, and F. Lévy. 1994. Electrical and optical properties of TiO₂ anatase thin films. *J. Appl. Phys.* 75: 2042–2047.

148. M. Dolata, P. Kedzierzawski, and J. Augustynski. 1996. Comparative impedance spectroscopy study of rutile and anatase TiO₂ film electrodes. *Electrochim. Acta* 41: 1287–1293.

149. D. Duonghong, J. Ramsden, and M. Grätzel. 1982. Dynamics of interfacial electron-transfer processes in colloidal semiconductor systems. *J. Am. Chem. Soc.* 104: 2977–2985.

150. A. C. Arango, S. A. Carter, and P. J. Brock. 1999. Charge transfer in photovoltaics consisting of interpenetrating networks of conjugated polymer and TiO₂ nanoparticles. *Appl. Phys. Lett.* 74: 1698–1700.

151. A. C. Arango, L. R. Johnson, V. N. Bliznyuk, Z. Schlesinger, S. A. Carter, and H.-H. Hörhold. 2000. Efficient titanium oxide/conjugated polymer photovoltaics for solar energy conversion. *Adv. Mater.* 12: 1689–1692.

152. B. Kraeutler and A. J. Bard. 1978. Heterogeneous photocatalytic decomposition of saturated carboxylic acids on TiO₂ powder: Decarboxylative route to alkanes. *J. Am. Chem. Soc.* 100: 5985–5992.

153. T. Watanabe, A. Fujishima, O. Tatsuoki, and K. Honda. 1976. pH-dependence of spectral sensitization at semiconductor electrodes. *Bul. Chem. Soc. Jpn.* 49: 8–11.

154. D. Fitzmaurice. 1994. Using spectroscopy to probe the band energetics of transparent nanocrystalline semiconductor films. *Sol. Energy Mater. Sol. Cells* 32: 289–305.

155. A. J. Nozik and R. Memming. 1996. Physical chemistry of semiconductor-liquid interfaces. *J. Phys. Chem.* 100: 13061–13078.

156. L. Sun, S. Zhang, X. W. Sun, and X. He. 2010. Effect of the geometry of the anodized titania nanotube array on the performance of dye-sensitized solar cells. *J. Nanosci. Nanotechnol.* 10: 1–10.

157. E. Hendry, M. Koeberg, B. O'Regan, and M. Bonn. 2006. Local field effects on electron transport in nanostructured TiO₂ revealed by terahertz spectroscopy. *Nano Lett.* 6: 755–759.

158. V. Kytin, Th. Dittrich, J. Bisquert, E. A. Lebedev, and F. Koch. 2003. Limitation of the mobility of charge carriers in a nanoscaled heterogeneous system by dynamical coulomb screening. *Phys. Rev. B* 68: 195308.

159. H. G. Agrell, G. Boschloo, and A. Hagfeldt. 2004. Conductivity studies of nanostructured TiO₂ films permeated with electrolyte. *J. Phys. Chem. B* 108: 12388–12396.

160. E. Enache-Pommer, J. E. Boercker, and E. S. Aydil. 2007. Electron transport and recombination in polycrystalline TiO₂ nanowire dye-sensitized solar cells. *Appl. Phys. Lett.* 91: 123116.

161. D. Kim, A. Ghicov, and P. Schmuki. 2008. TiO₂ nanotube arrays: Elimination of disordered top layers (Nanograss) for improved photoconversion efficiency in dye-sensitized solar cells. *Electrochem. Commun.* 10: 1835–1838.

162. R. Hahn, T. Stergiopoulus, J. M. Macak et al. 2007. Efficient solar energy conversion using TiO₂ nanotubes produced by rapid breakdown anodization—A comparison. *Phys. Stat. Sol. (RRL)* 1: 135–137.

163. K. Shankar, G. K. Mor, M. Paulose, O. K. Varghese, and C. A. Grimes. 2008. Effect of device geometry on the performance of TiO₂ nanotube array-organic semiconductor double heterojunction solar cells. *J. Non-Cryst. Sol.* 354: 2767–2771.

164. K. Zhu, T. B. Vinzant, N. R. Neale, and A. J. Frank. 2007. Removing structural disorder from oriented TiO₂ nanotube arrays: reducing the dimensionality of transport and recombination in dye-sensitized solar cells. *Nano Lett.* 7: 3739–3746.

165. D. Kim, A. Ghicov, S. P. Albu, and P. Schmuki. 2008. Bamboo-type TiO₂ nanotubes: Improved conversion efficiency in dye-sensitized solar cells. *J. Am. Chem. Soc.* 130: 16454–16455.

166. O. K. Varghese, D. Gong, M. Paulose, K. G. Ong, E. C. Dickey, and C. A. Grimes. 2003. Extreme changes in the electrical resistance of titania nanotubes with hydrogen exposure. *Adv. Mater.* 15: 624–627.

167. S. P. Albu, A. Ghicov, S. Aldabergenova et al. 2008. Formation of double-walled TiO₂ nanotubes and robust anatase membranes. *Adv. Mater.* 20: 4135–4139.

168. J. M. Macak, S. P. Albu, and P. Schmuki. 2007. Towards ideal hexagonal self-ordering of TiO₂ nanotubes. *Phys. Stat. Sol. (RRL)* 1: 181–183.

3 Progress and Challenges of Photovoltaic Applications of Silicon Nanocrystalline Materials

Jatin K. Rath

CONTENTS

3.1 INTRODUCTION

Nanocrystalline silicon (nc-Si) is a very important material for the fabrication of thin film solar cells, as this material combines the advantages of amorphous silicon (a-Si) and crystalline silicon (c-Si), namely, deposition over a large area, low-temperature processing, low material consumption, and high stability. The deposition rate of nc-Si has received much attention in recent years due to its importance as a rate-limiting step in the fabrication of multijunction solar cells using nc-Si as one of the bottom cells in combination with a-Si as the top cell. It is a scientific as well as technological challenge to obtain high-efficiency nc-Si cells at a deposition rate ≥ 5 nm/s. There are three areas where nc-Si faces competition: (1) the c-Si wafer that is getting thinner and thinner ($<100 \mu$m); (2) a-Si and, especially, its alloy a-SiGe; and (3) non-silicon materials, such as copper indium gallium diselinide (CuInGaSe$_2$ or CIGS) and cadmium telluride (CdTe). CdTe-based modules with a system price of ~1\$/W$_p$ certainly seem to be the most attractive solar cells at present [1]. CIGS is drawing a lot of attention due to its high efficiency, which touches almost 20% [2] on the laboratory scale.

However, silicon is still the most widely used material because of its abundance in nature and nontoxic character. Crystalline Si modules still continue to dominate the market share of around 90% [3]. Moreover, due to rapid progress in the sawing process, wafers of less than 100 microns are now available, and efforts are in progress in utilizing these thin wafers in the normal homojunction structure or in the heterojunction structure. In this regard, in the latter category, HIT (heterojunction with intrinsic thin layer) type of cells from Sanyo, Japan, which uses a-Si as a passivation layer between the wafer and the emitter, have been made with efficiencies of around 21.4% using a very thin wafer of ~85 microns and have shown high performance down to a 70 μm thickness [4]. Thus, nc-Si has to show that it can deliver what a thin silicon wafer cannot. This, of course, means that all the advantages of a thin film compared to those of a bulk material, namely, deposition over a large area, wide range of substrate materials, low material consumption, and small payback time, have to win over the relatively low efficiencies. All these advantages can be realized if the homogeneity and deposition rate demands are met on large-size substrates without compromising on the high efficiency. The size of the glass substrate has reached 5.7 m² (Applied Materials, Santa Clara, California) [5]. The deposition process used for such substrates is the standard radio frequency plasma-enhanced chemical vapor deposition (St. RF PECVD) at 13.56 MHz to make a-Si layers and cells. Though, recently, nc-Si cells on such large substrates have been demonstrated with high homogeneity [5], the deposition rate is still an issue. It has to be seen whether a very-high-frequency (VHF) PECVD process can be employed on such large areas without sacrificing homogeneity, because VHF PECVD at present looks most promising to meet the demands of both high efficiency and high deposition rate. The equipment manufacturer Oerlikon [6], which has the best expertise on VHF PECVD technology, has still limited its substrate size to 1.4 m² and the frequency is still at the lower side (40.68 MHz), though at the laboratory scale the frequencies between 60 and 100 MHz have shown the best results.

3.2   DEPOSITION OF NANOCRYSTALLINE SILICON

Since the first report on nc-Si deposition by Veprek and Marecek [7] more than 40 years back, a lot of efforts have been put into understanding the growth process and finding ways to make better nc-Si materials. It has also become clear that the nc-Si deposition regime is determined using two distinct parameters space: (1) the gas-phase condition and (2) the thermodynamics in the growth region. Information on the growth process has been obtained using various in situ diagnosis tools to probe (1) the gas-phase condition and its changes during the growth period and (2) the evolution of the films. Modeling and simulations are employed to analyze the growth process.

3.2.1   GROWTH PROCESS

The nc-Si films are made at temperatures around 200°C, much below the melting temperature of silicon and the temperature at which c-Si ingots are made. Hence, the nucleation process and the growth of the crystalline phase at low temperature have been a sort of puzzle for the researchers and varieties of models have been proposed. The models have basically taken two routes: (1) the reactions at the growing material and (2) the gas phase. These two aspects are, however, related to each other in some way or the other. The very first model by Veprek et al. [8] tried to link the growth of the crystalline phase to the gas-phase condition that is at equilibrium due to deposition and etching. The subsequent models gave more emphasis to the reactions in the material, whether at the surface or in the bulk, without getting into the specifics of the gas-phase condition. In a surface reaction model [9], a silicon species on the growing surface has to have a high diffusion length so that it can find an appropriate site to react, gain energy, and create nucleation. Thus, the diffusion length of the species is the most important parameter in this concept and not the sort of species reaching from the gas phase, though a species such as SiH_3 is more preferred due to its low reactive nature [10]. Moreover, a high flux of hydrogen is beneficial, as it increases the hydrogen coverage of the growing surface [11] and also provides extra energy through exothermic reactions [12,13]; both of these aspects increase the diffusion length. On the other hand, hydrogen also penetrates into the bulk of the growing film, where it modifies the structure [14]. Layer-by-layer (LBL) growth has shown clearly that the hydrogen changes bulk amorphous to crystalline [15], and a molecular model has been proposed to explain this phase transition [16]. According to this model, hydrogen atoms diffusing through the disordered matrix modify it and leave the region in an orderly state. In a more phenomenological way, Street explains the bulk reaction using his equilibrium model [17], according to which the hydrogen chemical potential basically forces a material to a structurally ordered state that is beyond the physical limit of an amorphous matrix and the material changes the phase to crystalline.

The growth models now differentiate between the nucleation and further growth of the crystallites. The free energy of the c-Si phase is nearly 0.1 eV lower than that of the amorphous phase; however, the amorphous phase has to surmount an energy barrier for nucleation to move the structure to the crystalline phase [18]. Once in the crystalline phase, the material continues to be in this phase without using any extra energy. However, the growth of the crystalline phase around the nucleus needs a certain precursor flux condition to drive the local epitaxy. Hence, a high hydrogen dilution is necessary in the beginning to create nucleation, and after that, the dilution can be reduced to a level that is sufficient to drive the local epitaxy [19]. This principle is used in what is known as grading or hydrogen profiling [20]. However, the use of a seed layer is a simple solution to grow nc-Si without the amorphous incubation phase [21], and it has been proved from in situ ellipsometry studies that a thin film silicon layer made at a low-hydrogen-dilution condition on a seed layer made at high hydrogen dilution grows with a considerable crystalline fraction, whereas the same layer deposited on glass is purely amorphous [22]. In a solar cell, the nc-Si-doped layer can function as a seed layer (Figure 3.1). The need for a low hydrogen dilution is to decrease the defect density in the nc-Si, which could partly be attributed to lower ion energies at lower-hydrogen-dilution conditions [23]. However, hydrogen profiling is also used in ion-less depositions, such as HWCVD (hot-wire chemical vapor deposition) [24], because even in this growth process, nc-Si made at low hydrogen dilution is less defective, which could be attributed to lower nucleation density. On the other hand, if the hydrogen dilution is too low, there is an epitaxy breakdown, and the material changes from the crystalline to the amorphous phase [19]. This has been observed in solar cells, where the initial growth on n-type nc-Si, which acts as a seed layer, is in the crystalline phase, but the material changes halfway of the i-layer growth to the amorphous phase and grows further in the amorphous phase. In the case of HWCVD, the filament aging during the deposition is also considered to be one of the causes of breakdown of the crystalline phase, and an inverse grading [25], i.e., gradually increasing dilution, is used to improve the i-layer quality and the solar cell performance. The disadvantage of an ion-less deposition is more clearly visible at low deposition temperature. For a-Si

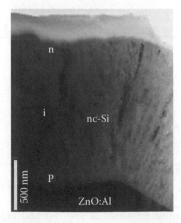

FIGURE 3.1   XTEM image of an nc-Si silicon p-i-n cell on texture-etched ZnO:Al TCO.

deposition, it has been found that at the comparable deposition conditions, HWCVD materials are more porous compared to the PECVD-made samples and, for HWCVD, more hydrogen dilution is needed than that in the case of PECVD to attain the same compactness as the material made at 200°C. We expect similar effect in the growth of the nc-Si material.

Recently, there has been more interest in the gas-phase conditions to explain the amorphous-to-crystalline phase transition. The amorphous-to-crystalline phase transition is determined by the ratio of fluxes of atomic hydrogen and silane species in the gas phase. This is generally determined by a threshold value of H_α/Si* intensity ratio, which is obtained using optical emission spectroscopy (OES), where H_α is the atomic hydrogen Balmer line and Si* is the excited state of atomic silicon species. The ratio is of course specific to the type of reactor and the processing temperature. However, recent studies have shown that this threshold ratio, t (for the same reactor), rises to a high value for the high-pressure [26], high-power regime that is used for the deposition of nc-Si at a high rate. Figure 3.2 shows this difference in threshold values at low- and high-deposition-rate conditions. This shift of the threshold is attributed to increasing abstraction reaction in the gas phase, which increases the concentration of silicon species (for example, $SiH_4 + H \rightarrow SiH_3 + H_2$) and decreases atomic hydrogen flux. Due to this abstraction reaction, the H_α/Si* underestimates the real ratio of radical fluxes of hydrogen and silicon species, which is the determining factor for the phase transition. Hence, to use the H_α/Si* ratio as a monitoring tool for a-Si to nc-Si transition, a calibration factor is needed that correlates the H_α/Si* ratio with the ratio of fluxes of atomic hydrogen and silicon species. Another report says that the concentration of silane in the plasma (c_p), which is determined by the input silane concentration and the fraction of silane depletion, should be considered as the key to determine the phase of growth [27], because the ratio of radical fluxes of hydrogen and silicon species is shown to be correlated with this silane concentration. The silane concentration in the plasma (c_p) is given by the expression $c_p = (1-D)c$, where c is the input silane concentration and D is the depletion fraction given by the relation $c = (p_{SiH_4}^0 - p_{SiH_4})/p_{SiH_4}^0$, where p_{SiH_4} and $p_{SiH_4}^0$ are the partial pressures of silane with and without discharge, respectively. The implication of this concept is that the nc-Si material can be made with pure silane, subject to a high depletion condition of silane [28]. A similar argument linking the degree of depletion with the phase transition is also put forward in another study, in which a parameter called gas use parameter, defined as $c_d = r_d/f_{SiH_4}$, where r_d is the deposition rate and f_{SiH_4} is the silane flow rate, is considered to represent the degree of depletion [90]. This gas use parameter, c_d, shows better correlation with the growth of the crystalline phase at high-growth-rate conditions than the optical emission parameter, H_α/Si*. It has also been noted that the concentration of silane in the plasma is in a dynamic phase due to back diffusion of silane from the passive region of the reactor into the discharge zone, caused by silane depletion, due to which a transient plasma condition occurs [28,29]. Figure 3.3 shows a schematic of the flows of silane and hydrogen from the active to the inactive region and

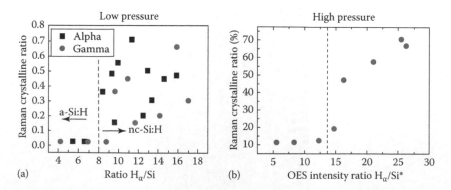

FIGURE 3.2 Amorphous-to-crystalline transition at low- (a) and high-pressure (b) conditions detected using OES.

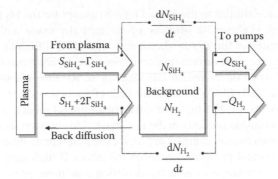

FIGURE 3.3 Schematic of the rate of change of hydrogen and silane concentration in the background (inactive region of the reactor) gas.

back diffusion. The consequence of this is that a large amorphous incubation phase occurs in the initial stages of growth. Moreover, it has been noted that this incubation phase is bigger in case of high deposition rate. This could partially explain the deterioration of the solar cell performance at high deposition rate, in addition to the increasing defect density.

Starting the plasma with pure hydrogen before letting in silane for deposition is a proposed way to get rid of the incubation phase [30]. One more result that has come into picture is that the standard procedure of varying the hydrogen flow for grading is not applicable at high-depletion conditions, which is generally encountered at high-deposition-rate conditions. Hence, a silane profiling should be instead used to control the crystalline fraction during growth [29].

3.2.2 DEPOSITION TECHNIQUES

Deposition techniques for nc-Si have basically followed those used for a-Si deposition. Thus, one finds the same deposition systems used for a-Si as well as nc-Si. For example, all sorts of CVD processes, such as PECVD at standard RF (13.56 MHz) [31] or VHF [32] excitations, which have been used for a-Si deposition, have also been proved to be excellent for nc-Si deposition. This is possible because of a simple underlying principle: a single parameter space of the ratio of hydrogen to silane in the gas phase can determine whether the material is amorphous or nanocrystalline. The other techniques that have been subsequently developed, such as HWCVD [33], photo CVD [34], expanding thermal plasma (ETP) [35], low-energy plasma-enhanced chemical vapor deposition (LEPECVD) [36], microwave PECVD (MWCVD) [37], electron cyclotron resonance chemical vapor deposition (ECR CVD) [38], and surface wave plasma CVD (SW PECVD) [39], also use the same principle, although each of these techniques have other parameters that also have a strong influence on phase transition. More information on some of these techniques is given below.

The very first nc-Si deposition was made through chemical transport of silicon particles after etching of a silicon target [7]. However, neither this technique nor the similar process of the sputtering technique have been successful in obtaining device-quality nc-Si material, and hence, will not be discussed in detail. PECVD was the first direct deposition technique to be used to make nc-Si layers. In this case, St. RF PECVD at 13.56 MHz was used [31]. In a PECVD process, the grounded electrode (also called anode) is separated from the RF electrode (called also cathode) at which power is fed from an RF power supply. A gas mixture of silane and hydrogen is fed into the chamber. The gas between the electrodes gets electrical discharge, and the gas species produced through electron impact dissociation and subsequent secondary reactions, diffuse (radicals) and drift (ions) to the walls (substrate) and get deposited. In these depositions, the gas is normally fed from the sides of a chamber. The typical power density and pressure are ~0.1 W/cm^2 and ≤1 mbar, respectively. The deposition rates with these types of deposition conditions are below 0.1 nm/s. Most of the plasma conditions are similar to those used for a-Si except for the H_2 to SiH_4 ratio, which, for a typical nc-Si deposition, is above 10, whereas a-Si is normally made with pure silane. However, it was not until the beginning of the 1990s when a new type of RF PECVD, namely, VHF PECVD, was used for nc-Si deposition, that device-quality intrinsic nc-Si layer could be made and tested in solar cell devices [32]. Further development was made in St. RF PECVD to reach the same quality of the nc-Si i-layer, as in VHF PECVD and solar cell, efficiencies reached up to more than 10% [40]. The new developments are the high-pressure condition (>1 mbar), which decreases the ion energies due to multiple collisions of ions in the gas phase. This concept is particularly very useful for St. RF PECVD, in which the ion energies are high enough to create defects in the material if the depositions are made at normal pressure conditions (≤1 mbar). If high power is applied to increase the deposition rate and to remain at or near the depletion condition, pressure is also substantially increased to keep the ion energy low enough not to create defects. This is the concept behind the high-pressure-depletion (HPD) condition [41]. Typical pressures used now for high rate depositions are 2–10 mbar, and power densities are above 1 W/cm². One more development that became necessary for deposition near the depletion condition is the showerhead, without which large inhomogeneity in the film in the lateral direction is obtained. Innovative showerhead designing with optimum distance between the holes and arrangement of the holes to distribute the gas has taken place. One of the special cases of the showerhead design is the so-called hollow cathode design [42]. The diameter of the holes has been optimized in such a way that for a certain pressure condition, intense plasma occurs at the mouth of the holes; as a result, the deposition rate is increased.

VHF PECVD, a special type of PECVD process, in which the frequency of the applied power is typically between 20 and 150 MHz, is now widely used. A higher frequency is preferred because of the advantage of a higher deposition rate caused by high electron density [43] and the growth of less defective material due to softer ion bombardment [44]. However, a practical limit to high frequency is encountered due to (1) the inhomogeneity in the layer when the length of the standing wave is in the range of the size of the substrate and (2) increasing loss of RF power at cables and interconnects at high frequencies. In fact, the earlier studies had put 70 MHz to be the upper limit, where the maximum deposition rate was obtained [43]. However, subsequent experiments confirmed that this limit is imposed due to the power loss at higher frequencies. Using better power feeding and reducing all the parasitic losses, depositions at frequencies higher than 100 MHz have been made with a high deposition rate [45]. For industrial processing, where large-size substrates above 1 m² are being used and the size is ever increasing as fabrication processes used for displays are adapted to solar cell depositions, the inhomogeneity related issue to the wavelength of the applied power is a concern. However, a number of methods have been employed to mitigate such effects, namely, sharing the power feeding [46] and the curved RF electrode surface [47], and these developments dispel the concerns over dimension-related inhomogeneity. At the laboratory scale, VHF PECVD has been the most successful method to make device-quality materials, and the best solar cell performances have been reported using this technique (nip [48], pin [49–51]), especially for the i-layers made at high deposition rates (see Figure 3.4). However, an efficient use of VHF power and reproducibility of the deposition process need a careful estimation and monitoring of the delivered power to the plasma and knowledge of the plasma characteristics. This is obtained by the use of the V–I impedance probe, which in addition also gives information whether the plasma is capacitive or resistive.

There are a number of other deposition techniques that have been tried for nc-Si deposition with varying degrees of success. The plasma-based techniques that contain high-energy particles and ions, such as DC plasma and sputtering, are not suitable to produce device-quality nc-Si. At the other extremes, the deposition techniques in which the ion energies are very low or there are no ions, are promising. Considering the facts that lower ion energies are beneficial for the growth of less defective materials and that VHF PECVD delivers better electronic quality material than St. RF PECVD, an extrapolation to lower ion energies should have given us excellent materials. However, that has not really taken place. The ion energies in ETP, MWCVD, LEPECVD, ECR CVD, etc., are in the range of ~1 eV, and these processes can be grouped together as "low-energy plasma

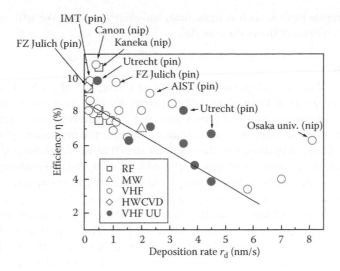

FIGURE 3.4 Efficiency versus deposition rate of the best reported performance of nc-Si solar cells made using various techniques in p-i-n and n-i-p configurations.

processes." Without sufficient defect density data on these films, it is hard to evaluate precisely the quality of these films; however, the solar cells data so far have been poor. Only HWCVD, which is an ion-less technique, has been successful to some extent, in the sense that at a low deposition rate of 0.1 nm/s, efficiency very near to that has been achieved using PECVD processes, i.e., slightly below 10% has been obtained [52]. However, at high deposition rate, it has so far not made any significant progress. The reason could be that HWCVD lacks the advantage enjoyed by PECVD processes of providing extra energy to the growing surface so long as the energies are below the threshold of making electronic defects. Hence, the energy for the surface reactions has to be provided by the heated substrate, and thus a higher substrate temperature than PECVD is needed to get good electronic quality. This reasoning holds good also for the low-energy plasma processes stated above. This requirement of high substrate temperature is usually met in the case of HWCVD even without heating the substrate holder from the back side, due to the heat from the filaments from the front side of the substrate. A high temperature has a deleterious effect on the stability of the TCO (transparent conducting oxide), the doped layers, etc., and interdiffusion of the various species, such as boron, phosphorous, and hydrogen, can damage the cell performance. Moreover, filament heating and filament aging [53] through silicidization have been some of the main drawbacks of HWCVD. Whereas the former can be mitigated by external cooling of the substrate from the back side, filament aging is still a major issue, though various schemes to increase filament life, such as choice of filament (Ta filament instead of W filament [53]) and decontamination by preheating the filaments in hydrogen ambient before and after deposition [54], have been proposed.

3.3 MATERIAL CHARACTERIZATION

3.3.1 NC-SI MATERIAL CHARACTERISTICS

An nc-Si material has many specific characteristics that help to identify the material and differentiate it from an amorphous or polycrystalline type of material. Although, in the literature, especially in the former days, this material was many times called polycrystalline and in a way treated like a small grain version of poly-Si, research on these materials and more so in device structures has revealed that nc-Si materials behave very differently than what is expected if one extrapolates the behavior of poly-Si to the low grain-size limit [55]. This also has a consequence on how this material responds to the external probes, such as light, heat, and electric power. We will discuss below some of the main characteristics of the nc-Si material.

3.3.1.1 Structure

nc-Si does not grow directly, but evolves out of another phase, namely, a-Si. This has been extensively studied by the group at Pennsylvania State University [56]. In a chemical vapor deposition process, at first, an amorphous material grows, and after a certain thickness, called the incubation phase, the crystalline phase starts with a nucleation. Because the crystalline phase grows around the nucleus but in the growth direction, the resultant structure is of a conical shape (Figure 3.5). These cones will ultimately touch each other, after which the crystalline columns grow parallel to each other in the growth direction, leaving very little space for the amorphous phase. We use the term crystalline columns here instead of crystal columns because close study of the cones as well as the columns reveals an ensemble of small grains, rather than one single cone or column. In some rare cases, almost single columns have been observed [21]. The importance of hydrogen in the nucleation process is evident, as one could see a monotonic increase of the incubation phase with a decrease of the hydrogen to silane flow ratio in the gas phase [56]. However, this ideal growth model and phase diagram that are derived from the simulation results using ellipsometry data are not always observed in the structural study. The Julich group, with the data obtained from XTEM (cross-sectional transmission electron microscopy), has presented a very realistic schematic of the structure and its changes with the hydrogen to silane ratio in the gas phase [57]. A lot of distributed crystalline grains are observed in the transition phase from amorphous to crystalline growth, instead of simply regular conical regions. Moreover, the cones and the columns are packed with crystallites and there are many macroscopic defects within these columns. Another study revealed that the crystallites are formed with various shapes; small spherical shapes, elongated nanocrystals, small cones, elongated columns, etc., depending on the hydrogen dilution [58]. All these variations in the growth process make the nc-Si material very complex, and it is now rightly said that there is nothing like "the nc-Si material," but a "whole range of nc-Si materials," as the crystals come in various shapes and sizes and orientations. A consequence of the evolution type of growth, which makes nc-Si so distinct in character, is the inhomogeneous structure in the growth direction. It starts with an amorphous phase, and at some thickness a two-phase growth starts, and after that the crystalline fraction grows in a quadratic way due to the conical nature, and after some thickness the crystalline fraction reaches its maximum and saturates. To have a homogeneous structure, grading of the hydrogen dilution or hydrogen profiling has been proposed. The idea is to have a fast nucleation, starting with a high hydrogen dilution, followed by a decreasing hydrogen dilution in steps, whether in a linear [59] or a more exotic (quadratic [60]) manner. One of the ways to get rid of the incubation amorphous phase is to use a seed layer. The silicon film on top of this seed layer grows immediately as nanocrystalline if the hydrogen to silane ratio is above a threshold for

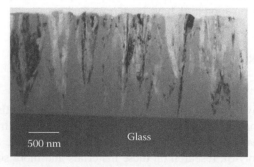

FIGURE 3.5 XTEM bright-field image of a typical nc-Si layer deposited through HWCVD on glass substrates. The conical nature of growth of the crystalline phase is clearly seen in this figure.

the local epitaxy. It has to be noted that this threshold may change with increasing thickness and, if the H_2/SiH_4 ratio is too low, an epitaxial breakdown occurs at a certain thickness. In this case, an inverse grading is employed. The epitaxy breakdown is particularly easy to encounter when a stiff grading sequence is used. Hence, a very careful grading sequence has to be used in which the H_2/SiH_4 ratio at increasing thickness stays just above the threshold for the local epitaxy.

As it is mentioned above, the nc-Si material is made with a whole variety of characteristics. Hence, for any device application, it has to be decided which nc-Si is to be used. In a thin film transistor (TFT), an nc-Si material of almost 100% crystallinity without any incubation phase has to be used, as the electronic path is in the lateral direction and the channel region in a few nm range. On the other hand, for solar cells, the nc-Si material is made at ~50% crystallinity. At this condition, the material has the best electronic quality [57,61]. Hence, a deposition plan has to be made to keep the crystalline fraction around 50% throughout its thickness. To achieve this, deposition starts on a seed layer to obtain the required crystalline fraction, followed by hydrogen grading to maintain this crystalline fraction. Raman spectrum taken from the front and back sides of the sample confirms whether the crystallinity has changed from bottom to top [60].

3.3.1.2   Electronic Characteristics

The electronic defects in the nc-Si materials are microscopic in nature. Dangling bond types of defects dominate their photoelectronic properties. The structural defects, such as dislocations, stacking faults, and twin boundaries, are not as dominant as in the case of c-Si or poly-Si. Electronically, nc-Si behaves like a-Si; the density of states (DOS) of an nc-Si material [62] is considered to be similar to a-Si, i.e., a continuous DOS in the sub-bandgap, comprising dangling bond states in the middle of the gap and tail states at the conduction and valence band edges. However, the local nature of defects and the mechanism of defect creation seem to differ from a-Si. In the case of a-Si, the dangling bond defects are homogeneously distributed in the bulk and a numerous models have been proposed to link between the disorder in the material, manifested by the weak bond tail-state distribution and the midgap defect density. No such correlation between the tail states and midgap states has been established in the case of the nc-Si material. Unlike a-Si, the dangling bond defects in nc-Si materials are inhomogeneously distributed; they are predominantly located at the grain boundaries and column surfaces [63]. Hence, the origin and behavior of the defect states are similar to the c-Si surface states, and specific types of dangling bond centers, typical of c-Si surface defects, such as P_b centers, have been identified in nc-Si [64]. Moreover, due to close proximity of defects at thin intercolumnar regions, defect pairs can form in special cases [65]. Nevertheless, the similarities of the bulk behavior of defect states of nc-Si to that of a-Si allow nc-Si to be characterized in the same way as in the case of a-Si, for which a very large range of characterization techniques are available. The reduction of defect density has been one of the main targets of nc-Si research. In fact, for a long time since the first deposition of nc-Si [7], it was considered to be not suitable for solar cell applications due to high density of defects, as detected by PDS (photothermal deflection spectroscopy) and ESR (electron spin resonance) techniques. In the beginning of the 1990s, a substantial improvement in the quality of the nc-Si material took place, mainly at IMT (Neuchatel) employing the VHF PECVD technique, and a high photosensitivity allowed photoconductivity techniques such as CPM (constant photocurrent method) to measure the sub-gap absorption accurately [32]. For the first time, an a-Si-like sub-bandgap absorption could be observed, in which an exponential region, Urbach tail, could be discerned from the midgap defect absorption. Further studies revealed that the standard CPM or PDS experiment overestimates the absorption coefficient due to the contribution from internal and surface scattering [66]. Methods have been proposed to subtract the contribution from scattering and obtain the real absorption coefficient. The density defects in the nc-Si material obtained using this method showed much lower values. Subsequently, it became clear that the total density of defects does not convey the real potential of a material for a solar cell. As the electrical transport is in the perpendicular direction and the growth of nc-Si is columnar in nature, the defects in the transport path are more relevant [67]. Moreover, the nc-Si i-layer material in a solar cell grows differently than that on the glass, due to the presence of the underlying doped nc-Si layer and also the presence of a textured surface. Hence, the determination of sub-bandgap absorption of the nc-Si material in the real cell structure became important. A defect density of $<10^{15}\,cm^{-3}$ and an Urbach tail parameter of 35 meV have been obtained for an nc-Si i-layer of a high-efficiency state-of-the-art nc-Si solar cell [68]. In fact, the earliest successful implementation of the nc-Si material for a solar cell is attributed to a low defect density. Moreover, the characteristics specific to device-quality material have also been identified.

3.3.2 Characterization Techniques

The special characteristics of an nc-Si material allow it to be probed by a wide range of techniques, with the analysis procedure directed toward finding the physical properties of the nc-Si layers. The important techniques are roughly divided into structural, electrical (and electronic defects) and optical characterizations.

3.3.2.1 Structure

Probably, the most widely used technique to characterize an nc-Si material is Raman spectroscopy. It uses the difference in Si–Si transverse-optic vibrational frequencies in c-Si and a-Si matrices, which are at 520 and $480\,cm^{-1}$, respectively. The estimation of the crystalline fraction is a simple ratio of the intensity of the crystalline part to the total intensity of the TO vibration spectrum. However, the TO spectrum is more complex than the simple summation of the above-said crystalline and amorphous peaks. A simple deconvolution of the Raman spectrum of the TO modes gives in addition to the modes around 480 and $520\,cm^{-1}$, an intermediate peak around $510\,cm^{-1}$ [69]. In other words, without a third peak, it is not possible to fit the TO vibration spectrum. This intermediate vibrational mode has been attributed to the grain boundaries or small grains. The complication arises because, unlike the a-Si peak, the crystalline peak varies strongly with the grain size. In fact, the deconvolution of a Raman spectrum using a set of peaks does not always give unique fittings because of the uncertainties of the number of peaks and their line shapes and positions. Simple methods have been proposed to avoid these errors in fittings. One of the methods is to scale the Raman spectrum of a standard a-Si to the amorphous part of the Raman spectrum of the nc-Si sample and subtract the amorphous part from the total spectrum [70]. The remaining part is from the crystalline region, without distinguishing which is from big crystals and which is from small grain or grain boundaries. The ratio of the intensity of this crystalline part to the total intensity of the TO region is used as an estimate of the crystalline fraction. This method ignores the fact that the Raman spectrum of a-Si changes with the structure, confirmed from numerous papers showing changes in full width at half maximum (FWHM) of the peaks, related to structural disorder, the peak position, the ratio of TA and TO line intensities, and the effect of the medium range order on the intensity of the longitudinal modes. Another assumption that the amorphous part in the nc-Si is independent of the growth condition or that this does not differ from a pure amorphous material is also without any basis. However, this method has been used because of its simplicity. A more accurate method [71] is to first fit the amorphous part $(<480\,cm^{-1})$ of the Raman spectrum of nc-Si to a set of Gaussians of TA, LA, LO, and TO peaks and then subtract this amorphous part from the total spectrum (Figure 3.6). In this method, no assumptions are made about the amorphous spectrum, which could vary depending on the growth condition of the nc-Si layer. One of the weak points of the use of Raman spectroscopy for nc-Si characterization is that one gets a ratio of intensities, also called the Raman crystalline ratio, which is not the real crystalline fraction unless a calibration factor is used to compensate the difference in the Raman cross sections between amorphous and crystalline matrices. The size of the grains is used to calculate this calibration term [78]. The crystalline fraction is given by $v_c = I_c/(I_c + mI_a)$, where m is the calibration factor given by $m(L) = 0.1 + e^{-(L/250)}$, where L is the size of the grains obtained from XRD (x-ray diffraction) peaks. In the large grain-size limit (for example, poly-Si), the calibration factor reaches 0.1.

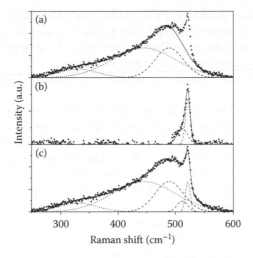

FIGURE 3.6 (a) Raman spectrum of a thin nc-Si layer (scattered data) of which the amorphous contribution is fit in the range <480 cm^{-1} with three Gaussians for the LA and LO peaks at 330 and 440 cm^{-1}, respectively (dotted lines), and the amorphous contribution to the TO peak at 480 cm^{-1} (dashed). (b) The crystalline contribution to the TO peak (scattered data) fit to two Gaussians at 510 and 520 cm^{-1} (dashed line). (c) The fitted peaks from (a) and (b) (LA and LO dotted; TO dashed) and the sum (solid line) compared to the measured data (scattered data).

Another simple method that gives very useful information on the structure is infrared (IR) spectroscopy. Fourier transform infrared spectroscopy (FTIR) is a widely used technique to identify the various vibrational modes. The absorption due to rocking–wagging modes at ~640 cm^{-1} is used to estimate the bonded hydrogen concentration just as it is used in the case of a-Si. High hydrogen content would ensure better probability of passivation of grain boundary defects and also neutralize oxygen-related donor states [72]. The presence of oxygen in the film is identified with the absorption around 1050 cm^{-1}. The most interesting region of the IR spectrum is the hydride stretching modes around 2000 cm^{-1}. For a long time it was accepted that whereas a-Si has these vibrational modes predominantly centered at 2000 cm^{-1}, for nc-Si the vibration shifts to 2100 cm^{-1}, and this was used as a fingerprint to know whether the material is amorphous or nanocrystalline [73]. However, it was later realized [74] that an nc-Si material with hydrogen at compact sites also shows the vibrational mode predominantly at 2000 cm^{-1}, whereas a porous structure shows the typical double peak at 2100 cm^{-1} attributed to Si–H bonds at the grain surface (Figure 3.7). It was proposed that 2000 cm^{-1} mode is a fingerprint of a device-quality nc-Si. Further studies have confirmed this hypothesis [75,76].

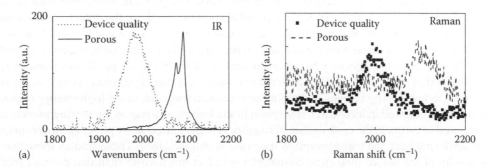

FIGURE 3.7 Si:H vibrational modes of nc-Si. (a) FTIR spectrum and (b) Raman spectrum. The device-quality material shows vibration predominantly at 2000 cm^{-1}, whereas for the porous material, mainly the peaks around 2100 cm^{-1} are observed.

XRD is also a widely used technique. From the line widths of the XRD peaks, using Scherer's formula [77], it is possible to obtain the grain size, which is used in the calibration term explained above for the determination of the crystalline fraction from the Raman spectrum [78]. Moreover, XRD also tells about the orientation of the grains: a (220)-oriented crystalline structure is considered to be beneficial for the solar cell characteristics [79], as such a structure allows better transport of the photogenerated carriers.

The third method, which is less widely employed due to the time-consuming process of sample preparation but, otherwise, probably the most important technique for structural characterization, is the XTEM and high-resolution electron microscopy (HREM). The XTEM image gives direct information on the evolution of the crystalline phase. The selective area diffraction pattern (SADP) gives information on structural ordering, bond lengths, and orientation of grains. The void structure and void density can be obtained from the TEM image in under- and over-focused modes. HREM allows looking at the microscopic details of the grains and the structural faults [21].

Ellipsometry is also an important technique to obtain the information on the structure, how it varies across the thickness of the sample; void fraction, and crystalline volume fraction [15,22,56]. The pseudo-dielectric functions obtained in this experiment are simulated using a structural model that contains at least three layers; an incubation phase, a nanocrystalline bulk, and a roughness layer. The roughness layer is modeled as 50% voids and 50% polycrystalline. The incubation phase consists of amorphous fraction f_a and void fraction f_v, whereas the nanocrystalline bulk consists of polycrystalline fraction f_c, amorphous fraction f_a, and void fraction f_v. Effective medium approximation is used to simulate the dielectric function of the mixed phase system. There are basically two types of setups used for the ellipsometry: rotating analyzer (RAE) and phase modulation (PME).

3.3.2.2   Density of States

The estimation of defect density of an nc-Si material is more complex than that of an a-Si material. It is not a simple extension of the techniques employed for the a-Si materials, because the analysis of the data is more complicated and subject to errors. The ESR technique has been used to estimate defect density, as the dangling bond defects have almost the same characteristics as in the case of a-Si with a g-value of 2.0055. However, many other lines with g-values close to this value, representing midgap defects, have also been reported [62]. The drawback of this method is that the nanocrystalline material easily attains an n-type character, when there is unintentional doping, presumably due to oxygen. Hence, the defect density is grossly underestimated, as ESR detects only neutral dangling bonds.

The sub-bandgap absorption techniques, such as PDS and CPM, which are very successful in obtaining the DOS and the estimation of midgap defect density for a-Si, have been very cautiously used for nc-Si. For a long time, it was debatable whether the DOS diagram of nc-Si should be represented by a sharp band edge and discrete midgap states, or a continuous DOS model as is used for a-Si. The latter has gained acceptance in recent years. The problem associated with the sub-bandgap absorption spectrum is that it has a broad absorption, masking the defect-related absorption. However, a marked improvement in the quality of the nc-Si layer made it possible to obtain the CPM of the nc-Si material with the distinct tail region and the midgap defect-related absorption region [32], just as in the case of a-Si. Detailed studies by the CPM on nc-Si materials revealed that the absorption spectrum has other contributions such as scattering (surface as well as internal) that affects the low-energy regions and amorphous content that affects the high-energy region. A polished surface showed indeed lower absorption in the low-energy region. Scattering models have been proposed to estimate the contribution of roughness to the optical absorption, and after subtracting this scattering part, the true absorption spectrum is obtained [66]. These studies show that the defect density of the device-quality nc-Si material is in fact much lower than that earlier expected from CPM or PDS. Nowadays, the trend is to use techniques that can obtain the defect density of the i-layer in a solar cell configuration and defects that take part in the recombination/trapping process in a solar cell operation, instead of finding the global defect density, as estimated from PDS and

ESR. The techniques that have been used for this purpose are dual-beam photoconductivity (DBP) [80,81] and Fourier transform photocurrent spectroscopy (FTPS) [71,82], both of which are basically spectral response measurements. Out of these, FTPS is a faster and more versatile measurement technique to record absorption to very low energies (even below 0.5 eV) with very low noise, as it uses a Fourier technique.

3.4   SOLAR CELL DEVICES

One of the most important developments that have taken place is the choice of the i-layer used in nc-Si cells. As it was pointed out in the preceding section, nc-Si materials are obtained in a wide range of structures; hence, it is necessary to pick the nc-Si material with the right physical properties. Following the reports from KFA (Julich) [57] and IMT [61], and confirmed by various groups, it is now more or less accepted that the material made at the amorphous-to-nc-Si transition, or the so-called "transition material," "edge material," or "optimum phase material (OPM)," with around a 50% crystalline fraction is best suited for nc-Si cells. Solar cells with the highest possible open circuit voltage (V_{oc}), keeping an nc-Si character, can be made with this transition material because V_{oc} is linearly dependent on the crystalline fraction [83]. Any further decrease of the crystalline fraction changes the character of the cell to amorphous type, and V_{oc} is in fact used as a probe to determine whether the cell is nc-Si or a-Si type. The consequence of the sensitivity of V_{oc} to the crystalline fraction is that any reduction of crystallinity along the growth direction due to an epitaxial breakdown can be easily detected. To avoid such a case, the materials are made at a slightly higher side of the nc-Si phase and grading is precisely planned to avoid receding into the amorphous phase. As it was mentioned above, a reverse grading is sometimes used to bring the material back to the nc-Si phase.

3.4.1   ON GLASS SUBSTRATES

Most of the developments in nc-Si solar cells and modules have taken place on glass substrates. The very first reported nc-Si cell was on a glass substrate [32]. The advantage of a glass substrate is its robust nature, so that substrate handling is simple. Moreover, glass acts as a very good encapsulation shield and protects the cell from moisture. It is easy to attach the glass substrate to a substrate holder and make the cell in a batch or in-line process. Moreover, from an industrial point of view, the glass substrate gives a huge advantage in laser scribing and monolithic integration. However, unlike in the case of a-Si cells, where the choice of the glass substrate is predominantly due to commercially available TCO glasses (SnO$_2$:F-coated glass) from Asahi Glass Co., Nippon Sheet Glass Co., etc., for nc-Si cells, such a choice does not exist. SnO$_2$:F is not a preferred TCO material for nc-Si solar cells, because this TCO degrades under strong atomic hydrogen ambient during the nc-Si p-layer and subsequent nc-Si i-layer deposition. It should be noted that for an nc-Si cell, an nc-Si p-layer instead of an amorphous p-layer is used to avoid the amorphous incubation phase for the i-layer, and the p-layer has the function of a seed layer. Thus, out of necessity, another type of TCO layer, namely, doped zinc oxide (ZnO), was developed. At present, textured zinc oxide layers can be made using various methods such as (1) texture etching [84] and (2) as-grown texture [85]. As it will be discussed below, the requirement of an optimum texture for an nc-Si cell is very different from that of an a-Si cell, because of the lower absorption strength of an nc-Si material compared to an amorphous material and the difference in the bandgaps. The TCOs used for nc-Si cells have a higher *rms* roughness (around 100 nm) [86] than that of Asahi U-type TCOs (around 40 nm) [87], and with the high-roughness TCOs, a high current density can be obtained. For p-i-n nc-Si cells on textured TCO glass, current slightly exceeding 25 mA/cm^2 has been obtained [50]. This TCO development has allowed wide use of glass substrates for nc-Si cells in superstrate configuration. It should be however noted that for n-i-p nc-Si cells on a rough Ag/ZnO:Al substrate, current density near 30 mA/cm^2 has been reached [88]. Figure 3.8 shows the SEM images of the textured substrates used for p-i-n (texture-etched ZnO:Al) and n-i-p (rough Ag/ZnO:Al) cells.

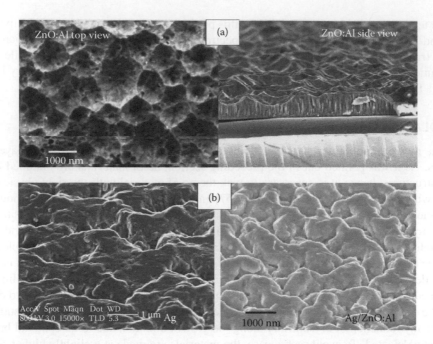

FIGURE 3.8 Scanning electron microscope images of the rough substrates used in solar cells. (a) Texture-etched ZnO:Al, used as front TCO for p-i-n-type solar cell; and (b) rough Ag/ZnO:Al, used as back reflector in n-i-p-type solar cells.

Although most of the nc-Si cells on glass have been made in superstrate configuration, taking advantage of the flat surface and robust nature of glass, cells have also been made in substrate configuration. One of the highest recorded efficiency for nc-Si is in fact in n-i-p configuration on glass in what is called a STAR structure [79]; $\eta = 10.87\%$ at a deposition rate of 0.5 nm/s for the i-layer has been achieved by the Kaneka corporation, made by St. RF PECVD at 13.56 MHz. However, all other high-efficiency cells on glass have been reported for p-i-n configuration, by three groups [49–51] from Europe, reaching an efficiency of around 10%. The deposition rates in these cases are ≤1 nm/s.

The deposition rate has been receiving much attention in recent years due to its importance in industrial production, which has either to do with throughput in in-line and batch process production or the belt size in a roll-to-roll production. Here again, the highest deposition rate has been reported for n-i-p cell configuration, 8 nm/s with a cell efficiency of 6.3% [89]. Two other groups have pioneered in the high-deposition-rate area, Utrecht University [90] and AIST [91]. Both of these groups have excelled in the technique of depositing at high-pressure, high-power conditions. High pressure demands small interelectrode distance, to keep gas-phase reactions and powder formation to minimum. Powder formation has been a lingering issue in the high-deposition-rate regime. The gas-phase condition shifting to a dusty regime by a phenomenon called α to γ' transition at the high-pressure regime takes place. Amplitude modulation and high gas flows are some of the techniques employed to retard the dust formation process. In any case, increasing the deposition rate monotonically decreases the efficiency of nc-Si solar cells (Figure 3.4). This is predominantly due to the increasing concentration of defects in the nc-Si i-layer with an increase in the deposition rate (Figure 3.9) [68]. A detailed analysis of I–V characteristics and defects in the i-layer has revealed that solar cell characteristics are dominated by the defect density in the i-layer, which basically controls the diode characteristics [68]. A good correlation between diode quality factor and defect density in the i-layer has been obtained. There are several reasons for the increase in the defect density at a high deposition rate. One of the reasons could be that the species on the growing surfaces

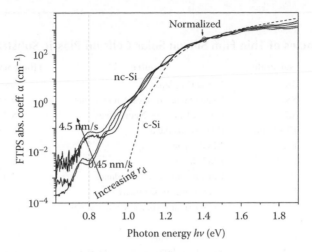

FIGURE 3.9 Optical absorption coefficient of the nc-Si i-layers of solar cells, obtained using FTPS. The deposition rates of the nc-Si layers are indicated in the figure.

do not get sufficient time to diffuse to an appropriate position due to the high flux of precursors reaching the surface. However, there is one more reason: high ion energies. At high deposition rates, high plasma powers are used. Though the pressure is increased to mitigate the ion energies, the energies may still be high enough to create defects. Decreasing ion energies by using other external effects can decrease the defect density if this proposition is valid. Applying an external bias to the RF electrode indeed showed a decrease in the defect density in the i-layer [90]. This was accompanied by an improvement in fill factor (FF), diode quality factor, and reverse saturation current, all correlating well with the decrease in defect density. This method allowed an improvement of 10% in the efficiency of the solar cell.

3.4.2 On Flexible Substrates

Flexible electronics is gaining momentum due to its advantages of flexible and unbreakable products, cheap substrates, easy handling and transport. Various types of devices ranging from displays, sensors, and markers to solar cells are being made on flexible substrates. Each device has an optimum process temperature that depends on the various constituents of the device and, most crucially, the photoactive material that is the central piece of such devices. For thin film silicon-based solar cells, the silicon layers, i.e., a-Si and nc-Si materials, are generally made at 200°C, the optimum temperature at which the best electronic qualities (low defect density, high structure order, high photosensitivity, etc.) are obtained. This is related to the optimum diffusion length of growth precursors, such as SiH$_3$, on the growing surface [9]. On the other hand, a-Si films deposited at lower temperatures, for example, at 100°C, are more porous and disordered. However, the structural quality of the films can be improved by providing extra energy through the flux of atomic hydrogen and ions to the growing surface [92]. The same argument would hold good also for nc-Si; however, no systematic work on this has been reported, though the nc-Si films made at low temperature are more defective and disordered compared to high-temperature samples [93]. In this sense, plasma-mediated growth processes, such as PECVD, have a definite advantage and VHF PECVD in combination with high hydrogen dilution can deliver device-quality materials at a substrate temperature as low as 100°C [94,95]. At low temperature, the deposition rate becomes even more critical, both for a-Si and nc-Si, due to slow diffusion of physisorbed species on the growing surface. Hence, the deposition process has to be adapted to meet the temperature limit that the substrate imposes. For flexible substrates, the processing temperature and other

TABLE 3.1
Reported Efficiencies of Thin Film Silicon Solar Cells on Plastic Substrates

Cell Type	Substrate	Source	Efficiency (%)	Reference
n-i-p (nc-Si)[a] D	SS	Canon Co., Japan	10.87	[48]
n-i-p (nc-Si)[a] D	SS	United Solar Ovonics, United States	9.51	[88]
n-i-p (nc-Si)[b] D	SS	Utrecht Univ., the Netherlands	8.7	[25]
n-i-p (nc-Si)[a] D	LT plastics	IMT, Switzerland	8.7	[97]
n-i-p (nc-Si)[a] D	E/TD	AIST, Japan	6	[98]
n-i-p (nc-Si)[a] D	LCP	AIST, Japan	8.1	[99]
n-i-p (nc-Si)[a] D	PET	Utrecht Univ., the Netherlands	2.6	[94]
n-p (mono-Si) T	Plastics	Univ. Stuttgart, Germany	14.6	[100]
n-i-p (a-Si/a-SiGe/ nc-Si)[a,c] D	SS	United Solar Ovonics, United States	15.4	[101]
n-i-p (a-Si/nc-Si)[a] D	LT plastics	IMT, Switzerland	10.9	[97]
p-i-n (a-Si/nc-Si)[a] T	Polyester	Univ. Utrecht/Nuon, the Netherlands	8.1	[102]
p-i-n (a-Si/nc-Si)[c] T	Polyester	Univ. Utrecht/Nuon, the Netherlands	9.4	[96]

Note: E/DT, tetracyclododecene copolymer; LCP, liquid crystal polymer; PEN, polyethylene naphthalate; PET, poly-ethylene terephthalate; Kapton, polyimide; *AM0 light condition; D, direct deposition; T, transfer method; DT, direct deposition+transfer; SS, stainless steel; LT, low temperature.

[a] VHF PECVC.
[b] HWCVD.
[c] St. RF PECVD.

conditions, mostly related to thermal stress, depend on whether the processing is (1) directly on the permanent substrate or (2) on a temporary substrate and later transferred to the permanent substrate. In the first case, the glass transition temperature (or the deformation temperature) of the substrate is the deciding factor. Whereas, cell processing directly on polyimide (also called Kapton) plastics or metal foils, such as stainless steel foil type of substrates, can be made at the standard processing temperature (~200°C) of thin film silicon cells, the deposition on cheap plastics, such as polyethylene terephthalate (PET), needs a much lower processing temperature due to its glass transition temperature of 70°C. There are, however, a number of plastic materials to choose that can withstand intermediate temperatures: polyester sulfonate (PES), polyethylene naphthalate (PEN), thermoplastic ethylene–tetracyclododecene copolymer (E/TD), liquid crystal polymer, etc. (see Table 3.1 and references therein). In the transfer type of processing, the processing temperature is not a critical issue as the cell processing takes place on a substrate that can resist heat above the standard processing temperature of the cell, i.e., 200°C. The transfer of the cell from a temporary to a permanent substrate is the critical issue here. The advantage of this processing is that almost any permanent substrate can be employed, as the transfer processing takes place at room temperature. An example of such processes (see Table 3.1 and references therein) is the deposition of thin film silicon on a metal foil (Al) and transfer to a plastic substrate (the Helianthos concept). Figure 3.10 shows a picture of a thin film silicon solar cell minimodule made using the Helianthos concept. An efficiency of 9.6% for an a-Si/nc-Si tandem cell has been achieved using the St. RF PECVD process in a tandem structure [96]. In order to increase the deposition rate with an aim to decrease the processing time, a-Si/nc-Si solar cells have also been fabricated through VHF PECVD using again the same Helianthos processing. An efficiency of 8% has been reached. Notwithstanding these achievements, efficiency-wise, solar cells made using this process have to compete with the thin monocrystalline silicon n-p junction solar cells fabricated using a lift-off process from a c-Si wafer and transfer to plastic [100].

FIGURE 3.10   Picture of a thin film silicon minimodule on a plastic substrate (polyester) made using the Helianthos technique. (From Rath, J.K. et al., *Thin Solid Films*, 517, 4758, 2009. With permission.)

3.5   CONCLUSION

nc-Si single-junction cells have reached stabilized efficiencies of ~10%, which is slightly higher than the state-of-the-art stabilized efficiency achieved for a-Si-type solar cells. The deposition rate of the nc-Si solar cell has reached 8 nm/s with a cell efficiency above 6%. However, single-junction nc-Si cell module fabrication through the direct deposition of nc-Si material is yet to be realized. However, solid-phase-crystallized nc-Si modules have shown the potential of nc-Si. On the other hand, nc-Si in combination with a-Si and its alloy to make multijunction has shown great industrial viability, and such multijunction solar cell production is at various pilot stages. Initial efficiencies of 15.4% in triple junction and 14.7% in tandem cells have been reached. nc-Si-based solar cells on plastics have received much attention nowadays, and in this category of devices, an efficiency of 9.6% has been achieved on a polyester substrate using the Helianthos transfer method. On the other hand, nc-Si-based solar cells directly deposited on plastics are still at a very preliminary stage of development.

REFERENCES

1. M. Gloeckler, *Technical Digest of the International PVSEC-17*, Fukuoka, Japan (2007) p. 116.
2. NREL Press release, http://www.nrel.gov/news/press/2008/574.html
3. T. Tomita, *Technical Digest of the International PVSEC-17*, Fukuoka, Japan (2007) p. 1.
4. D. Ide, M. Taguchi, Y. Yoshimine, T. Baba, and T. Kinoshita, *Proceedings of the 33rd IEEE PVSC*, San Diego, CA (2008); Y. Tsunomura, Y. Yoshimine, M. Taguchi, T. Kinoshita, H. Kanno, H. Sakata, E. Maruyama, and M. Tanaka, *Technical Digest of the International PVSEC-17*, Fukuoka, Japan (2007) p. 387.
5. http://www.appliedmaterials.com/products/assets/brochures/solar_product_brochure.pdf.
6. J. Meier, U. Kroll, S. Benagli, T. Roschek, A. Huegli, J. Spitznagel, O. Kluth et al., *Mater. Res. Soc. Symp. Proc.*, **989** (2007) 545.
7. S. Veprek and V. Marecek, *Solid State Electron.*, **11** (1968) 683.
8. S. Veprek, Z. Iqbal, H.R. Oswald, F.A. Sarrott, and J.J. Wagner, *J. Phys.*, **42** (1981) 251; K. Ensslen and S. Veprek, *Plasma Chem. Plasma Process.*, **7** (1987) 139.
9. A. Matsuda, *J. Non-Cryst. Solids*, **59/60** (1983) 767.
10. A. Matsuda, *Jpn. J. Appl. Phys.*, **43** (2004) 7909.
11. K. Nomoto, Y. Urano, J.L. Guizot, G. Ganguly, and A. Matsuda, *Jpn. J. Appl. Phys.*, **29** (1990) L1372.
12. A. Matsuda, K. Nomoto, Y. Takeuchi, A. Suzuki, A. Yuuki, and J. Perrin, *Surf. Sci.*, **227** (1990) 50.
13. J. Perrin, Y. Takeda, N. Hirano, Y. Takeuchi, and A. Matsuda, *Surf. Sci.*, **210** (1989) 114.

14. J.K. Rath, L.A. Klerk, A. Gordijn, and R.E.I. Schropp, *Sol. Energy Mater. Sol. Cells*, **90** (2006) 3385.
15. P.I. Cabarrocas, N. Layadi, T. Heitz, B. Drevillon, and I. Solomon, *Appl. Phys. Lett.*, **86** (1995) 3609.
16. S. Srinivasan, S. Agarwal, E.S. Aydil, and D. Marouda, *Nature*, **418** (2002) 62.
17. R.A. Street, *Phys. Rev. B*, **43** (1991) 2454.
18. J. Roberson, *J. Appl. Phys.*, **93** (2003) 731.
19. L. Houben, C. Scholten, M. Luysberg, O. Vetterl, F. Finger, and R. Carius, *J. Non-Cryst. Solids*, **299** (2002) 1189.
20. B. Yan, G. Yue, Y. Yan, C.-S. Jiang, C.W. Teplin, J. Yang, and S. Guha, *Mater. Res. Soc. Symp. Proc.*, **1066** (2008) 61.
21. J.K. Rath, F.D. Tichelaar, H. Meiling, and R.E.I. Schropp, *Mater. Res. Soc. Symp. Proc.*, **507** (1998) 879.
22. P.A.T.T. van Veenendaal, G.W.M. van der Mark, J.K. Rath, and R.E.I. Schropp, *Thin Solid Films*, **430** (2003) 41.
23. A. Verkerk, J.K. Rath, M. Brinza, R.E.I. Schropp, W.J. Goedheer, V.V. Krzhizhanovskaya, Y.E. Gorbachev, K.E. Orlov, E.M. Khilkevitch, and A.S. Smirnov, *Mater. Sci. Eng. B*, **159–160** (2009) 53.
24. J.E. Bouree, *Thin Solid Films*, **395** (2001) 157.
25. H. Li, R.L. Stolk, C.H.M. van der Werf, R.H. Franken, J.K. Rath, and R.E.I. Schropp, *J. Non-Cryst. Solids*, **352** (2006) 1941.
26. J.K. Rath, A. Verkerk, Y. Liu, M. Brinza, W.J. Goedheer, and R.E.I. Schropp, *Mater. Sci. Eng. B*, **159–160** (2009) 38–43.
27. B. Strahm, A.A. Howling, L. Sansonnens, Ch. Hollenstein, U. Kroll, J. Meier, Ch. Ellert, L. Feitknecht, and C. Ballif, *Sol. Energy Mater. Sol. Cells*, **91** (2007) 495.
28. M.N. van den Donker, B. Rech, F. Finger, L. Houben, W.M.M. Kessels, and M.C.M. van de Sanden, *Prog. Photovolt. Res. Appl.*, **15** (2007) 291.
29. Verkerk, Plasma deposition of thin film silicon at low substrate temperature and at high growth rate, PhD thesis, Utrecht University, Utrecht, the Netherlands (2009).
30. A. Verkerk, J.K. Rath, and R.E.I. Schropp, *Phys. Status Solidi A*, **1–5** (2010), DOI: 10.1002/pssa.200982883.
31. A. Matsuda, S. Yamasaki, K. Nakagawa, H. Okushi, K. Tanaka, S. Iizima, M. Matsumura, and H. Yamamoto, *Jpn. J. Appl. Phys.*, **19** (1980) L305.
32. J. Meier, R. Fluckiger, H. Keppner, and A. Shah, *Appl. Phys. Lett.*, **65** (1994) 860.
33. J.K. Rath, F.D. Tichelaar, and R.E.I. Schropp, *Solid State Phenom.*, **67–68** (1999) 465.
34. Y. Zhao, S. Miyajima, Y. Ide, A. Yamada, and M. Konagai, *Jpn. J. Appl. Phys.*, **41** (2002) 6417.
35. C. Smit, E.A.G. Hamers, B.A. Korevaar, R.A.C.M.M. van Swaaij, and M.C.M. van de Sanden, *J. Non-Cryst. Solids*, **299–302** (2002) 98.
36. M. Acciarri, S. Binetti, M. Bollani, A. Comotti, L. Fumagalli, S. Pizzini, and H. von Känel, *Sol. Energy Mater. Sol. Cells*, **87** (2005) 11.
37. S.J. Jones, R. Crucet, X. Deng, D.L. Williamson, and M. Izu, *Mater. Res. Soc. Symp. Proc.*, **609** (2000) A15.1.1; H. Jia, H. Kuraseko, H. Fujiwara, and M. Kondo, *Sol. Energy Mater. Sol. Cells*, **93** (2009) 812.
38. V.L. Dalal, S. Kaushal, E.X. Ping, J. Xu, R. Knox, and K. Han, *Mater. Res. Soc. Symp. Proc.*, **377** (1995) 137.
39. Y. Hotta, H. Toyoda, and H. Sugai, *Thin Solid Films*, **515** (2007) 4983.
40. K. Yamamoto, T. Suzuki, M. Yoshimi, and A. Nakajima, *Jpn. J. Appl. Phys.*, **36** (1997) L569.
41. L. Guo, M. Kondo, M. Fukawa, K. Saitoh, and A. Matsuda, *Jpn. J. Appl. Phys.*, **37** (1998) L1116.
42. C. Niikura, M. Kondo, and A. Matsuda, *Sol. Energy Mater. Sol. Cells*, **90** (2006) 3223.
43. H. Curtins, N. Wyrsch, M. Favre, and A.V. Shah, *Plasma Chem. Plasma Process.*, **7** (1987) 267.
44. J. Dutta, U. Kroll, P. Chabloz, A. Shah, A. Howling, J.-L Dorier, and Ch. Hollenstein, *J. Appl. Phys.*, **72** (1992) 3220.
45. M. Heintze, *Mater. Res. Soc. Symp. Proc.*, **467** (1997) 471.
46. H. Takatsuka, M. Noda, Y. Yonekura, Y. Takeuchi, and Y. Yamauchi, *Sol. Energy*, **77** (2004) 951.
47. L. Sansonnens and J. Schmitt, *Appl. Phys. Lett.*, **82** (2003) 182; Patent U.S. 6228438.
48. K. Saito, M. Sano, A. Sakai, R. Hayasi, and K. Ogawa, *Technical Digest of the International PVSEC-12*, Jeju, Korea (2001) p. 429.
49. A. Gordijn, J.K. Rath, and R.E.I. Schropp, *Prog. Photovolt Res. Appl.*, **14** (2006) 305.
50. Y. Mai, S. Klein, R. Carius, H. Stiebig, L. Houben, X. Geng, and F. Finger, *J. Non-Cryst. Solids*, **352** (2006) 1859.
51. J. Bailat, D. Dominé, R. Schlüchter, J. Steinhauser, S. Faÿ, F. Freitas, C. Bücher et al., *Proceedings of the 4th WCPEC*, Waikoloa, HI (2006) p. 1533.

52. S. Klein, F. Finger, R. Carius, T. Dylla, B. Rech, M. Grimm, L. Houben, and M. Stutzmann, *Thin Solid Films*, **430** (2003) 202.
53. P.A.T.T. van Veenendaal, O.L.J. Gijzeman, J.K. Rath, and R.E.I. Schropp, *Thin Solid Films*, **395** (2000) 194.
54. D. Knoesen, C. Arendse, S. Halindintwali, and T. Muller, *Thin Solid Films*, **516** (2008) 822.
55. J. Werner, *Technical Digest of the 13th Sunshine Workshop on Thin Film Solar Cells*, M. Konagai (Ed.), NEDO, Tokyo, Japan (2000) p. 41.
56. R.W. Collins, A.S. Ferlauto, and C.R. Wronski, *Sol. Energy Mater. Sol. Cells*, **78** (2003) 143.
57. O. Vetterl, F. Finger, R. Carius, P. Hapke, L. Houben, O. Kluth, A. Lambertz, A. Muck, B. Rech, and H. Wagner, *Sol. Energy Mater. Sol. Cells*, **62** (2000) 97.
58. E. Vallat-Sauvain, U. Kroll, J. Meier, and A. Shah, *J. Appl. Phys.*, **87** (2000) 3137.
59. G. Yue, B. Yan, G. Ganguly, J. Yang, S. Guha, C.W. Teplin, and D.L. Williamson, *Conference Record of the 2006 IEEE 4th World Conference on Photovoltaic Energy Conversion*, Waikoloa, HI, vol. 2 (2007) p. 1588.
60. A. Gordijn, Microcrystalline silicon for thin-film solar cells, PhD thesis, Utrecht University, Utrecht, the Netherlands (2005).
61. J. Meier, E. Vallat-Sauvain, S. Dubail, U. Kroll, J. Dubail, S. Golay, L. Feitknecht, P. Torres, S. Fay, D. Fischer, and A. Shah, *Sol. Energ. Mater. Sol. Cells*, **66** (2001) 73.
62. J.K. Rath, *Sol. Energy Mater. Sol. Cells*, **76** (2003) 431.
63. S. Hiza, A. Yamada, M., and Konagai, *Jpn. J. Appl. Phys.*, **47** (2008) 6222.
64. M. Kondo, S. Yamasaki, and A. Matsuda, *J. Non-Cryst. Solids*, **266–269** (2000) 544.
65. J.K. Rath, A. Barbon, and R.E.I. Schropp, *J. Non-Cryst. Solids*, **266–269** (1999) 548.
66. A. Poruba, A. Fejfar, Z. Remes, J. Springer, M. Vanecek, J. Kocka, J. Meier, P. Torres, and A. Shah, *J. Appl. Phys.*, **88** (2000) 148.
67. J.K. Rath, F.A. Rubinelli, M. van Veghel, C.H.M. van der Werf, Z. Hartman, and R.E.I. Schropp, *Proceedings of the 16th European Photovoltaic Conference and Exhibition*, Glasgow, Scotland (2000) p. 462.
68. A. Gordijn, L. Hodakova, J.K. Rath, and R.E.I. Schropp, *J. Non-Cryst. Solids*, **352** (2006) 1868.
69. T. Kaneko, M. Wakagi, K. Onisawa, and T. Minemura, *Appl. Phys. Lett.*, **64** (1994) 1865.
70. C. Smit, R. Van Swaaij, H. Donker, A. Petit, W.M.M. Kessels, and M.C.M. Van de Sanden, *J. Appl. Phys.*, **94** (2003) 3582.
71. A. Gordijn, J. Löffler, W.M. Arnoldbik, F.D. Tichelaar, J.K. Rath, and R.E.I. Schropp, *Sol. Energy Mater. Sol. Cells*, **87** (2005) 445.
72. Y. Nasuno, M. Kondo, and A. Matsuda, *Appl. Phys. Lett.*, **78** (2001) 2330.
73. F. Finger, R. Carius, P. Hapke, K. Prasad, and R. Flueckiger, *Mater. Res. Soc. Symp. Proc.*, **283** (1993) 471.
74. J.K. Rath, R.E.I. Schropp, and W. Beyer, *J. Non-Cryst. Solids*, **266–269** (1999) 190.
75. S. Klein, F. Finger, R. Carius, H. Wagner, and M. Stutzmann, *Thin Solid Films*, **395** (2001) 305.
76. A.H.M. Smets, T. Matsui, and M. Kondo, *Appl. Phys. Lett.*, **92** (2008) 033506.
77. H.P. Klug and L.E. Alexander, *X-Ray Diffraction Procedure*, Wiley, New York (1974).
78. E. Bustarret, M.A. Hachichia, and M. Brunel. *Appl. Phys. Lett.*, **52** (1988) 1675.
79. K. Yamamoto, M. Yoshimi, Y. Tawada, Y. Okamoto, and A. Nakajima, *J. Non-Cryst. Solids*, **266–269** (2000) 1082.
80. J.K. Rath, A. Barbon, and R.E.I. Schropp, *J. Non-Cryst. Solids*, **227–230** (1998) 1277.
81. V.L. Dalal and A. Madhavan, *J. Non-Cryst. Solids*, **354** (2008) 2403.
82. M. Vanecek and A. Poruba, *Appl. Phys. Lett.*, **80** (2002) 719.
83. C. Droz, E. Vallat-Sauvain, J. Bailat, L. Feitknecht, J. Meier, and A. Shah, *Sol. Energy Mater. Sol. Cells*, **81** (2004) 61.
84. J. Hupkes, B. Rech, O. Kluth, T. Repmann, B. Zwaygardt, J. Muller, R. Drese, and M. Wuttig, *Sol. Energy Mater. Sol. Cells*, **90** (2006) 3054.
85. S. Faÿ, L. Feitknecht, R. Schlüchter, U. Kroll, E. Vallat-Sauvain, and A. Shah, *Sol. Energy Mater. Sol. Cells*, **90** (2006) 2960.
86. T. Oyama, M. Kambe, N. Taneda, and K. Masumo, *Mater. Res. Soc. Symp. Proc.* **1101** (2008) 71; J. Springer, A. Poruba, and M. Vanecek, *J. Appl. Phys.*, **96** (2004) 5329.
87. K. Sato, Y. Gotoh, Y. Wakayama, Y. Hayashi, K. Adachi, and H. Nishimura, *Rep. Res. Lab. Asahi Glass Co. Ltd.*, **42** (1992) 129.
88. G. Yue, B. Yan, L. Sivec, J.M. Owens, S. Hu, X. Xu, J. Yang, and S. Guha, *Proceedings of the 34th IEEE PVSC*, Philadelphia, PA (2009).

89. Y. Sobajima, M. Nishino, T. Fukumori, M. Kurihara, T. Higuchi, S. Nakano, T. Toyama, and H. Okomoto, *Sol. Energy Mater. Sol. Cells*, **93** (2009) 980.

90. A. Gordijn, M. Vanecek, W.J. Goedheer, J.K. Rath, and R.E.I. Schropp, *J. Appl. Phys.*, Part 1: Regular papers and short notes and review papers, **45** (2006) 6166.

91. T. Matsui, M. Kondo, and A. Matsuda, *Jpn. J. Appl. Phys. Part 2 Lett.*, **42** (2003) L901.

92. J.K. Rath, R.E.I. Schropp, P. Roca i Cabarocas, and F.D. Tichelaar, *J. Non-Cryst. Solids*, **354** (2008) 2652.

93. M. Brinza, J.K. Rath, and R.E.I. Schropp, *Phys. Status Solidi C*, **1–4** (2010), DOI: 10.1002/pssc.200982879.

94. J.K. Rath and R.E.I. Schropp. *Sol. Energy Mater. Sol. Cells*, **93** (2009) 680.

95. P.C.P. Bronsveld, J.K. Rath, and R.E.I. Schropp, *Proceedings of the 20th EUPVSEC*, Barcelona, Spain (2005) p. 1675.

96. A. Gordijn, M.N. van den Donker, F. Finger, E.A.G. Hamers, G.J. Jongerden, W.M.M. Kessels, R. Bartl et al., *Conference Record of the IEEE Fourth World Conference on Photovoltaic Energy Conversion*, Waikoloa, HI (2006) p. 1716.

97. F.-J. Haug, T. Soderstrom, M. Python, V. Terrazzoni-Daudrix, X. Niquille, and C. Ballif, *Sol. Energy Mater. Sol. Cells*, **93** (2009) 884.

98. H. Mase, M. Kondo, and A. Matsuda, *Sol. Energy Mater. Sol. Cells*, **74** (2002) 547.

99. T. Takeda, M. Kondo, and A. Matsuda, *Proceedings of the Third World Conference on Photovoltaic Energy Conversion*, B, Osaka, Japan (2003) p. 1580.

100. C. Berge, M. Zhu, W. Brendle, M.B. Schubert, and J.H. Werner, *Sol. Energy Mater. Sol. Cells*, **90** (2006) 3102.

101. B. Yan, G. Yue, J.M. Owens, J. Yang, and S. Guha, *Conference Record of the 2006 IEEE Fourth World Conference on Photovoltaic Energy Conversion*, Waikoloa, HI, vol. 2 (2007) p. 1461.

102. J.K. Rath, M. Brinza, Y. Liu, A. Borreman, and R.E.I. Schropp, *Sol. Energy Mater. Sol. Cells* (2010), DOI: 10.1016/j.solmat.2010.01.013; J.K. Rath, Y. Liu, M. Brinza, A. Verkerk, C. van Bommel, A. Borreman, and R.E.I. Schropp, *Thin Solid Films*, **517** (2009) 4758.

4 Semiconductive Nanocomposite Films for Clean Environment

Joe H. Hsieh

CONTENTS

4.1 NANOCOMPOSITE THIN FILMS

4.1.1 NANOCRYSTALLINE MATERIALS

Nanocrystalline materials are single-phase or multiphase polycrystalline materials having grain sizes smaller than 100 nm. Nanocrystalline materials may contain crystalline, quasicrystalline, or amorphous phases, and can be metals, ceramics, or composites (nanocomposites). Because of the extremely small dimensions, a large volume fraction of the atoms is located at the grain boundaries, and this brings special properties to these materials. The properties of nanocrystalline materials are very often superior to those of conventional polycrystalline coarse-grained materials. Nanocrystalline materials exhibit superior mechanical properties, such as high strength/hardness, enhanced diffusivity, improved ductility/toughness, reduced density, and reduced elastic modulus, and also have higher electrical resistivity, higher thermal expansion coefficient, and superior magnetic properties in comparison with conventional coarse-grained materials. The concepts of nanocomposites are being investigated with certain emphasis on semiconductor composites with nano-sized metal particles embedded in the matrix [1,2]. Overall, there appears to be a great potential for the nanocrystalline materials to be applied in almost every area. The extensive investigations on structure–property correlations in nanocrystalline materials have begun to unravel the complexities of these materials and have provided many important results as well as many new concepts [3].

During the past decade, nanocomposites have emerged as the new research area for constructing light-energy-harvesting assemblies. Organic and inorganic hybrid structures that exhibit improved selectivity and efficiency toward catalytic processes have been designed. Size-dependent properties, such as size quantization effects in semiconductor nanoparticles and quantized charging effects in metal nanoparticles, provide the basis for developing new and effective systems [4]. According to Kamat [4], there are three major ways in which one can utilize nanocomposite structures for the design of solar energy conversion devices. The first is to mimic photosynthesis with donor–acceptor molecular assemblies and clusters. The second is the semiconductor-assisted photocatalysis to produce fuels such as hydrogen. The third is the nanostructure-based solar cells. Kamat's paper summarizes some of the highlights on the recent development in nanocomposite materials for a cleaner environment [4].

4.1.2 Metal Nanoparticles

Metal nanoparticles, usually with a crystalline structure and with a characteristic size of the order of a few nanometers (<10 nm), show very special properties that include quantum size effect, weak localization effect that makes them interesting for many scientific and industrial applications. Figure 4.1 shows the variation of resistivity versus temperature. This figure exhibits clearly the transition of metal-like silver to semiconductor-like silver due to quantum size effect. In addition, due to the large surface-to-volume ratio (for instance, the fraction of interfaces is ~50% for the nano-materials with a grain size of 5 nm) of nanoparticles, the influence of the interfaces on the electron transport characteristics cannot be neglected, as has been done in conventional polycrystalline materials. Therefore, some special features involving optical, electrochemical, and electrical behaviors may appear, and are closely associated with their characteristic size.

4.1.3 Nanocomposite Materials

Nanocomposite materials are a new form of nanocrystalline materials and have been investigated tremendously because these may have superior properties due to the size effect and especially due to the mixing of different nano-phases with different properties. This material represents a new class of film, powder, or bulk materials that are, in the simplest case, composed of two phases with a grain size below 100 nm, typically about 10 nm or smaller. This is illustrated in Figure 4.2.

At present, many studies have been conducted with a special emphasis on ceramic–metal composites. Ceramic–metal composites are composed of two or more immiscible components with nano-sized metal particles dispersed in a dielectric or semiconductor matrix. They can also be called "cermets," such as $Cu-Al_2O_3$ [1], Co-TiN [6], Ag-TiN [7], Co-ZnO [8], and so on (see Figure 4.3). These kinds of nanocomposite materials may have superior properties due to the mixing of various phases with different properties. So far, most efforts have been made to investigate the optical [1,9], magnetic [6,8], mechanical [10,11], as well as optoelectronic properties.

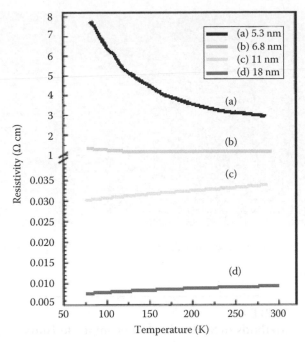

FIGURE 4.1 Variation of resistivities of nano-sized Ag with different grain sizes as a function of temperature. (Unpublished drawing.)

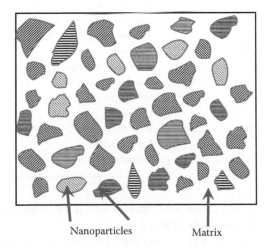

FIGURE 4.2 Schematic diagram of a nanocomposite material. (From Musil, J. and Vicek, J., *Mater. Chem. Phys.*, 54, 116, 1998.)

It has been proved that the properties of cermets can be altered in a wide range by changing the metal volume fraction. The composite's characteristic behavior may change from insulator/semiconductor-like to metal-like as the metal volume fraction increases to a certain extent [14].

4.1.4 SYNTHESIS OF NANOCOMPOSITE FILMS

As mentioned, increased activity on the synthesis of nanocomposite films in recent years is due to the combining advantages of thin films and nanocomposite structures. So far, a number of techniques

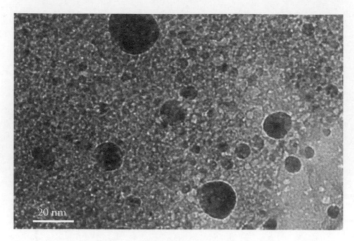

FIGURE 4.3   TEM image of Ag particles embedded in TaON. (From Wang, D. et al., *Mater. Chem. Phys.*, 81, 341, 2003.)

TABLE 4.1
Methods to Synthesize Nanocomposite Films

Starting Phase	Techniques
Vapor	Co-sputtering
	Co-evaporation
	Pulse laser deposition (PLD)
	Chemical vapor deposition (CVD)
	Combustion CVD (CCVD)
Liquid	Electrodeposition
	Electrophoretic deposition
	Rapid solidification
	Sol–gel process
	Casting
Solid	Mechanical alloying/milling
	Spark erosion
	Powder processing

have been used to synthesize nanocomposite films, in which the starting material can be either in the solid, the liquid, or the gaseous state (Table 4.1).

In principle, any method capable of producing films can be used to produce nanocomposite films. If a phase transformation is involved, e.g., liquid to solid or vapor to solid, then certain steps have to be taken to control the nucleation and growth rates in order to control the composite's structure and properties. The particle's second-phase size, morphology, and texture can be varied by suitably modifying or controlling the process variables in each of these methods.

4.2   APPLICATION IN WATER SPLITTING

Although solar cells have many obvious desirable features, they have some restrictions including reliable storage and stable light source. On the other hand, hydrogen is an ideal energy carrier mainly because hydrogen can be converted into a useful form of energy more efficiently than any other fuel [15]. Furthermore, hydrogen can be produced from water by solar irradiation through a photo-electrochemical (PEC) process, without using other forms of energy, especially electricity. Therefore, an efficient PEC process to directly convert sunlight into hydrogen becomes much promising.

Hydrogen production using a PEC process basically requires a stable semiconductive material as a photoanode or photocathode to absorb sunlight. This process is initiated by light absorption with light energy equal to or greater than the semiconductor bandgap. During the process, electrons (e^-) can be excited to jump from the valence band (VB) to the conduction band (CB), and leave holes (h^o) in the VB. Prior to the recombination of electrons and holes, hydrogen can be generated due to a reduction process initiated by free electrons, while holes can induce an oxidation reaction. These reactions can be illustrated as in Figure 4.4 [16] and are expressed as the following equations:

$$2H^+ + 2e^- \rightarrow H_2 \quad E^o = 0 \text{ eV versus SHE}$$

$$H_2O + 2h^o \rightarrow \frac{1}{2}O_2 + 2H^+ \quad E^o = 1.23 \text{ eV versus SHE}$$

Figure 4.5 shows a typical experimental setup for water splitting.

According to the reactions, the material-property requirements for water splitting can be summarized as follows:

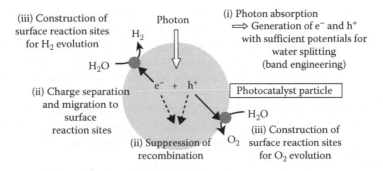

FIGURE 4.4 Photocatalytic process for water splitting. (From Kudo, A. et al., *Chem. Lett.*, 33, 1534, 2004.)

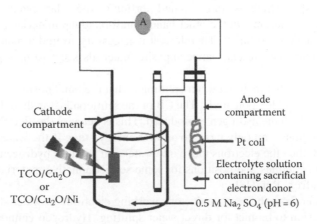

FIGURE 4.5 A twin-compartment PEC cell for water splitting. (From Somasundaram, S. et al., *Int. J. Hydrogen Energy*, 32, 4661, 2007.)

1. The energy bandgap must be greater than 1.23 eV.
2. The potential of CB must be more negative than the potential required to reduce water (H_2 evolution).
3. The VB should be located more positively than the O_2/H_2O level so that the oxidation power is strong enough to oxidize water.

To satisfy these requirements, semiconductive oxides or nitrides seem to be the most suitable candidates.

4.2.1 PROGRESS ON ELECTRODE MATERIALS

The principle of semiconductor-based photocatalysis is presented in Figure 4.4. The efficiency of the electrode is seen depending on the optical bandgap of the electrode. For example, anatase TiO_2 has a bandgap of 3.2 eV. According to the equation

$$E = \frac{hc}{\lambda} = \frac{6.626 \times 10^{-34} \times 3 \times 10^8 \ (J)}{\lambda \ (m)} \times \frac{10^9 \lambda \ (m)}{\lambda \ (nm)} \times \frac{1 \ eV}{1.602 \times 10^{-19} (J)} = \frac{1240 \ (eV)}{\lambda \ (nm)}$$

TiO_2 can only absorb light with a wavelength shorter than 380 nm, which is in the ultraviolet range. Any light with a wavelength longer than 380 nm is unlikely to create electron–hole pairs in TiO_2.

Photoelectrode materials can be either n-type or p-type semiconductor materials, either in bulk or in thin-film form. Till date, most of the photoelectrode materials are n-type semiconductors, although some studies show that p-type photoelectrodes can also be used to produce hydrogen. However, the reducing reaction occurring on the electrode surface has been thought as a major problem. In Section 4.2.2, some of the developed and developing materials for hydrogen production are discussed.

As mentioned earlier, the efficiency of a photoelectrode depends on its ability to absorb sunlight. Only the solar radiation having energy greater than the bandgap of the semiconducting material is absorbed. Hence, the bandgap of the semiconductor should be such that it utilizes the maximum of the solar spectrum in energy conversion.

The stability of the semiconductor in an aqueous solution is another important factor that determines the success of any material as a photoanode. Materials with bandgaps higher than 3 eV, such as ZnO, WO_3, and TiO_2, which use only a small portion of the solar spectrum in photolysis, are stable in an electrolyte solution. On the other hand, small-bandgap materials, such as GaP and Si, are not stable in an aqueous solution. Therefore, it is necessary to find a material that can satisfy both these requirements. At present, nanocomposite materials seem to be a good answer to this question.

Another condition needed in direct water splitting is that the band position of the semiconductor should overlap the water redox potentials. The CB of the semiconductor should be higher in energy than the potential at which the hydrogen is produced. This is because the electrons in the CB, which are excited by the photons, can only move to a lower energy state. If the water reduction potential is lower than the CB, then the electrons move into water and produce hydrogen gas [15]. Figure 4.6 presents the bandgap energies and positions for some semiconductive materials. Further information can be obtained from Ref. [18].

The charge transfer from the semiconductor's working electrode to water should occur quickly. This is another criterion to be met for direct water splitting. Hydrogen cannot be produced if the charge cannot move into the water. If the electrons do not move into water quickly, they will accumulate on the semiconductor's CB. This will cause a shift in the Fermi level in the negative direction, which may not be favorable for the oxidation of water [15].

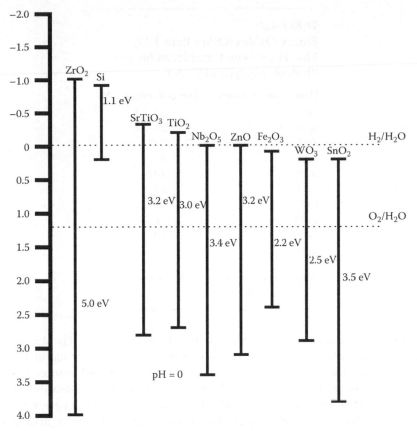

FIGURE 4.6 Bandgaps of some semiconductive materials. (From Somasundaram, S. et al., *Int. J. Hydrogen Energy*, 32, 4661, 2007.)

4.2.1.1 TiO₂

The first PEC splitting of water using an n-type TiO_2 electrode was demonstrated by Fujishima et al. in 1969. In 1972, this was reported in *Nature* by Fujishima and Honda [19]. This paper soon attracted much attention from scientists, especially after the first energy crisis. This is commonly thought to be the starting point for the rapid developing of semiconductive photocatalysts. Although TiO_2 is an inexpensive and stable material, its overall hydrogen production efficiency is still not high enough due to its wide bandgap nature. This wide bandgap nature makes the material only work under UV light irradiation. In order to make it work under visible light irradiation, many studies doped the material with N, C, F, and P elements or some transition metals. A thorough review in this regard can be found in Ref. [20]. Depending on the concentrations of various dopants, many studies show that the efficiency of the doped TiO_2 can be improved under visible light irradiation. In addition to this, many efforts have been devoted to the development of visible-light-induced photocatalytic materials in the past years [21,22].

4.2.1.2 Visible-Light Photocatalytic Materials

As mentioned, visible-light-activated TiO_2 materials have been obtained by doping with either transition metals or nonmetal anions. Anionic doping provides advantages of improved photocatalytic efficiency, dopant stability, and easiness of doping. Since Asahi et al. [23] reported the visible-light photo-activity of TiO_2 with nitrogen doping, many research groups demonstrated that anion doping of TiO_2 extended the optical absorbance of TiO_2 into the visible-light region. The reported anionic dopants include N, C, S, F, and P [20]. Despite the remarkable advances in absorption shift toward the visible-light region of the solar spectrum, the photocatalytic activity of these doped materials is

TABLE 4.2
Binary Oxides (Other than TiO_2)
That Have Been Considered for the
Photoelectrolysis of Water

Oxide Semiconductor	Energy Bandgap (eV)
WO_3	2.5–2.8
Fe_2O_3	2.0–2.2
ZnO	3.37
SnO_2	3.5
NiO	3.47
CdO	2.3
PdO	0.8
Cu_2O	2.0–2.2
CuO	1.7
Bi_2O_3	2.8

Source: Rajeshwar, K., J. Appl. Electrochem., 37, 765, 2007.

rather low owing to a low quantum yield under visible light activation [24]. There remains an urgent need for engineers and scientists to design some photocatalysts with higher sensitivity.

Recently, Sakatani et al. [25] successfully synthesized metal ion and nitrogen co-doped TiO_2. They suggested that the photocatalyst prepared with Sr^{2+} and nitrogen exhibits the highest activity among a group of studied metal ions. Apparently, these endeavors can open up new possibilities for the development of efficient solar-induced photocatalytic materials. However, until now, there are only a few reports on metal ion and anion co-doped TiO_2.

Besides TiO_2-based materials, other visible-light-induced photocatalytic materials for use in water splitting are developed. Some of them show much better performance than TiO_2. However, the cost may happen to be a major problem. The studied materials include binary and ternary oxide as well as oxy-nitride. The examples include Cu_2O, CuO, $InMO_4$ (M = Ta^{5+}, Nb^{5+}, V^{5+}) [26–28], and TaON [29,30]. Tables 4.2 and 4.3 list some of the examples adapted from Ref. [21]. Zou et al. and Ye et al. [26–28] reported that $InMO_4$ could be a good visible-light photocatalyst. The photocatalysts belonging to this category have shown a good water-splitting efficiency. $InTaO_4$ and $InNb_4$ belong to the category of the ABO_4 compound. They have a $A^{3+}B^{5+}O_4$ wolframite structure. The addition of some minor transition metals can further increase its efficiency. $InVO_4$ belongs to the category of an orthorhombic structure. It has a bandgap of 1.9 eV, which can efficiently absorb visible light. The addition of 1.0 wt.% NiO provides a water-splitting efficiency of 5.0 μmol/g·cat·h, while the addition of Pt can increase the efficiency to 14.0 μmol/g·cat·h. Figure 4.7 shows the bandgap diagram of $InMO_4$ photocatalysts. Figure 4.8 shows the comparison of light absorption for various $InMO_4$ compounds. Another recently developed material is TaON. TaON is categorized as a visible-light-inducible photocatalyst [29,30]. Figure 4.9 shows the band structure of TaON, Ta_2O_5, and Ta_3N_5. Figure 4.10 shows the comparison of light absorption for TiO_2 and TaON. Apparently, TaON should also be a good candidate for direct water splitting.

4.2.2 ENHANCED EFFICIENCY BY NANOCOMPOSITES

4.2.2.1 Metal–Semiconductor Nanocomposites

A previous study [32] shows that metal–semiconductor nanocomposites may have an increased efficiency for water splitting. For example, Pt, Ru, and Ag are doped to photocatalysts in order to

TABLE 4.3
Ternary Oxides with the General Formula ABO$_3$ That Have Been Examined for Water Splitting

Oxide Semiconductor	Energy Bandgap (eV)
SrTiO$_3$	3.2
BaTiO$_3$	2.8
FeTiO$_3$	2.16
YFeO$_3$	2.58
LuRhO$_3$	2.2
BaSnO$_3$	3.0
CaTiO$_3$	3.6
KNbO$_3$	3.1
Ba$_{0.8}$Ca$_{0.2}$TiO$_3$	3.25
KTaO$_3$	3.5
CdSnO$_3$	1.77
LaRhO$_3$	1.35
NiTiO$_3$	1.6
LaMnO$_3$	1.1

Source: Rajeshwar, K., *J. Appl. Electrochem.*, 37, 765, 2007.

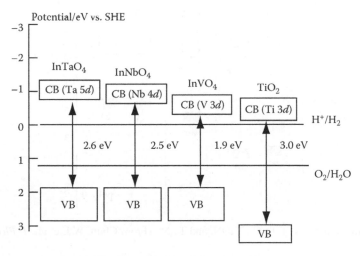

FIGURE 4.7 Band structures of InTaO$_4$, InNbO$_4$, InVO$_4$, and TiO$_2$. (From Ye, J. et al., *J. Photochem. Photobiol. A Chem.*, 148, 79, 2002.)

extend the light absorption range and increase the electron–hole pair separation rate. Eventually, the photocatalytic performance can be improved.

In general, the advantages of these semiconductor composites can be summarized as follows:

1. They act as a type of sensitizer and increase light absorption [33].
2. They enhance the electron–hole pair separation, and/or to prohibit the recombination of electrons and holes [34].
3. They act as the base for hydrogen reduction [35].

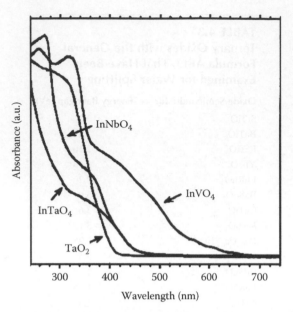

FIGURE 4.8 Absorbance of $InTaO_4$, $InNbO_4$, and TiO_2. (From Ye, J. et al., *J. Photochem. Photobiol. A Chem.*, 148, 79, 2002.)

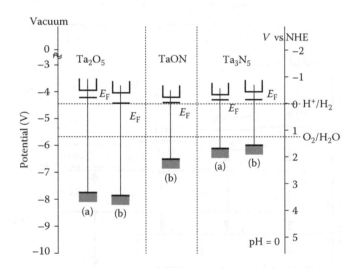

FIGURE 4.9 Band structures of Ta_2O_5, TaON, and Ta_3N_5. (From Chun, W.J. et al., *J. Phys. Chem. B*, 107, 1798, 2003.)

Shan et al. [36] proposed that the Schottky barrier formed at the metal–semiconductor interface can serve as an efficient electron trap to avoid electron–hole recombination in a normal photocatalytic process. A metal island contact with a semiconductor leads to the formation of a Schottky barrier, as illustrated in Figure 4.11. The electron migration from the semiconductor to the metal occurs until the two Fermi levels are aligned, since the metal has a work function (ψ_m) higher than that of the semiconductor (ψ_s). The surface of the metal acquires an excess negative charge, while the semiconductor exhibits an excess positive charge as a result of electron migration away from the barrier region. A Schottky barrier is then formed at the metal–semiconductor interface. Figure 4.12 illustrates the attached noble metals in storing and shuttling photo-generated electrons from the semiconductor in a photocatalytic process [37]. The photo-induced electrons in the CB

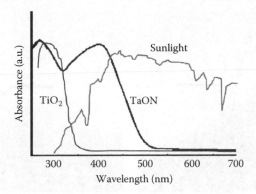

FIGURE 4.10 Absorbance spectra of TaON and TiO$_2$ in comparison to sunlight spectrum. (From Chun, W.J. et al., *J. Phys. Chem. B*, 107, 1798, 2003.)

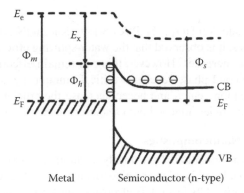

FIGURE 4.11 Schematic diagram of Schottky barrier. (From Shan, Z. et al., *J. Phys. Chem. C*, 112, 15423, 2008.)

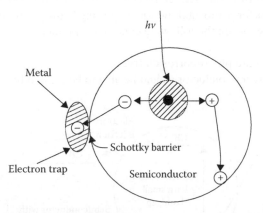

FIGURE 4.12 Schematic diagram for charge separation in metal–semiconductor nanocomposite. (From Shan, Z. et al., *J. Phys. Chem. C*, 112, 15423, 2008.)

of the semiconductor are believed to readily transfer to the metal, which facilitates the separation of the photo-induced electron–hole pairs and effectively inhibits their recombination. Hence, the efficiency is increased. Furthermore, the metal attached on the semiconductor surface can shift the Fermi level to the more negative potential direction to improve the energy system of the composite and the efficiency of the interfacial charge-transfer process [37,38].

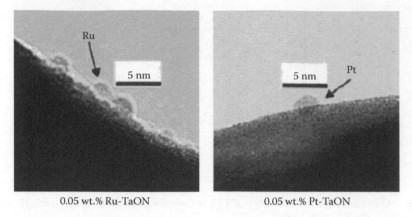

0.05 wt.% Ru-TaON   0.05 wt.% Pt-TaON

FIGURE 4.13 TEM images of Ru-TaON and Pt-TaON. (From Yamasita, D. et al., *Solid State Ionics*, 172, 591, 2004.)

More examples that include Ag/TiO_2 [36], Pt/CdS [39], Ru/TaON [40], and Pt/TaON [40] have been studied. In these studies, it is observed that the water-splitting efficiency would increase as the content of the second phase increases. However, this can remain effective up to a certain degree. When the content of the second phase reaches a critical amount, the second phase may become the electron–hole recombination center, and hence decrease the efficiency [41]. Figure 4.13 shows micrographs of metal nanoparticles attached onto a semiconductor matrix [40].

4.2.2.2 Semiconductor Nanocomposites

Semiconductor composites are mentioned previously as another answer to utilize visible light for hydrogen production. When a large-bandgap semiconductor is coupled with a small-bandgap semiconductor with a more negative CB level, CB electrons can be injected from the small-bandgap semiconductor to the large-bandgap semiconductor. Thus, a wide electron–hole separation is achieved, as shown in Figure 4.14 [42]. This process is similar to dye sensitization. The difference is that electrons are injected from one semiconductor to another semiconductor. Successful coupling of the two semiconductors for photocatalytic water-splitting hydrogen production under visible light irradiation can be achieved when the following conditions are met:

1. The semiconductors are photo-corrosion free.
2. The small-bandgap semiconductor should be able to be excited by visible light.

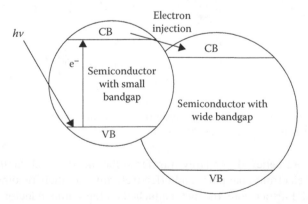

FIGURE 4.14 Electron injection in composite semiconductor. (From Ni, M. et al., *Renew. Sustain. Energ. Rev.*, 11, 401, 2007.)

3. The CB of the small-bandgap semiconductor should be more negative than that of the large-bandgap semiconductor.
4. The CB of the large-bandgap semiconductor should be more negative than $E_{H_2H_2O}$.
5. Electron injection should be fast as well as efficient.

It has been reported that coupling of CdS (2.4 eV bandgap) with SnO_2 (3.5 eV bandgap) could produce hydrogen under visible light irradiation [43]. Electrons excited to the CB (−0.76 eV) of CdS are injected to the CB (−0.34 eV) of SnO_2, resulting in wide electron–hole separation. So et al. [44] conducted photocatalytic hydrogen production using CdS–TiO_2 composite semiconductors. The optical absorption spectra analysis showed that the CdS–TiO_2 composite could absorb photons with a wavelength up to 520 nm. Under visible light illumination, CdS–TiO_2 composite semiconductors produced hydrogen at a higher rate than CdS and TiO_2 used separately. In another case, CdS–ZnS [45] composite semiconductors have also been studied with Si doping. Under solar irradiation, the addition of n-Si enhanced hydrogen production. This was due to the smaller bandgap of n-Si together with its more negative CB. When exposed to a radiation with a wavelength longer than 520 nm, electrons were excited from the VB of n-Si to the CB of n-Si, and then transferred to the CB of CdS sequentially, resulting in higher solar radiation utilization.

Besides coupling with small-bandgap semiconductors, TiO_2 coupled with a large-bandgap semiconductor has also been investigated and proven to be more efficient under UV irradiation. Keller and Garin [46] coupled TiO_2 with WO_3 (2.7 eV bandgap) and SiC (3.0 eV bandgap). As the CB of SiC was more negative, electron transfer to the CB of TiO_2 is more efficient. On the other hand, the CB of WO_3 was less negative than that of TiO_2, and thus electrons were transferred from the CB of TiO_2 to that of WO_3, resulting in a wide electron–hole separation. These composite semiconductors were found to be more effective than TiO_2 alone for charge separation. However, although WO_3 coupling with TiO_2 could enhance photocatalytic oxidation, it could not reduce protons, since the CB of WO_3 was not negative enough. Comparatively, the electrons transferred from the CB of SiC to the CB of TiO_2 were more negative than $E_{H_2H_2O}$. Therefore, SiC coupling was more suitable for hydrogen production under UV illumination.

Li et al. [47] developed a novel photocatalyst by doping nitrogen into composite semiconductors, as shown in Figure 4.15. Nitrogen-doped ZnO was coupled with WO_3, V_2O_5, and Fe_2O_3. By doping with nitrogen, ZnO could respond with visible spectrum. Although N-doped ZnO-WO_3 and ZnO-V_2O_5 work better under visible light irradiation for acetaldehyde decomposition, they are not suitable for hydrogen production since the CB of both WO_3 and V_2O_5 are not negative enough. It is expected that an N-doped composite semiconductor with a CB level more negative than $E_{H_2H_2O}$, such as SiC–TiO_2, may serve as an efficient photocatalyst for hydrogen production under visible light irradiation. More examples include RuO_2/TiO_2 [48] and $InVO_4$–TiO_2 [49–51].

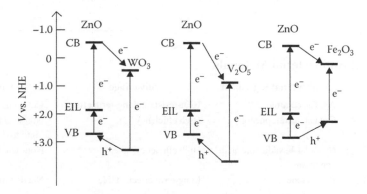

FIGURE 4.15 Schematics of charge separation for N-doped MO_x–ZnO composite semiconductors (M represents W, V, and Fe). (From Li, D. et al., *Catal. Today*, 93–95, 895, 2004.)

Overall, semiconductor nanocomposites have been developed in recent years to improve the photocatalytic activity of TiO_2 and other semiconductors under visible light irradiation. An increased water-splitting efficiency could be expected. A more detailed review can be found in Ref. [42].

4.3 APPLICATION IN BACTERICIDE AND WATER DISINFECTION

4.3.1 BACTERIA

Bacteria exist in every corner of the natural world. They belong to the group of prokaryotic microorganisms, which do not contain a nucleus as distinguished from the eukaryotes. Their DNA (deoxyribonucleic acid) molecules do not have a nuclear membrane to separate them from the cytoprism. Their sizes range from 0.5 to 5 μm. According to their shape, they can be classified as co-cocus (spherical), bacillus (rod), and spirella (spiral). Based on the different staining effects on cell wall components by the Gram stain, bacteria can be divided into gram-positive and gram-negative organisms. *Staphylococcus aureus* is a representative of gram-positive bacterium, while *Escherichia coli* (*E. coli*) is a typical Gram-negative bacterium. *E. coli* are the most popular model bacteria for antibacterial testing in the lab. These bacteria have a thin peptidoglycan layer (2–6 nm) in the cell wall. The outer layer is a membrane (6–18 nm) accounting for 90% of the cell wall (50% lipopolysaccharides, 35% phospholipids, and 15% lipoproteins) [52].

4.3.2 ANTIBACTERIAL MATERIALS

It has been long since human beings learned to use natural antibacterial materials. For example, milk was stored in silverware to avoid becoming rotten. At present, antibacterial materials can be classified as inorganic, organic, and natural materials. Each has its pros and cons. In general, organic and natural materials cannot sustain heat. Some of them are not workable. This leaves inorganic materials as the best choice for antibacterial purpose. Table 4.4 lists the pros and cons of different antibacterial materials.

4.3.3 APPLICATIONS OF ANTIBACTERIAL MATERIALS

Antibacterial materials are a kind of functional materials that can inactivate and kill bacteria. These can be used almost everywhere to prevent infection and maintain a good and healthy environment. Certainly, they are needed largely in medical-related areas. Recently, they have been applied in many kinds of biomaterials such as surgical implant and catheter [53]. Table 4.5 lists some of the common applications.

TABLE 4.4
Pros and Cons of Antibacterial Materials

Classification	Antibacterial Ingredient	Advantage	Disadvantage
Inorganic materials	Ag, Cu, ceramic	Heat-resistant, long-term effect	Color change, oxidation
Organic materials	Organic metal, phenol	High efficiency	Non-heatable, dissociates into toxic materials
Natural materials	(*E,E*)-2,4-hexadienoic acid chitosan	High efficiency, safe	Non-heatable, non-workable
Polymeric materials	Polystyrene	Long-term effect, stable	Needs further study

Source: Singleton, P., *Bacteria*, 6th edn., Wiley, Chichester, U.K., 2004.

TABLE 4.5
Common Applications of Antibacterial Materials

Applications	Products
Convenient goods	Brush, razor blade, scissors, containers
Electrical appliances	Telephone, handles, washing machine, air cleaner, air conditioner, water fountain
Fabrics	Pillow, quilt, socks, shoe pad, mask, surgical gown
Kitcheware and bath accessories	Toilet bowl, kitchen knife, handle, washbasin or sink
Medical supplies	Catheter, surgical knife, public utilities

4.3.4 Mechanisms of Bactericide

4.3.4.1 Bactericide by Metal Ions

It is widely known that materials containing silver or copper may have the most promising antibacterial performance against some bacteria [54–56]. When Cu or Ag atoms transform into ions and enter the environment, the multiplication of bacteria may be stopped, probably due to their engagement with the bacteria's DNA structure [57]. More importantly, these positively charged ions may interact with the cell wall membrane (slightly negative), leading to the breakage of the cell wall [57]. The SEM images of *E. coli*, before and after being exposed to Ag ions, are shown in Figure 4.16. The term "antibacterial performance" is normally defined as the effectiveness of the antibacterial materials to stop the multiplication of the bacteria and to kill the bacteria.

The use of metal ions so far is the most popular inorganic approach to kill bacteria. They can be used in many different forms and in different areas. Although there is no doubt about the effect of metal ions, some are concerned with their cytotoxicity. In the following, the antibacterial efficiency and cytotoxicity for some critical metal ions are compared:

Antibacterial efficiency [53]

$$As^{2+}, Sb^{2+}, Cd^{2+}, Se^{2+} > Hg^{2+} > Ag^+ > Cu^{2+} > Zn^{2+}$$

Cytotoxicity effect on human cells [53]

$$As^{2+}, Sb^{2+}, Cd^{2+}, Se^{2+} > Hg^{2+} > Zn^{2+} > Cu^{2+} > Ag^+$$

According to the ranking, it could be concluded that a small amount of Zn, Cu, and Ag is beneficial to the human body, but is detrimental to many bacteria. On the safe side, it can be said that copper

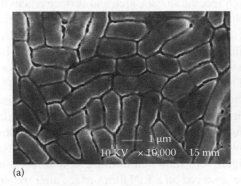 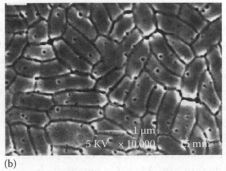

(a) (b)

FIGURE 4.16 SEM micrographs of *E. coli* before (a) and after (b) contact with Ag ions. (From Sondi, I. and Salopek-Sondi, B., *J. Colloid Interface Sci.*, 275, 177, 2004.)

TABLE 4.6
Antibacterial Efficiency for Some of the Metal Ions

Metal Ions		
Strong Antibacterial Effect	Medium Antibacterial Effect	Weak Antibacterial Effect
Ag	Cd	Co
Bi	Pt	Cr
Cu	Si	Mg
Mo	Ti	Mn
Tl	Al	Zn

Source: Hiramatsu, S. and Fukuzaki, J., *Surf. Technol.*, 52, 58, 2001 (in Japanese); Kobayashi, D., *Surf. Technol.*, 49, 433, 1998 (in Japanese).

and silver ions are the best choice as inorganic antibacterial materials. Other metal ions are too toxic to human cells. According to the metals' antibacterial efficiency, Table 4.6 classifies some of the common metal ions.

4.3.4.2 Bactericide by Photocatalysis

Several mechanisms have been proposed on the killing of bacteria by photocatalysts [60–62]. Sokmen et al. [63] and the Blake group [64,65] have proposed that photocatalysis may lead to the leakage of cell contents. From the results of *E. coli* testing, they suggest that photocatalysts degrade the micro-organisms' cell wall lipids via lipid peroxidation. This directly causes the peroxidation of the unsaturated phospholipid components of the lipid membrane with simultaneous losses of respiratory activities, although this is debated [66]. Another possible mechanism could be the detrimental effects of the photocatalytic effect on DNA molecules. Dunford et al. [67] conducted in vitro studies on supercoiled plasmid DNA with TiO_2, and found that the plasmid DNA was first converted to a relaxed form and then to a linear form, suggesting the occurrence of strand breakage. This lethal event could be the cause of cell death. Recently, photocatalytic peroxidation and oxidation of biomolecules such as nucleic acid and phosphatidylethanolamine lipid have also been observed [68]. Regardless of these proposed mechanisms, a comprehensive understanding on the mechanisms of photocatalytic bactericide and disinfection is still not reached [69].

4.3.5 NANOCOMPOSITE THIN FILMS FOR BACTERICIDE

4.3.5.1 Enhanced Dissolution of Metal Ions

Nano-sized metal particles are known to have a high specific surface area and surface reactivity. As a result, these particles would have a higher dissolution rate comparing with the same material but in a bulk form. In the case when the antibacterial silver or copper nanoparticles are embedded in a ceramic matrix, the constructed nanocomposite thin films would have varied antibacterial behaviors, depending on the metal particle size and the total exposed amount. The enhanced dissolution rate due to nano-sized particles was also reported by Cai et al. [70]. Figure 4.17 shows the effect of particle size and the effect of galvanic coupling on the dissolving rate of Cu. Besides showing the effect of particle size, this figure also implies that a galvanic enhancement may accelerate the dissolution of Cu. The Cu powder sample with an average particle size of 25 nm shows the most obvious galvanic effect. Furthermore, Figure 4.17 also shows that the galvanic-couple-enhanced dissolution could be enhanced by the size of Cu particles. Figure 4.18 shows an example of a TaN–Cu nanocomposite thin film. Many Cu nanoparticles can be observed on a TaN film surface after the sample is annealed using a rapid thermal annealing (RTA) system. Cu particles, in this case, are electrically connected with a TaN film, which means that a galvanic couple is formed. According to Figure 4.17,

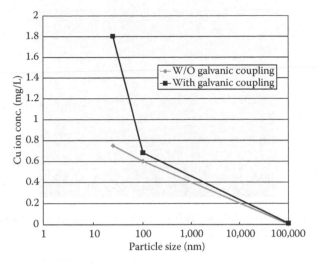

FIGURE 4.17 Cu ion concentration versus particle size, with and without TaN galvanic coupling. The immersion time was 3 h. (From Liu, P.C. et al., *Thin Solid Films*, 517, 4956, 2009.)

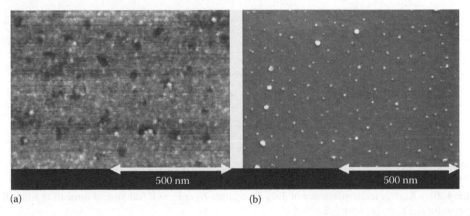

FIGURE 4.18 SEM micrographs of the annealed sample with 11 at.% Cu (TaN-11% Cu) (a) before annealing (plane view) and (b) after annealing (plane view). (From Liu, P.C. et al., *Thin Solid Films*, 517, 4956, 2009.)

the dissolution of the Cu particles surfaced on annealed TaN films will be accelerated, particularly when the size is at the nano level due to the proved synergistic effect.

Therefore, it is possible to create an antibacterial film having a short-term effect by reducing the particle size of Ag or Cu. On the other hand, one can also produce an antibacterial film showing a long-term effect by using large metal particles (lower dissolution rate). Both of these conditions can be reached by optimizing particle size, distribution, and density. In the case of TaN-Cu or TaN-Ag, proper annealing plus precise control of Cu or Ag contents is needed.

4.3.5.2 Metal–Ceramic Nanocomposites

It has been mentioned previously that materials containing copper or silver show antibacterial property. However, in many applications, antibacterial property is not the only requirement. Many medical- or bio-related devices or instruments require both antiwear and antibacterial properties. Surgical tools, catheters, hospital equipment, and kitchenware are some of the typical examples. Recently, it was reported that the mechanical properties of Ag- or Cu-doped nitride films could be improved after RTA [71,72]. This is due to the formation of nano-sized metal particles in the nitride matrix. This type of coating is categorized as an nc-MeN/soft-metal (nc, nanocrystalline;

FIGURE 4.19 SEM micrographs of TaN-Ag (10.0 at.%) after annealing (cross-sectional view). (From Hsieh, J.H. et al., *Surf. Coat. Technol.*, 202, 5586, 2008.)

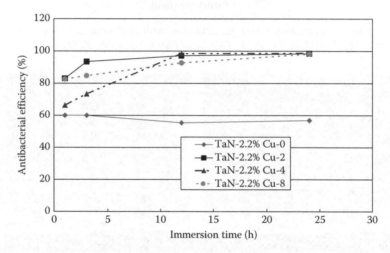

FIGURE 4.20 Antibacterial efficiency of TaN-Cu against *E. coli*, as a function of annealing time (RTA: 0, 2, 4, 8 min) for the samples with 2.2 at.% Cu.

MeN, metal nitride) nanocomposite thin film. The nitride phases can be TiN, ZrN, TaN, or other hard nitrides, while the soft metals can be Cu, Ag, or Ni. Although metals are usually soft and ductile, it has been shown that the mechanical properties of these metal–ceramic composites can be improved significantly when the size of the embedded metal particles is reduced to less than a few nanometers [73]. In addition to the improved mechanical properties, the friction coefficient is found to be lower due to the incorporation of soft metals [74,75]. Besides the immiscibility between the MeN phase and the metal phase, there are other critical requirements that need to be satisfied in order to enhance the films' mechanical properties. These requirements include (1) controlled metal particle size, (2) controlled metal concentration, and (3) controlled nitride structure [74,76].

Figures 4.18 and 4.19 show the emergence of Cu and Ag particles on the surface of TaN films. Figures 4.20 and 4.21 show the antibacterial efficiency of TaN-Cu and TaN-Ag, respectively. Apparently, in addition to the forming material of nanoparticles, the particle size as well as the exposed amount will affect the antibacterial behaviors.

4.3.5.3 Metal–Photocatalyst Nanocomposites

Ag is so far the most popular antibacterial metal being doped in oxide-based photocatalysts. As reported previously, adding extra elements into TiO_2 or other oxides is one way to improve oxides'

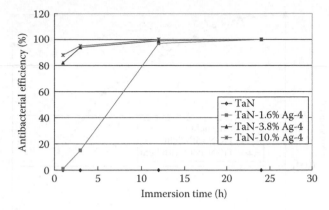

FIGURE 4.21 Antibacterial efficiency of TaN-Ag against *E. coli*, as a function of Ag contents. All samples were annealed (RTA) for 4 min. (From Hsieh, J.H. et al., *Surf. Coat. Technol.*, 202, 5586, 2008.)

photocatalytic behavior. Recently, more and more studies are focusing on embedding Ag nanoparticles into these oxide films. Ag is a nontoxic, precious metal with remarkable catalytic activity [77,78]. In addition, Ag particles may have promising size- and shape-dependent optical properties under visible light irradiation [79,80].

Overall, adding Ag into oxide-based photocatalysts may have the following advantages:

1. The bandgap may be narrowed [81].
2. Ag can be an efficient antibacterial material with or without light irradiation [82].
3. Ag doping may increase the light absorption intensity due to the plasmonic effect [83].
4. Ag can serve as an electron-trapping site due to the Schottky barrier effect [84,85]. It can also facilitate electron excitation by creating a local electric field.

4.3.5.4 Semiconductor–Semiconductor Nanocomposites

As described in Section 4.2.1.2, semiconductive nanocomposites will have the advantage of double bandgaps. Accordingly, the light absorption will be enhanced. Li et al. [86] reported a visible-light photocatalyst based on palladium oxide nanoparticles dispersed on nitrogen-doped TiO_2 (PdO/TiON). This composite has demonstrated a much faster photocatalytic disinfection rate on *E. coli* under visible light illumination than TiON (Figure 4.22). Figure 4.23 shows the comparison of the light absorption of TiO_2, TiON, and PdO/TiON. In the study of Li et al. [86], it is found that when PdO and TiON are in contact, an optoelectronic coupling between PdO nanoparticles and the TiON semiconductor would occur. This will promote the charge carrier separation in TiON and results in the chemical reduction of PdO to Pd. While the separation of the charge carrier greatly enhances the visible-light photocatalytic killing of bacteria, a "memory" antibacterial effect will result from the catalytic effect of metal Pd when the light is not irradiated. The synergistic effect of this type of semiconductor–semiconductor nanocomposite has opened up new possibilities, such as continuous solar-powered disinfection during daytime and at night. Besides PdO, Wu [69] also proposed that $Ag_2O/TiON$ would have a similar function.

4.4 SUMMARY AND RECENT DEVELOPMENTS

Since the finding of the catalytic function of TiO_2 in 1969, applications of many different semiconductor-based photocatalytic materials have been used in various areas that need a cleaner environment [87]. Since then, the photocatalytic materials have been significantly improved

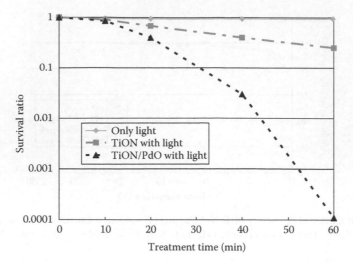

FIGURE 4.22 Survival ratios of *E. coli* after various treatments. (From Li, Q. et al., *Adv. Mater.*, 20, 3717, 2008.)

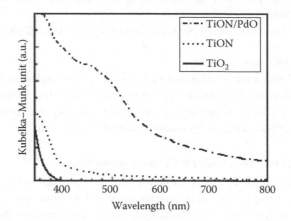

FIGURE 4.23 Optical absorbance of TiON/PdO (dots), TiON (dash), and TiO$_2$ (solid). (From Li, Q. et al., *Adv. Mater.*, 20, 3717, 2008.)

though several approaches. Also, many new theories and applications have been proposed. All of these can be summarized as follows:

1. To improve the quantum yield. It was found that the crystal structure [88], presence of hydroxyl groups on the surface [89] and the presence of oxygen deficiencies in the oxide-based photocatalysts affect the photocatalytic activities [90] including water splitting, bactericide, and disinfection.

2. To enhance the photocatalytic activity by extending light absorption from the UV region into the visible region. A considerable increase in the photocatalytic activity in the visible region has been observed in nitrogen-doped oxides (e.g., TaON and TiON) [22].

3. To suppress the recombination of electron–hole pairs in oxide semiconductors, by creating oxide/noble metal nanocomposites. The embedded particles may act as electron traps aiding electron–hole separation [91,92].

4. To extend the range of light absorption and enhance photocatalytic efficiency, the creation and use of new types of nanocomposites is a plausible approach.

5. Metal–semiconductor nanocomposites can be applied in bactericide in case Ag or Cu is in the form of nanoparticles. More interestingly, short- or long-term effects can be designed by controlling mainly the size of metal particles.

To create a cleaner environment, the last three approaches are thought to have unlimited possibilities, and are easy to improve the currently developed materials.

Awazu et al. [81] recently proposed another concept based on the new term plasmonic photocatalysis. The idea of plasmonic photocatalysis uses the theory of plasmon resonance that is known to enhance light absorption. For the anatase TiO_2 phase to be activated, a UV irradiation with a wavelength shorter than 380 nm needs to be used to excite electron–hole pairs. Embedded Ag nanoparticles could show a very intense absorption band in the near-UV region due to the influence of plasmon resonance [93]. This is associated with a considerable enhancement of the electric near-field in the vicinity of the Ag nanoparticles. It is thought that this enhanced near-field could boost the excitation of electron–hole pairs in TiO_2 and, thus, increase the efficiency of the photocatalysis. Similar ideas were already outlined in the past [94,95]. Figure 4.24 shows a cross-sectional TEM micrograph that shows embedded Ag nanoparticles in the matrix of TiO_2. The enhancement of light absorption at a wavelength of 420 nm is presented in Figure 4.25. It is known that the resonance wavelength depends on the size and shape of the nanoparticles as well as on the particle-forming material. Therefore, it is possible to create a structure similar to that in Figure 4.26. With this structure, the range of light absorption could be extended further. The photocatalytic efficiency in this case could be enhanced without any question, although more detailed studies should be carried out.

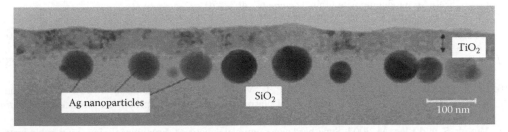

FIGURE 4.24 Cross-sectional TEM view of Ag nanoparticles embedded in TiO_2. (From Awazu, K. et al., *J. Am. Chem. Soc.*, 130, 1676, 2008.)

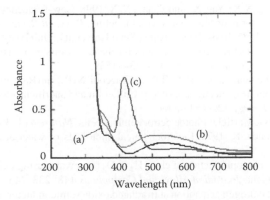

FIGURE 4.25 Optical absorption spectra. (a) TiO_2 thin film, (b) Ag nanoparticles embedded in TiO_2, and (c) Ag nanoparticles covered with SiO_2 layer embedded in TiO_2. (From Awazu, K. et al., *J. Am. Chem. Soc.*, 130, 1676, 2008.)

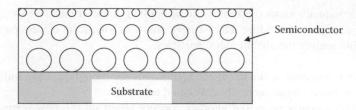

FIGURE 4.26 A concept of using embedded size-varying particles in a semiconductor-based photocatalyst matrix to extend the absorption range.

REFERENCES

1. Serna, R., Babonneau, D., Suarez-Garcia, A., and Afonso, C. N. 2002. Effect of oxygen pressure on the optical and structural properties of $Cu:Al_2O_3$ nanocomposite films. *Physical Review B* 66: 205402.
2. Malzbender, J., den Toonder, J. M. J., Balkenende, A. R., and de With, G. 2002. Measuring mechanical properties of coatings: A methodology applied to nano-particle-filled sol–gel coatings on glass. *Materials Science and Engineering R* 36: 47–103.
3. Suryanarayana, C. 1995. Nano-materials. *International Materials Reviews* 40: 41–84.
4. Kamat, P. V. 2007. Meeting the clean energy demand: Nanostructure architectures for solar energy conversion. *Journal of Physical Chemistry C* 111: 2834–2860.
5. Qin, X. Y., Zhang, W., Zhang, L. D., and Jiang, L. D. 1997. Low-temperature resistance and its temperature dependence in nanostructured silver. *Physical Review B* 56: 10596–10604.
6. Chen, C. C., Shi, J., and Hashimoto, M. 2002. Preparation of Co–Ti–N nanocomposite films. *Surface and Coatings Technology* 151–152: 59–62.
7. de los Arcos, T., Oelhafen, P., Aebi, U., Hefti, A., and Duggelin, M. 2002. Preparation and characterization of TiN–Ag nanocomposite films. *Vacuum* 67: 463–470.
8. Viart, N., Richard-Plouet, M., Muller, D., and Pourroy, G. 2003. Synthesis and characterization of Co/ZnO nanocomposites: Towards new perspectives offered by metal/piezoelectric composite materials. *Thin Solid Films* 437: 1–9.
9. Dalacu, D. and Martinu, L. 2000. Spectroellipsometric characterization of plasma-deposited Au/SiO_2 nanocomposite films. *Journal of Applied Physics* 87: 228–235.
10. Li, Z. G., He, J. L., Matsumoto, T., Mori, T., Miyake, S., and Muramatsu, Y. 2003. Preparation of nanocomposite thin films by ion beam and plasma based sputtering processes. *Surface and Coatings Technology* 174–175: 1140–1144.
11. Veprek, S. and Jilek, M. 2002. Superhard nanocomposite coatings. From basic science toward industrialization. *Pure and Applied Chemistry* 74: 475–482.
12. Musil, J. and Vlcek, J. 1998. Magnetron sputtering of films with controlled texture and grain size. *Materials Chemistry and Physics* 54: 116–122.
13. Unpublished results.
14. Pakhomov, A. B., Wong, S. K., Yan, X., and Zhang, X. X. 1998. Low-frequency divergence of the dielectric constant in metal-insulator nanocomposites with tunneling. *Physical Review B* 58: R13375–R13378.
15. Prakasam, H. E. 2008. PhD thesis, Pennsylvania State University, University Park, PA.
16. Kudo, A., Kato, H., and Tsuji, I. 2004. Strategies for the development of visible-light-driven photocatalysts for water splitting. *Chemistry Letters* 33: 1534–1539.
17. Somasundaram, S., Chenthamarakshan, C. R. N., de Tacconi, N. R., and Rajeshwar, K. 2007. Photocatalytic production of hydrogen from electrodeposited $p-Cu_2O$ film and sacrificial electron donors. *International Journal of Hydrogen Energy* 32: 4661–4669.
18. Gratzel, M. 2001. Review article: Photoelectrochemical cells. *Nature* 414: 338–344.
19. Fujishima, A. and Honda, K. 1972. Electrochemical photolysis of water at a semiconductor electrode. *Nature* 238: 37–43.
20. Ekambaram, S. 2008. Photoproduction of clean H_2 or O_2 from water using oxide semiconductors in presence of sacrificial reagent. *Journal of Alloys and Compounds* 448: 238–245.
21. Rajeshwar, K. 2007. Hydrogen generation at irradiated oxide semiconductor–solution interfaces. *Journal of Applied Electrochemistry* 37: 765–787.
22. Aroutiounian, V. M., Arakelyan, V. M., and Shahnazaryan, G. E. 2005. Metal oxide photoelectrodes for hydrogen generation using solar radiation-driven water splitting. *Solar Energy* 78: 581–592.

23. Asahi, R., Morikawa, T., Ohwaki, T., Aoki, K., and Taga, Y. 2001. Visible-light photocatalysis in nitrogen-doped titanium oxides. *Science* 293: 269–271.

24. Li, Q. 2000. PhD thesis, University of Illinois at Urbana-Champaign, Urbana and Champaign, IL.

25. Sakatani, Y., Nusoshige, J., Ando, H., Okusako, K., Koike, H., Takata, T., Kondo, M., Hara, J. N., and Domen, K. 2003. Photocatalytic decomposition of acetaldehyde under visible light irradiation over La^{3+} and N Co-doped TiO$_2$. *Chemistry Letters* 32: 115611–115619.

26. Ye, J., Zou, Z., Arakawa, H., Oshikiri, M., Shimoda, M., Matsushita, A., and Shishido, T. J. 2002. Correlation of crystal and electronic structures with photophysical properties of water splitting photocatalysts InMO$_4$. *Journal of Photochemistry and Photobiology A: Chemistry* 148: 79–83.

27. Zou, Z., Ye, J., Sayama, K., and Arakawa, H. 2002. Photocatalytic hydrogen and oxygen formation under visible light irradiation with M-doped InTaO$_4$ (M=Mn, Fe, Co, Ni and Cu) photocatalysts. *Journal of Photochemistry and Photobiology A: Chemistry*. 148: 65–69.

28. Ye, J., Zou, Z., Oshikiri, M., Matsushita, A., Shimoda, M., Imai, M., and Shishido, T. 2002. A novel hydrogen-evolving photocatalyst InVO$_4$ active under visible light irradiation. *Chemical Physics Letters* 356: 221–226.

29. Hara, M., Hitoki, G., Takata, T., Kondo, J. N., Kobayashi, H., and Domen, K. 2003. TaON and Ta$_3$N$_5$ as new visible light driven photocatalysts. *Catalyst Today* 78: 555–560.

30. Hitoki, G., Takata, T., Kondo, J. N., Hara, M., Kobayashi, H., and Domen, K. 2002. Ta$_3$N$_5$ as a novel visible light-driven photocatalyst ($\lambda < 600$ nm). *Chemistry Letters* 31: 736–737.

31. Chun, W. J., Ishikawa, A., Fujisawa, H., Takata, T., Kondo, J. N., Hara, M., Kawai, M., Matsumoto, Y., and Domen, K. 2003. Conduction and valence band positions of Ta$_2$O$_5$, TaON, and Ta$_3$N$_5$ by UPS and electrochemical methods. *Journal of Physical Chemistry B* 107: 1798–1803.

32. Kato, H. and Kudo, A. 2003. Photocatalytic water splitting into H$_2$ and O$_2$ over various tantalate photocatalysts. *Catalysis Today* 78: 561–569.

33. Maruthamutha, P. and Ashokkumar, M. 1988. Doping effects of transition metal ions on the photosensitization of WO$_3$ particles. *Solar Energy Materials* 17: 433–438.

34. Maruthamutha, P. and Ashokkumar, M. 1988. Hydrogen generation using Cu(II)/WO$_3$ and oxalic acid by visible light. *International Journal of Hydrogen Energy* 13: 677–680.

35. Maruthamutha, P. and Ashokkumar, M. 1989. Hydrogen production with visible light using metal loaded-WO$_3$ and MV^{2+} in aqueous medium. *International Journal of Hydrogen Energy* 14: 275–277.

36. Shan, Z., Wu, J., Xu, F., Huang, F. Q., and Ding, H. 2008. Highly effective silver/semiconductor photocatalytic composites prepared by a silver mirror reaction. *Journal of Physical Chemistry C* 112: 15423–15428.

37. Linsebigler, A. L., Lu, G., and Yates, J. T. 1995. Photocatalysis on TiO$_2$ surfaces: Principles, mechanisms, and selected results. *Chemical Reviews* 95: 735–758.

38. Subramanian, V., Wolf, E. E., and Kamat, P. V. 2004. Catalysis with TiO$_2$/gold nanocomposites. Effect of metal particle size on the Fermi level equilibration. *Journal of the American Chemical Society* 126: 4943–4950.

39. Taqui Khan, M., Bhardwaj, R. C., and Bhardwaj, C. 1988. Photodecomposition of H$_2$S by silver doped cadmium sulfide and mixed sulfides with ZnS. *International Journal of Hydrogen Energy* 13: 7–10.

40. Yamasita, D., Takata, T., Haraa, M., Kondoa, J. N., and Domen, K. 2004. Recent progress of visible-light-driven heterogeneous photocatalysts for overall water splitting. *Solid State Ionics* 172: 591–595.

41. Ashokkumar, M. 1998. An overview on semiconductor particulate systems for photoproduction of hydrogen. *International Journal of Hydrogen Energy* 23: 427–428.

42. Ni, M., Leung, M. K. H., Leung, D. Y. C., and Sumathy, K. 2007. A review and recent developments in photocatalytic water-splitting using TiO$_2$ for hydrogen production. *Renewable and Sustainable Energy Reviews* 11: 401–425.

43. Gurunathan, K., Maruthamuthu, P., and Sastri, V. C. 1997. Photocatalytic hydrogen production by dye-sensitized Pt/SnO$_2$ and Pt/SnO$_2$/RuO$_2$ in aqueous methyl viologen solution. *International Journal of Hydrogen Energy* 22: 57–62.

44. So, W. W., Kim, K. J., and Moon, S. J. 2004. Photo-production of hydrogen over the CdS–TiO$_2$ nanocomposite particulate films treated with TiCl$_4$. *International Journal of Hydrogen Energy* 29: 229–234.

45. De, G. C., Roy, A. M., and Bhattacharya, S. S. 1996. Effect of n-Si on the photocatalytic production of hydrogen by Pt-loaded CdS and CdS/ZnS catalyst. *International Journal of Hydrogen Energy* 21: 19–23.

46. Keller, V. and Garin, F. 2003. Photocatalytic behavior of a new composite ternary system: WO$_3$/SiC-TiO$_2$. Effect of the coupling of semiconductors and oxides in photocatalytic oxidation of methylethylketone in the gas phase. *Catalysis Communications* 4: 377–383.

47. Li, D., Haneda, H., Ohashi, N., Hishita, S., and Yoshikawa, Y. 2004. Synthesis of nanosized nitrogen-containing MOx–ZnO (M = W, V, Fe) composite powders by spray pyrolysis and their visible-light-driven photocatalysis in gas-phase acetaldehyde decomposition. *Catalysis Today* 93–95: 895–901.

48. Sakata, T., Hashimoto, K., and Kawai, T. 1984. Catalytic properties of ruthenium oxide on n-type semi-conductors under illumination. *Journal of Physical Chemistry* 88: 5214–5221.

49. Ge, L., Xu, M., and Fang, H. 2007. Synthesis of novel photocatalytic InVO4–TiO2 thin films with visible light photoactivity. *Materials Letters* 61: 63–66.

50. Ge, L. and Xu, M. 2006. Influences of the Pd doping on the visible light photocatalytic activities of InVO4–TiO2 thin films. *Materials Science and Engineering B* 131: 222–229.

51. Ge, L., Xu, M., and Fang, H. 2006. Photo-catalytic degradation of methyl orange and formaldehyde by Ag/InVO4–TiO2 thin films under visible-light irradiation. *Journal of Molecular Catalysis A: Chemical* 258: 68–76.

52. Singleton, P. 2004. *Bacteria*, 6th edn., Wiley, Chichester, U.K.

53. Kim, T. Z. 2004. *Inorganic antibacterial Materials*, Chemical Industry Publisher, Beijing, China (in Chinese).

54. Shimamura, K., Matsumoto, Y., and Matsunaga, T. 2002. Some applications of amorphous alloy coatings by sputtering. *Surface and Coatings Technology* 50: 127–133.

55. Kim, J. Cho, M., Oh, B., Choi, S., and Yoon, J. 2004. Control of bacterial growth in water using synthesized inorganic disinfectant. *Chemosphere* 55: 775–780.

56. Dowling, D. P. J., Betts, A., Pope, C., McConnell, M. L., Eloy, R., and Arnaud, M. N. 2003. Anti-bacterial silver coatings exhibiting enhanced activity through the addition of platinum. *Surface and Coatings Technology* 163–164: 637–640.

57. Sondi, I. and Salopek-Sondi, B. 2004. Silver nanoparticles as antimicrobial agent: A case study on *E. coli* as a model for Gram-negative bacteria. *Journal of Colloid and Interface Science* 275: 177–183.

58. Hiramatsu, S. and Fukuzaki, J. 2001. The antibacterial activity of plated coatings. *Surface Technology* 52: 58–64 (in Japanese).

59. Kobayashi, D. 1998. Antibacterial metal surface treatment. *Surface Technology* 49: 433–441 (in Japanese).

60. Ohtani, B., Okugawa, Y., Nishimoto, S., and Kagiya, T. 1987. Photocatalytic activity of titania powders suspended in aqueous silver nitrate solution: Correlation with pH-dependent surface structures. *Journal of Physical Chemistry* 91: 3550–3555.

61. Ohtani, B., Okugawa, Y., and Nishimoto, S. 1997. Photocatalytic activity of amorphous–anatase mixture of titanium(IV) oxide particles suspended in aqueous solutions. *Journal of Physical Chemistry B* 101: 3746–3752.

62. Ohtani, B. and Nishimoto, S. 1993. Effect of surface adsorptions of aliphatic alcohols and silver ion on the photocatalytic activity of titania suspended in aqueous solutions. *Journal of Physical Chemistry* 97: 920–926.

63. Sokemen, M., Candan, F., and Sumer, Z. 2001. Disinfection of *E. coli* by the Ag-TiO2/UV system: Lipid peroxidation. *Journal of Photochemistry and Photobiology A* 143: 241–244.

64. Maness, P. C., Smolinski, S., Blake, D. M., Huang, Z., Wolfrum, E. J., and Jacoby, W. A. 1999. Bactericidal activity of photocatalytic TiO2 reaction: Toward an understanding of its killing mechanism. *Applied and Environmental Microbiology* 65: 4094–4098.

65. Zheng, H., Maness, P. C., Blake, D. M., Wolfrum, E., Smolinski, J. S., and Jacoby, W. A. 2000. Bactericidal mode of titanium dioxide photocatalysis. *Journal of Photochemistry and Photobiology A* 130: 163–170.

66. Cronan, J. and Rock, C. J. 1994. The presence of linoleic acid in *Escherichia coli* cannot be confirmed. *Journal of Bacteriology* 176: 3069–3071.

67. Dunford, R., Salinaro, A., Cai, L., Serpone, N., Horikoshi, S., and Hikada, H. 1997. Chemical oxidation and DNA damage catalysed by inorganic sunscreen ingredients. *FEBS Letters* 418: 87–90.

68. Horikoshi, S., Serpone, N., Yoshizawa, N., and Hikada, H. 1999. Photocatalyzed degradation of poly-mers in aqueous semiconductor suspensions: IV Theoretical and experimental examination of the pho-tooxidative mineralization of constituent bases in nucleic acids at titania/water interfaces. *Journal of Photochemistry and Photobiology A* 120: 63–74.

69. Wu, P. 2007. PhD thesis, University of Illinois at Urbana-Champaign, Urbana and Champaign, IL.

70. Cai, S., Xia, X., and Xie, C. 2005. Research on Cu^{2+} transformations of Cu and its oxides particles with different sizes in the simulated uterine solution. *Corrosion Science* 47: 1039–1047.

71. Liu, P. C., Hsieh, J. H., Li, C., Chang, Y. K., and Yang, C. C. 2009. Dissolution of Cu nanoparticles and antibacterial behaviors of TaN–Cu nanocomposite thin films. *Thin Solid Films* 517: 4956–4960.

72. Hsieh, J. H., Tseng, C. C., Chang, Y. K., Chang, S. Y., and Wu, W. 2008. Antibacterial behavior of TaN–Ag nanocomposite thin films with and without annealing. *Surface and Coatings Technology* 202: 5586–5589.
73. Zhang, S., Sun, D., Fu, Y. Q., and Du, H. J. 2004. Effect of sputtering target power on microstructure and mechanical properties of nanocomposite nc-TiN/a-SiN$_x$ thin films. *Thin Solid Films* 447–448: 462–467.
74. Musil, J. and Vlcek, J. 2001. Magnetron sputtering of hard nanocomposite coatings and their properties. *Surface and Coatings Technology* 142–144: 557–566.
75. Mulligan, C. P. and Gall, D. 2005. CrN–Ag self-lubricating hard coatings. *Surface and Coatings Technology* 200:1495–1500.
76. Zeman, P., Cerstvy, R., Mayrhofer, P. H., Mitter, C., and Musil, J. 2000. Structure and properties of hard and superhard Zr–Cu–N nanocomposite coatings. *Materials Science and Engineering A* 289: 189–197.
77. Guin, D., Manorama, S. V., Latha, J. N. L., and Singh, S. J. 2007. Photoreduction of silver on bare and colloidal TiO$_2$ nanoparticles/nanotubes: Synthesis, characterization, and tested for antibacterial outcome. *Journal of Physical Chemistry C* 111: 13393–13397.
78. Arabatzis, I. M., Stergiopoulos, T., Bernard, M. C., Labou, D., Neophytides, S. G., and Falaras, P. 2003. Silver-modified titanium dioxide thin films for efficient photodegradation of methyl orange. *Applied Catalysis B* 42: 187–201.
79. Ohko, Y., Tatsuma, T., Fujii, T., Naoi, K., Niwa, C., Kubota, Y., and Fujishima, A. 2003. Multicolour photochromism of TiO$_2$ films loaded with silver nanoparticles. *Nature Materials* 2: 29–31.
80. Naoi, K., Ohko, Y., and Tatsuma, T. 2004. TiO$_2$ Films loaded with silver nanoparticles: Control of multi-color photochromic behavior. *Journal of the American Chemical Society* 126: 3664–3668.
81. Page, K., Palgrave, R. G., Parkin, I. P., Wilson, M., Savin, S. L. P., and Chadwick, A. V. 2007. Titania and silver–titania composite films on glass—Potent antimicrobial coatings. *Journal of Materials Chemistry* 17: 95–104.
82. Skorb, E. V., Antonouskaya, L. I., Belyasova, N. A., Shchukin, D. G., Mohwald, H., and Sviridov, D. V. 2008. Antibacterial activity of thin-film photocatalysts based on metal-modified TiO$_2$ and TiO$_2$:In$_2$O$_3$ nanocomposite. *Applied Catalysis B: Environmental* 84: 94–99.
83. Awazu, K., Fujimaki, M., Rockstuhl, C., Tominaga, J., Murakami, H., Ohki, Y., Yoshida, N., and Watanabe, T. 2008. A plasmonic photocatalyst consisting of silver nanoparticles embedded in titanium dioxide. *Journal of the American Chemical Society* 130: 1676–1680.
84. Zhao, G., Kozuka, H., and Yoko, T. 1996. Sol–gel preparation and photoelectrochemical properties of TiO$_2$ films containing Au and Ag metal particles. *Thin Solid Films* 277: 147–154.
85. Herrmann, J. M., Tahiri, H., Ait-Ichou, Y., Lassaletta, G., Gonzalez-Elipe, A. R., and Fernandez, A. 1997. Characterization and photocatalytic activity in aqueous medium of TiO$_2$ and Ag-TiO$_2$ coatings on quartz. *Applied Catalysis B* 13: 219–228.
86. Li, Q., Li, Y. W., Wu, P., Xie, R., and Shang, J. K. 2008. Palladium oxide nanoparticles on nitrogen-doped titanium oxide: Accelerated photocatalytic disinfection and post-illumination catalytic "Memory". *Advanced Materials* 20: 3717–3723.
87. Fujishima, A., Rao, T. N., and Tryk, D. N. 2000. Titanium dioxide photocatalysis. *Journal of Photochemistry and Photobiology C* 1: 1–21.
88. Zhang, H. and Banfield, J. F. 2000. Matrix isolation investigation of the interaction of SiH$_4$ with NH$_3$ and (CH$_3$)$_3$N. *Journal of Physical Chemistry* 104: 3481–3486.
89. Okamoto, K., Yamamoto, Y., Tanaka, H., and Itaya, A. 1985. Heterogeneous photocatalytic decomposition of phenol over TiO$_2$ powder. *Bulletin of the Chemical Society of Japan* 58: 2015–2022.
90. Nakajima, A., Koizumi, S., Watanabe, T., and Hashimoto, K. Effect of repeated photo-illumination on the wettability conversion of titanium dioxide. *Journal of Photochemistry and Photobiology A* 146: 12932.
91. Kraeutler, B. and Bard, A. J. 1978. Heterogeneous photocatalytic decomposition of saturated carboxylic acids on titanium dioxide powder. Decarboxylative route to alkanes. *Journal of the American Chemical Society* 100: 59855–59892.
92. Lee, J. and Choi, W. 2005. Photocatalytic reactivity of surface platinized TiO$_2$: Substrate specificity and the effect of Pt oxidation state. *Journal of Physical Chemistry B* 109: 7399–7406.
93. Kerker, M. 1985. The optics of colloidal silver: Something old and something new. *Journal of Colloid and Interface Science* 105: 297–314.
94. Hirakawa, T. and Kamat, P. V. 2004. Photoinduced electron storage and surface plasmon modulation in Ag@TiO$_2$ clusters. *Langmuir.* 20: 5645–5647.
95. Hirakawa, T. and Kamat, P. V. 2005. Charge separation and catalytic activity of Ag@TiO$_2$ core–shell composite clusters under UV–irradiation. *Journal of the American Chemical Society* 127: 3928–3934.

5 Thin Coating Technologies and Applications in High-Temperature Solid Oxide Fuel Cells

San Ping Jiang

CONTENTS

5.1 INTRODUCTION

A solid oxide fuel cell (SOFC) is an electrochemical device to convert the chemical energy of fuels, such as hydrogen and hydrocarbons, to electricity, with potential applications in transportation, distributed generation, remote power, defense, and many others [1–3]. They offer extremely high chemical-to-electrical conversion efficiencies because the efficiency is not limited by the Carnot

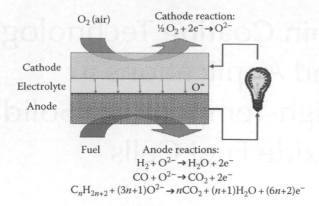

FIGURE 5.1 Schematic diagram of a solid oxide fuel cell. (From Jiang, S.P., *J. Mater. Sci.*, 43, 6799, 2008. With permission.)

cycle of a heat engine. Further energy efficiency can be achieved when the produced heat is used in combined heat and power, or gas turbine applications. Furthermore, the greenhouse gas emission from an SOFC is much lower than that from conventional power generation technologies. Due to its high operating temperature, an SOFC has a high tolerance to typical catalyst poisons, produces high-quality heat for reforming of hydrocarbons, and offers the possibility of direct utilization of hydrocarbon fuels.

An individual SOFC cell is composed of a porous anode or fuel electrode, a fully dense solid electrolyte, and a porous cathode or air electrode (see Figure 5.1) [4]. Driven by the differences in oxygen chemical potential between fuel and air compartments of the cell, oxygen ions migrate through the electrolyte to the anode where they are consumed by the oxidation of fuels such as hydrogen, methane, and hydrocarbons (C_nH_{2n+2}). Thus, the electrolyte must be dense in order to separate the air and fuel, must possess high oxygen ionic conductivity but negligible electronic conductivity, and must be chemically and structurally stable over a wide range of partial pressure of oxygen and temperatures [5]. On the other hand, the cathode and anode must be porous, chemically and thermally compatible with the electrolyte and interconnect, and be electrocatalytically active for the oxygen reduction and fuel oxidation reaction, respectively. In the case of hydrocarbon fuels, the anode must also possess certain tolerance toward sulfur and carbon deposition under SOFC operating conditions. Yttria-stabilized zirconia (YSZ) is the most commonly used solid electrolyte, while lanthanum strontium manganite (LSM) and nickel-YSZ often serve as the cathode and anode, respectively. Issues facing the development of electrode materials of SOFC have been reviewed [4,6,7].

Traditional SOFCs operate at high temperatures (900°C–1000°C) because of the low oxygen ion conductivity and high activation energy of oxide electrolytes such as YSZ. However, the lowering of the operating temperature of SOFCs brings both dramatic technical and economic benefits. The cost of an SOFC system can be substantially reduced by using less costly metal alloys as interconnect and compliant temperature gaskets [8]. Furthermore, as the operation temperature is reduced, thermodynamic efficiency, system reliability, and durability of cell performance increase. This increases the possibility of using SOFCs for a wide variety of applications, including residential and automotive applications. On the other hand, reduction in operation temperature results in a significant increase in the electrolyte and electrode resistivity and polarization losses. To compensate for the performance losses associated with a lower operating temperature, the thickness of electrolyte layer has to be reduced in order to lower the ohmic resistance of the cell. Using a thin electrolyte layer, the electrolyte can no longer mechanically support the cell. Thus, anode- or cathode-supported structures need to be employed. Anode-supported structure based on Ni/YSZ or Ni/GDC (Gd-doped ceria) cermets is the most popular one for the deposition of a thin electrolyte film on a thick, mechanically strong, and porous anode substrate.

The state-of-the-art anode-supported SOFC is based on porous Ni/YSZ cermets as a support. The anode support usually consists of a relatively thick porous supporting substrate and a thin and fine structured electrode layer, the anodic function layer (AFL). To reduce the electrolyte ohmic resistance and to enhance the cell efficiency, the electrolyte layer deposited should be as thin as possible. In addition, the thin electrolyte layer on the Ni/YSZ anode porous substrate/support should also have (1) high ionic conductivity, (2) high density and uniformity to minimize the fuel or oxidant crossover, (3) enough mechanical strength to resist high gas pressure gradient, and (4) good chemical and thermal compatibility with the anode as well as cathode without forming resistive phases at the electrode–electrolyte interface. Deposition techniques for the fabrication of thin and dense electrolyte films for SOFCs were reviewed by Will et al. [9].

Chromia-forming ferric stainless steel has been considered to be the primary candidate as the interconnect material for SOFCs for the intermediate temperature SOFCs, IT-SOFC, due to the economic and easy processing benefits. However, chromia-forming alloy forms a chromium oxide scale at the SOFC operation temperature. Without effective protective coating, the vaporization of chromium species poisons the cathode of SOFCs and seriously degrades the cell performance [10]. To reduce the growth rate of the oxide scale and the vaporization of chromium species, a thin and dense coating is commonly deposited on the metallic interconnect. In this case, the thin coating technique should not only deliver a protective film with high density and pore free or at least with no cross pores, but also with the targeted composition and stoichiometry to ensure the high electrical conductivity.

Micro-fuel cells are a potential replacement for high efficiency and high specific energy batteries in portable power generation. To date, miniaturized fuel cells utilizing proton exchange membranes and liquid methanol fuels (i.e., direct methanol fuel cells or DMFCs) have been the primary focus of interests. However, high loading of precious metal catalysts such as Pt and PtRu is required for DMFCs to obtain the beneficial energy output due to the CO poisoning, and under operating conditions methanol crossover is still a serious problem [11,12]. Due to the persistent challenges with polymer-based fuel cells, there is a growing interest in the development of micro-SOFCs or μ-SOFCs for portable power sources. With these systems, hydrocarbon fuels, in addition to hydrogen, can be used directly at the anode and this reduces the need for the reforming of fuels [13]. To develop μ-SOFCs, the design of the support structure and the deposition techniques for the thin-film electrolyte have been the areas of challenge [14]. This chapter will focus on the most commonly used thin-film techniques for the fabrication and development of thin-film components such as electrolyte, electrode, and protective coatings in SOFCs. The techniques are classified into chemical and physical methods based on the nature of the process. Examples are given for the preparation and characterization of electrolyte films as well as electrode and protective coating for the metallic interconnect of SOFCs.

5.2 CHEMICAL DEPOSITION METHODS

Chemical methods can be further divided into chemical vapor deposition and liquid precursor techniques. There are two main chemical vapor deposition techniques: chemical vapor deposition (CVD) and electrochemical vapor deposition (EVD). These methods make it possible to control chemical composition and to form a dense film. They are also known to be suitable for mass production. Chemical methods based on liquid precursor methods will be reviewed separately in Section 5.4.1.

5.2.1 CHEMICAL VAPOR DEPOSITION AND ATOMIC LAYER DEPOSITION

Chemical vapor deposition or CVD is a chemical process in which one or more gaseous precursors form a solid material by means of an activation process. CVD has been widely used for fabricating microelectronics. Therefore, the underlying processes are well understood.

A schematic diagram of a typical CVD apparatus is shown in Figure 5.2 [15]. Typically, fused-silica glass is used as the substrate material, which is heated to a deposition temperature of 600°C–1200°C depending on the reactivity of the precursors. For fabricating SOFC components,

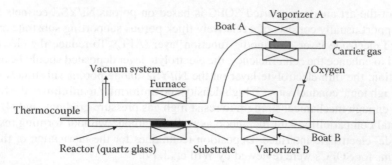

FIGURE 5.2 Schematic diagram of a CVD for the preparation of Y_2O_3-ZrO_2 films. (From Aizawa, M. et al., *Nippon Seramikkusu Kyokai Gakujutsu Ronbunshi—J. Ceram. Soc. Jpn.*, 101, 291, 1993. With permission.)

halogen compounds such as $ZrCl_4$ and YCl_3 [16,17] and metal organic compounds such as metal alkoxides [18] or β-diketones [15] have been used as precursor materials. The growth rates of the film thickness are in the range of 1–10 μm h^{-1}, depending on the evaporation rate and substrate temperature. The capital cost for the CVD equipment is relatively high [19].

Atomic layer deposition (ALD) is a modified CVD technique. In ALD, the substrate surface is exposed alternately to different vaporized precursors. Because gaseous precursors are strictly separated from each other during deposition and the precursors have self-limiting chemistry, one reaction cycle may produce only one atomic layer. For this reason, ALD can be an ideal technique to grow ultrathin oxide films because the composition of ALD films can be altered at each atomic layer with desired ratio. The growth rate for zirconia is 0.1–0.17 nm cycle^{-1} [20].

Prinz's group [21] fabricated freestanding ultrathin YSZ electrolyte films with a target stoichiometry, $(ZrO_2)_{0.92}(Y_2O_3)_{0.08}$, by ALD. For ZrO_2 and Y_2O_3, commercial tetrakis(dimethylamido)zirconium $(Zr(NMe_2)_4)$ and tris(methylcyclopentadienyl)-yttrium $(Y(MeCp)_3)$ were used as precursors with distilled water as oxidant. ALD YSZ films were grown on a Si_3N_4-buffered Si(100) wafer substrate. After ALD deposition, silicon nitride layer was removed by plasma-assisted chemical etching, leaving freestanding YSZ layers. Porous Pt layers were deposited using direct current sputtering as cathode and anode. A maximum power density of 270 mW cm^{-2} was reported for a cell based on an ultrathin YSZ film of 60 nm at 350°C. By using corrugated thin-film YSZ electrolyte design, the maximum power density was reported to increase to 677 mW cm^{-2} at 400°C [22]. However, the success rate of the cells would be limited by the possible electric shorts between electrodes through the nanoscale electrolyte.

5.2.2 Electrochemical Vapor Deposition

Electrochemical vapor deposition or EVD is a modified CVD process, originally developed by Siemens Westinghouse for the fabrication of thin YSZ electrolyte layer on tubular SOFCs [23]. EVD is a two-step process. The first step involves the pore closure by a normal CVD-type reaction between the steam (or oxygen), metal chloride, and hydrogen through the porous air electrode (phase I). These react to fill the air electrode pores with the yttria-stabilized zirconia electrolyte according to the following reaction:

$$2MeCl_y + yH_2O = 2MeO_{y/2} + 2yHCl \tag{5.1}$$

$$4MeCl_y + yO_2 + yH_2 = 4MeO_{y/2} + 4yHCl \tag{5.2}$$

where
  Me is the cation species (zirconium and/or yttrium)
  y is the valence associated with the cation

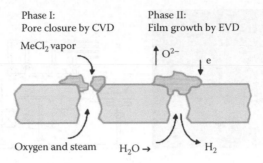

FIGURE 5.3   Schematic diagram of the EVD process. (From Pal, U.B. and Singhal, S.C., *J. Electrochem. Soc.*, 137, 2937, 1990. With permission.)

After the pores in the air electrode are closed, film growth then proceeds due to the presence of an electrochemical potential gradient across the deposited film. In this step, oxygen ions formed on the water vapor side of the substrate (i.e., the high oxygen partial pressure side) diffuse through the thin metal oxide layer to the metal chloride side (i.e., the low oxygen partial pressure side). The oxygen ions react with the metal chloride vapor to form the metal oxide products. The solid product or electrolyte is deposited as a thin film spreading over the internal pore surface in a desired region across the air electrode or membrane substrate. This second stage of the reaction is termed the electrochemical vapor deposition or EVD (phase II). Figure 5.3 shows a schematic diagram of the basic principles of the EVD process [23].

The growth of YSZ electrolyte film by the EVD process is parabolic with time and the rate determining step in the EVD process was found to be the electronic transport through the electrolyte film [23]. The EVD techniques were used by others for the fabrication of thin YSZ electrolyte films and thick Ru/YSZ cermet anodes [24,25]. The dissociated oxygen from metal oxide substrates, such as NiO, was also suggested as an oxygen source for the reaction instead of gaseous oxygen to form dense YSZ electrolyte films [26]. The effect of the NiO content of the substrate on the growth rate of YSZ film was studied in detail by Kikuchi et al. [27]. The high reaction temperature, the presence of corrosive gases, and relatively low deposition rates are some of the limiting factors in the application of EVD process in SOFCs.

5.3   PHYSICAL DEPOSITION TECHNIQUES

Physical deposition techniques reviewed in this chapter such as magnetron sputtering, laser ablation, and plasma spray have the common feature that atoms are brought to the gas phase through a physical process from a solid or molten target. The processes include evaporation, sputtering, laser ablation, and hybrid methods. The deposited films are typically polycrystalline with columnar structure, and the grain size can be tailored by varying the deposition conditions.

5.3.1   MAGNETRON SPUTTERING TECHNIQUES

Magnetron sputtering is one of the most common physical deposition techniques that is widely used to grow alloy and component films in which one or more of the constituent elements are volatile. Low-defect-density films of high melting point materials can be grown on unheated substrates because phase formation is mainly governed by kinetics, rather than by thermodynamics.

Radio frequency (RF) magnetron sputtering using an oxide target and DC reactive magnetron sputtering using metallic targets have been utilized to produce YSZ thin films of high quality. RF sputtering has frequently been utilized to deposit YSZ thin films in part because of the ability to use

(a)

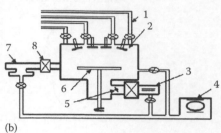
(b)

FIGURE 5.4 Outlook (a) and schematic diagram (b) of a magnetron sputtering system. In (b), 1, Gas feed line; 2, Magnetrons; 3, Cryogenic pump; 4, Rotary pump; 5, Butterfly valve; 6, Substrate holder; 7, Load lock; 8, Gate valve. (Courtesy of Dr. Liu Erjia, Nanyang Technological University, Singapore.)

either metallic or electrically insulating oxide target and the generally high quality of the deposit. In RF-sputtering deposition, an evacuated chamber is filled with the sputtering gas. A large negative voltage is applied to the cathode. The sputtering gas forms a self-sustained glow discharge. The physical sputtering of the target occurs when positive ions from the plasma that are accelerated across the gap strike the target surface. A metal oxide film is grown by sputtering a metal target in a discharge containing oxygen, usually in conjunction with a noble gas. The metal, metal oxide, and oxygen species that arrive at the substrate are adsorbed and ultimately incorporated into stable nuclei to form a continuous film. Oxygen is often included in the sputtering gas mixture as a means of controlling the metal-to-oxygen ratio in the target. However, RF-sputtering deposition rate decreases significantly with increased oxygen partial pressure [28]. Figure 5.4 shows the apparatus and a schematic diagram of a magnetron sputtering system.

DC magnetron sputtering has also been widely used to deposit thin YSZ electrolyte onto porous supports [29–31]. The target can be a single target consisting of an alloy of zirconium and yttrium or can be multiple targets of the pure metal, where the composition of the deposit is controlled by the relative exposed surface area of the targets. Barnett and coworkers described the formation of fully dense YSZ thin films on porous as well as dense electrode substrates by reactive DC magnetron sputtering with postdeposition annealing temperatures as low as 350°C [30,32]. Electrical conductivities of the YSZ deposit were similar to those films formed by other methods. The metal-to-oxygen ratio in the deposit was found to depend critically on the oxygen partial pressure in the sputter gas. Similar to RF sputtering, deposition rates are adversely affected by the oxygen partial pressure in the sputtering gas for DC reactive magnetron sputtering of YSZ films. This behavior was shown to be particularly apparent for oxygen partial pressures less than 10×10^{-4} mbar, as shown in Figure 5.5 [33]. The deposition rate dropped dramatically from ~700 nm h^{-1} at 7.5×10^{-4} mbar to ~100 nm h^{-1} when the oxygen partial pressure was higher than 10×10^{-4} mbar.

Surface morphology of the substrate such as pore size and pore size distribution is critical for the deposition of thin and high dense electrolyte film on the Ni/YSZ anode substrates as the morphology of the deposited film follows that of the substrate surface. Figure 5.6 shows the morphology of the YSZ film deposited on YSZ and Ni/YSZ cermet substrate by magnetron sputtering [6]. Thin YSZ film deposited is characterized by columnar structure and has the same grain and grain boundary pattern as that of the YSZ substrate (see Figure 5.6a), indicating that the surface morphology of the YSZ thin film follows closely the morphology of the substrate. The large pores on the substrate surface resulted in open pores on the YSZ electrolyte films deposited; see Figure 5.6b and c. Thus the pores diameter of the substrate surface needs to be in the order of the grain size of the deposited film to achieve dense and uniform YSZ electrolyte films; see Figure 5.6d and e. This can be done by using an interlayer or functional layer with fine microstructure and low porosity [34].

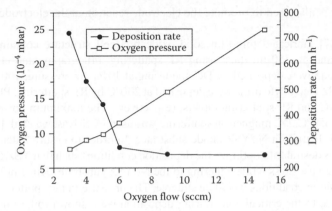

FIGURE 5.5   Relation of YSZ deposition rate by DC reactive magnetron sputtering to oxygen flow rate and oxygen partial pressure. (From Fedtke, P. et al., *J. Solid State Electrochem.*, 8, 626, 2004. With permission.)

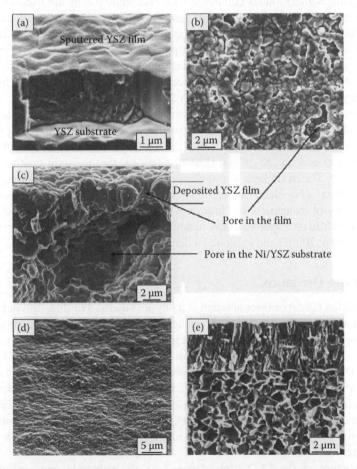

FIGURE 5.6   Morphology of the YSZ thin film deposited on YSZ and NiO/YSZ cermet anode substrate by magnetron sputtering. (a) YSZ film deposited on YSZ substrate, (b) and (c) YSZ film deposited on NiO/YSZ ceremt substrate, showing that the defects in the YSZ electrolyte film are due to the large pores on the anode substrate surface, (d) and (e) dense and uniform YSZ film on NiO/YSZ anode substrate with optimized surface morphology. (From Jiang, S.P. and Chan, S.H., *J. Mater. Sci.*, 39, 4405, 2004. With permission.)

The function layer is also used to promote the electrode reaction at the electrode–electrolyte interface [35,36].

Huang et al. [37] fabricated thin-film solid oxide fuel cell structure containing ultrathin YSZ electrolyte membrane 50–150 nm thick using RF sputtering, lithography, and etching. The porous Pt anode and cathode were deposited by DC sputtering at 10 Pa Ar pressure, 100 W, and room temperature. Dense YSZ electrolyte films were deposited at 200°C by RF sputtering. Pt and $Y_{0.16}Zr_{0.84}O_{1.92}$ were used as the DC and RF-sputtering targets, respectively. The maximum power density achieved is 200 mW cm^{-2} at 350°C. DC magnetron sputtering was used by Srivastava et al. [29] to deposit YSZ electrolyte films (5–16 μm) on Ni/YSZ anode substrates. In order to ensure a dense and impervious electrolyte layer, it is desirable to optimize the deposition conditions such that YSZ films would remain in a state of compressive stress while adhering to the Ni/YSZ substrate. This implies that only a narrow range of deposition conditions would be suitable. The pressure in the sputtering chamber and the flow of oxygen close to the critical zone during deposition were shown to be important parameters and required careful control [29]. Magnetron sputtering was also used to deposit a GDC protecting layer on YSZ electrolyte to prevent Sr^{2+} migration from $(La,Sr)(Co,Fe)O_3$ cathode toward the YSZ electrolyte [38].

Magnetron sputtering techniques combined with photolithography are used to produce unique, patterned electrodes such as LSM, Ni, platinum, and gold. The patterned, thin, dense, and uniformly structured electrodes allow the fundamental aspects of the reaction mechanism to be studied, which is not possible with conventional, heterogeneous, porous electrode structure. Horita et al. [39] studied the active site distribution of the O_2 reduction reaction on patterned LSM electrodes and demonstrated by SIMS that the O_2 reduction reaction occurs primarily at the three phase boundaries (TPB) where LSM cathode, oxygen reactant gas, and YSZ electrolyte meet. The linear relation between the length of TPB and the rate of the reaction was also found for the H_2 oxidation reaction on the patterned Ni anodes of SOFCs [40]. Sputtering method is also used to prepare protective coating for metallic interconnect. Lee and Bae [41] applied RF magnetron sputtering technique to deposit $La_{0.6}Sr_{0.4}CrO_3$ (LSCr) and $La_{0.6}Sr_{0.4}CoO_3$ (LSCo) as an oxidant-resistant coating on Fe-Cr ferric stainless steel SS430. The oxide layer shows dense structure and good adhesion to the SS430 alloy. LSCr-coated SS430 has a low electrical resistance as compared to LSCo-coated SS430 probably due to the low diffusivity of manganese or chromium in the LSCr layer. The RF magnetron-sputtered lanthanum chromite film on a stainless steel substrate formed the orthorhombic perovskite structure after being annealed at 700°C, and exhibited a dendritic microstructure [42].

5.3.2 PULSE LASER DEPOSITION

Pulse laser deposition (PLD) or laser ablation is a physical method of thin-film deposition in which a pulsed laser beam, usually of wavelength in the UV range, is employed to ablate a target composed of the desired thin-film composition, which is subsequently deposited onto a substrate. The usual range of laser wavelengths for thin film growth by PLD lies between 200 and 400 nm for most materials. In PLD, the temperature of the substrate is one of the main parameters affecting atomic surface mobility during the deposition process. PLD enables the fabrication of multicomponent stoichiometric films from a single target, and with an appropriate choice of the laser (e.g., Nd:YAG, KrF, XeCl), any material can be ablated and the growth can be carried out in a pressure of any kind of gas, reactive or not. Shown in Figure 5.7 is a schematic diagram of a pulse laser deposition process using KrF excimer laser [43].

The microstructure of the PLD films depends on the substrate temperature and pressure. Mengucci et al. [44] reported the formation of dense YSZ films by PLD at room temperature and oxygen pressure below 0.05 mbar. On the other hand, highly porous YSZ films can be obtained at high pressure of 0.3 mbar [45]. Infortuna et al. [46] studied in detail the characteristics of YSZ and GDC thin films by PLD and the microstructure of GDC and YSZ thin films deposited ranges from highly porous to dense depending on the substrate temperature and oxygen pressure during the

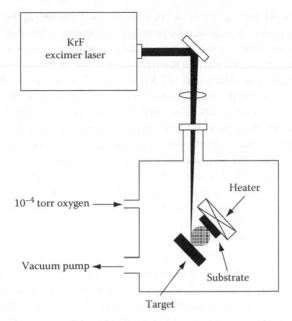

FIGURE 5.7   Schematic diagram of a pulse laser deposition process using KrF excimer laser. (From Suzuki, M. et al., *Solid State Ionics*, 96, 83, 1997. With permission.)

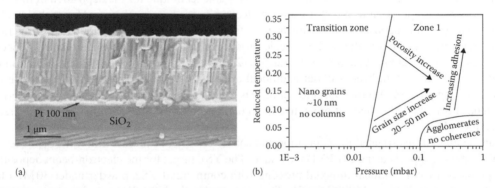

FIGURE 5.8   SEM micrograph of a dense YSZ thin film deposited at 800°C under 0.026 mbar oxygen pressure on a Si/SiO$_2$/TiO$_2$/Pt substrate (a); and structural map for GDC and YSZ thin film deposited by PLD (b). The normalized deposition temperature was calculated from the melting temperature of the materials (2500°C was used for both YSZ and GDC). (From Infortuna, A. et al., *Adv. Funct. Mater.*, 18, 127, 2008. With permission.)

deposition process. Figure 5.8 shows a typical SEM micrograph of a dense YSZ thin film deposited at 800°C and structure map for GDC and YSZ thin films deposited by PLD [46]. The electrical conductivity of the PLD films appears to be lower than that of the bulk materials. For example, the conductivity of a thin YSZ film deposited at 400°C in 0.026 mbar is 3×10^{-4} S cm^{-1} at 500°C, significantly lower than the conductivity value of $\sim10^{-3}$ S cm^{-1} for single YSZ crystal. Low electrical conductivity for YSZ thin films by PLD is also reported by Joo and Choi [47]. One potential problem is that PLD techniques result in crystalline microstructures that usually show columnar grains and texture, resulting in unfavorable anisotropic electrical conductivity [48]. Similar to the magnetron sputtering techniques, the surface artifacts of the function layer should also be smaller than the film thickness in order to avoid the defects on the YSZ film deposited by PLD technique [49].

PLD has been used to deposit multielement films such as LSM for SOFC [50]. Due to its unique method of dislodging atoms from the target, PLD provides some distinct advantages among thin-film techniques. Due to the weak interaction of lasers with gaseous species, ambient atmospheres can be used with little contamination, a significant advantage as compared to other deposition techniques based on ions or electrons. Thus, stoichiometric materials transfer from the target to the substrate could be achieved. Koep et al. [51] studied microstructure and electrochemical properties of LSM and LSC prepared by PLD. The deposition of LSM above 500°C results in thin films in the orthorhombic phase while the low-temperature-grown LSM films are amorphous. The stoichiometry of the film is the same as the target composition, and the conductivity of the LSM dense film is 21 S cm^{-1} at 700°C. Microfabrication process in the combination of photolithographic process is also used to produce well-defined three-dimensional geometries for the investigation of electrode reaction process and reaction sites at LSM electrode materials [52]. PLD is also used to fabricate cobalt and ferrite–based perovskite cathodes such as $La_{0.6}Ca_{0.4}Fe_{0.8}Ni_{0.2}O_{3-x}$ [53] and $(La,Sr)CoO_3$ (LSCo) [54] on YSZ electrolyte because the crystalline phase of the perovskite can be formed at a substrate temperature of 700°C to avoid the interfacial reaction between YSZ and lanthanum cobaltite–based cathodes. Reasonable area specific resistance (ASR) of 1.59 Ω cm^2 at 850°C was reported on the LSFN dense cathode layer by PLD [53]. Due to the dense structure of the PLD films, PLD method would be best suitable for the electrode application with high mixed ionic and electronic conducting oxides. For example, for the LSM cathode films prepared by PLD, the ASR can be as high as 32.1 Ω cm^2 at 850°C [55]. This is due to the fact that LSM is an electronic conductor with negligible oxygen ion conductivity. PLD is also used to deposit $La_{0.8}Sr_{0.2}Cr_{0.97}V_{0.03}O_3$ and $MnCr_2O_4$ protective films on Crofer 22APU interconnect at a substrate temperature of 750°C [56]. The deposited films were dense and significantly reduced the growth of the oxide scale of the alloy.

Another technique suitable for large-scale production for industrial purpose is electron beam-physical vapor deposition (EB-PVD). EB-PVD is a reliable technique for the deposition of thin film by the vacuum evaporation of a target using an electron-beam heating device. This technique has the advantages of the deposited films with an exact target composition, having high deposition rates at low temperatures, and realizing large-scale production for industrial application. The deposition rates of EB-PVD are typically in the range 1–2 μm min^{-1}. In addition, this method can make the deposited film crystallized without additional annealing, thus saving time and energy in the manufacturing process. Another variation of PLD is the large area filtered arc deposition (LAFAD), which has been applied to deposit protective thin coatings on Cr-Fe alloys using CoMn and CrAlY target alloys [57].

Jung et al. [58] investigated the deposition and characteristics of the YSZ thin films deposited on Ni/YSZ anode supports using EB-PVD technique. The YSZ target for the electron-beam deposition was prepared by conventional uniaxial pressing with commercial YSZ powder under 30 MPa followed by sintering in air at 1400°C for 5 h. The grain size of the YSZ film before heat treatment is 0.5–1 μm and grows to very dense columns with a grain size of 1–2 μm. Heat treatment improves the densification and the adhesion of the YSZ film with the anode functional layer. The results indicate that the electrical conductivity of EB-PVD YSZ electrolyte is smaller than that of the bulk YSZ at temperatures higher than 600°C [59]. The higher electrical resistance of the YSZ film deposited by EB-PVD could cause additional internal ohmic losses of the cell operating in the intermediate temperatures above 600°C. The maximum power density for an anode-supported YSZ thin electrolyte cell prepared by EB-PVD was 0.78 W cm^{-2} at 800°C. EB-PVD was also used to deposit GDC thin films on porous NiO/YSZ anode substrates [60]. The main dominating crystallite orientation of the GDC film repeats the characteristics of the used powder, and the crystalline size of the GDC thin film is influenced by the e-beam gun power.

5.3.3 PLASMA SPRAY DEPOSITION

The plasma spray process is a high-temperature process (up to 15,000 K for a typical DC torch operating at 40 kW). Figure 5.9 depicts the diagrams of DC and RF plasma spray processes [61]. The plasma spray process is based on the generation of a plasma jet consisting of argon or argon

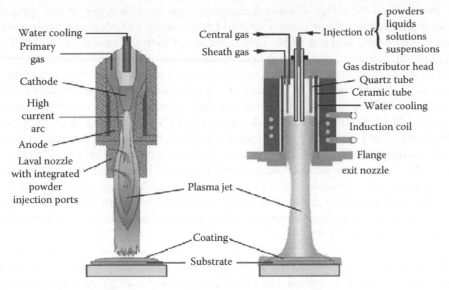

FIGURE 5.9 Schematic diagrams of DC and RF plasma spray processes. (From Pederson, L.R. et al., *Vacuum*, 80, 1066, 2006. With permission.)

with mixture of H_2 and He, which are ionized by a high-current arc discharge in a plasma torch. The powders to be sprayed are injected into the plasma where they are accelerated, melted, and finally projected onto a substrate. The coating is formed by solidification and flattening of the particles at impact on the substrate. This technique offers the possibility to deposit thin or thick layers in the millimeter range, which is hardly possible with other physical methods such as magnetron sputtering and PLD. With thermal spray technology, cells could be produced without using time- and energy-consuming sintering steps, reducing the production cost of SOFCs. The deposited layers often show the characteristics of anisotropic microstructure, microcracks due to thermal stress, and interlamellar porosity. The films produced by vacuum plasma spray (VPS) typically posses higher density than those produced by atmosphere plasma spray (APS) [62]. Thus VPS becomes particularly important in the fabrication of the SOFC electrolyte.

Tai and Lessing [63] used plasma spray to prepare porous LSM electrodes. The addition of pore formers, such as carbon, is necessary for spraying a porous coating. Coatings with a porosity of ~40% were deposited from LSM with a broad particle size distribution of 53–180 μm and 15 wt% solid Carbospheres as a pore former. The microstructure of the plasma-sprayed LSM coating is characterized by large agglomerates and coarse porous structure; see Figure 5.10.

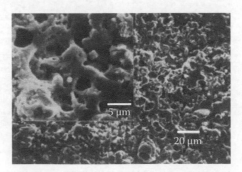

FIGURE 5.10 SEM micrograph of an annealed LSM coating plasma-sprayed from a mixture of broad-sized LSM and 15 wt% type-S130 Carbospheres as pore former. (From Tai, L.W. and Lessing P.A., *J. Am. Ceram. Soc.*, 74, 501, 1991. With permission.)

Li et al. [64] studied the effect of spray parameters such as plasma power and spray distance on the electrical conductivity of plasma-sprayed LSM coating and observed lower conductivity compared to that of the sintered LSM. This is explained by the formation of a lamellar structure of the plasma-sprayed LSM coating. The disadvantage of this method is that rapid heating and quenching involved induces nonstoichiometry and residual strain. Post treatment, e.g., at 1000°C in air for 2 h, was shown to be able to recover the crystallinity and stoichiometry of the plasma-sprayed LSM coating [65] and to increase the electrical conductivity [64]. Rambert et al. [66] prepared LSM/YSZ composite cathodes using the VPS technique. The performance of the LSM/YSZ composites is affected by the mixing process of the powders. Van herle et al. [67] prepared YSZ electrolyte by VPS and observed a strong anisotropy in the ionic conductivity of YSZ electrolytes. The cross-plane conductivities are several times lower than in-plane conductivities. This anisotropy can be eliminated by sintering the deposited YSZ layer at 1500°C for 2 h. The processing parameters such as the heat treatment, melting temperature of the raw materials, and particle size distribution have been shown to have a significant influence in the microstructure and conductivities of the films [68,69].

More recently, there have been attempts to fabricate SOFCs and LSM/YSZ composite cathodes on metallic substrates using APS, taking advantage of the relative simple and low cost of this process [70,71]. Vassen et al. [72] applied APS to fabricate porous NiO/YSZ coatings for anode, dense YSZ coatings for electrolyte, and functional coatings for reducing Cr-evaporation from interconnects. Figure 5.11 shows the cross section of a NiO/YSZ layer deposited on a tape-casted Chrofer APU22 metallic support using APS [72]. Some layer formation is still visible in the coating, indicating that the injection process needs to be further optimized. Under SOFC operation condition, the NiO will be reduced to Ni and additional pores will be formed. A dense YSZ layer was deposited onto this anode layer. In order to achieve the dense microstructure of the YSZ electrolyte layer, the velocity and temperature of the fused and crushed YSZ powder have to be high, the speed and temperatures are close to 320 m s^{-1} and ~3200°C. A cell with APS-deposited NiO/YSZ anode and YSZ electrolyte and screen-printed LSM cathode produced a power density of 0.8 W cm^{-2} at 800°C. By using high-speed plasma torches and specially designed nozzles, thin and gastight yttria- and scandia-stabilized ZrO_2 (YSZ and ScSZ) electrolyte layers (~30 μm in thickness and 1.5%–2.5% in porosity) and of porous electrode layers with high material deposition rates were fabricated by VPS techniques. The plasma-sprayed metal-supported cells showed good electrochemical performance and low internal resistances. Power densities of 300–400 mW cm^{-2} at 750°C–800°C were reported for plasma-sprayed cells [73,74].

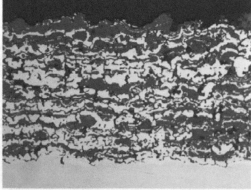

50 μm

FIGURE 5.11   SEM micrographs of the cross section of NiO/YSZ anode layer deposited on ferritic steel substrate (on the right reduced). (From Vassen, R. et al., *Surf. Coat. Technol.*, 202, 4432, 2008. With permission.)

5.4   COLLOIDAL AND CERAMIC POWDER TECHNIQUES

Colloidal and ceramic powder techniques are simple and cost-effective methods for the manufacturing of thin electrolyte films. The colloidal and ceramic powder methods include tape casting [75,76], slurry coating [77,78], spin coating [79,80], dip coating, screen-printing [81,82], etc.

5.4.1   COLLOIDAL TECHNIQUES

5.4.1.1   Slurry-Coating, Spin-Coating, and Dip-Coating Methods

In the case of slurry coating, spin coating and dip coating, additives such as dispersant and surfactants are added to the electrode or electrolyte powder in order to form a suitable and stable colloidal particle suspension. Moreover, in order to avoid cracks during the removal of the organic additives, careful drying and heat treatment procedures have to be adopted. Figure 5.12 shows the schematic diagram for spin-coating and dip-coating apparatuses. The process is simple and requires little investment. However, the coating process needs to be repeated several times to form a dense and pore-free thin electrolyte layer.

Chen et al. [83] prepared YSZ slurry mixed with ethyl cellulose and terpineol in a weight ratio of 25:3.4:71.6. The results indicate that heat treatment at 400°C for 10 min for each spin-coated layer is effective to remove the organic additives and to form a dense YSZ layer by repeated spin-coating process. A single cell with a 14 μm thick YSZ film and Sm-doped ceria (SDC)-impregnated LSM cathode yields a maximum power density of 634 mW cm^{-2} at 700°C in H$_2$/air. YSZ electrolyte layer as thin as 0.5 μm was prepared by the spin-coating method using YSZ precursors containing zirconia chloride hydroxide and yttrium chloride hydroxide as the source of Zr and Y with poly-vinylpyrrolidone (PVP) additive [84]. Dense and gastight YSZ films were obtained after repeating the spin-coating process three times (see Figure 5.13). Spin coating was also used for the preparation of thin porous Sm$_{0.5}$Sr$_{0.5}$CoO$_3$ (SSC) cathode for SOFCs, using a suspension solution made from SSC powder and ethyl cellulose in ethanol [85]. The suspension was applied to the doped ceria electrolyte surface and spun at 6000 rpm for 30 s to form a uniform layer and to remove the solvent. To form ~20 μm thick SSC cathode, 50 spin-coating cycles were performed [85]. However, by a combination of thick slurry coating and thinning and smoothing with a spinning step, a thin and dense YSZ electrolyte layer can be fabricated by a single coating–drying cycle [86].

Aerosol spray deposition was also used to fabricate YSZ thin electrolyte films on metal-supported structure from an isopropanol-based solution [87]. A metal-supported SOFC with thin YSZ electrolyte film achieved a maximum power density of 332 mW cm^{-2} with H$_2$/air at 700°C [88]. Wang et al. [89] described a fabrication process for the Ni/YSZ anode-supported thin-film YSZ cells based on tape casting and spray coating. The Ni/YSZ tape-cast green tapes were cut to desired sizes and annealed in air at ~1000°C for 1 h to obtain porous substrate with a thickness between 200 and 250 μm. The YSZ thin-film electrolyte was then deposited by spray coating a YSZ-water suspension on substrates under controlled conditions and sintered at 1400°C for 4 h. YSZ electrolyte film as thin

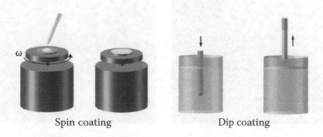

Spin coating   Dip coating

FIGURE 5.12   Schematic diagrams of spin-coating and dip-coating processes.

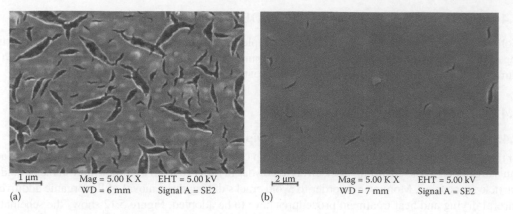

FIGURE 5.13   SEM micrographs of the surface of YSZ-coated samples after heat treatment at 600°C: One spin coating (a) and three spin coatings (b). (From Chen, Y.Y. and Wei, W.C.J., *Solid State Ionics*, 177, 351, 2006. With permission.)

as 3 μm can be obtained. The cell shows a very low ASR, 0.071 Ω cm^2 at 800°C, and has a power output of 0.85 W cm^{-2} at 800°C.

Slurry or dip coating has been used extensively for the fabrication of dense YSZ electrolyte [9]. Zhang et al. used the slip-casting method in combination with dip coating to fabricate anode-supported tubular SOFCs [90]. The Ni/YSZ anode tube support was prepared by the slip-casting method and presintered at 1000°C for 2h. The anode functional layer (AFL) and YSZ electrolyte thin film were prepared by dip coating. A suspension of 10 wt% NiO/YSZ (NiO:YSZ = 1:1 by weight) in isopropanol was dip coated onto the anode tubular substrate to form AFL. Similarly, a suspension of 10 wt% YSZ in isopropanol was dip coated onto the outer surface of the anode functional layer to form an electrolyte layer, followed by drying at room temperature. The process was repeated 10 times in order to prepare a dense and pore-free YSZ thin electrolyte layer. The YSZ-coated NiO/YSZ tubular substrate tubes were sintered at 1380°C for 2h to form a bilayer structure with a porous anode substrate and a dense electrolyte thin film. Multilayered cathodes of LSM, LSM/YSZ, and $Sm_{0.5}Sr_{0.5}CoO_3$ (SSC) were deposited on a thin YSZ electrolyte by the dip-coating method. Figure 5.14 shows SEM micrographs of an anode-supported tubular cell with a thin YSZ film and a multilayered cathode prepared by dip-coating method. With the impregnation of nano-sized GDC

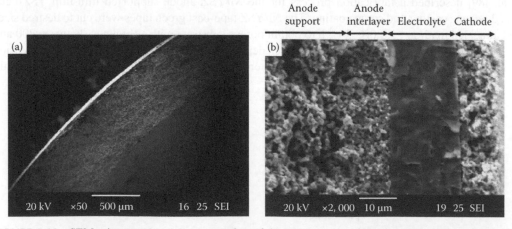

FIGURE 5.14   SEM micrographs of (a) an overview of the tubular cell and (b) the cross section of the tubular cell after testing, showing the thin YSZ electrolyte film and multilayer cathode prepared by dip-coating method. (From Zhang, L. et al., *J. Am. Ceram. Soc.*, 92, 302, 2009. With permission.)

particles into both anode supports and multilayer cathode, the cell achieved a peak power density of $1104\,mW\,cm^{-2}$ in H_2/air and $770\,mW\,cm^{-2}$ in CH_4/air at 800°C. Dip-coating method was also used by Yamaguchi et al. [91] to deposit GDC electrolyte on extruded anode-supported microtubular cells.

Dip-coating technique is also applied to deposit YSZ or GDC electrolyte films as thin as ~100 nm [92]. In this case, it is important to control the quality of the microstructure of the substrate as well as the YSZ and GDC sol containing nanoparticles (5–6 nm). For example, nano-structured zirconia sol can be produced by the controlled hydrolysis of $Zr(n-O_3H_7)_4$ and $Y(i-O_3H_7)_3$ in the presence of diethanol amine (DEA) as precursor modifier/polymerization inhibitor. DEA acts as a drying controlling additive and is important for the formation of nanoparticles in the synthesis process. In conventional slurry- or dip-coating process, as many as 10 sequential dip-drying-calcination steps are required to obtain sufficiently thick and dense film. These repeated dip-coating-drying-calcination processes are time consuming.

Wang et al. [93] showed that the composition of the dip-coating slurry has a significant influence on the quality of the YSZ thin films. Among the three commonly used composite solvents, trichloroethylene (TCE)/MEK, MEK/EtOH, and EtOH/TCE, YSZ suspension in MEK/EtOH solvents gave the best dispersibility and stability. Also MEK/EtOH has an evaporation temperature of 74.8°C, low enough for facilitating the formation of YSZ electrolyte thin films. Adding binder (PVB) and plasticizer (PEG and PHT) improves the stability. Thin and dense 16 μm thick YSZ films were prepared by dip coating twice, and the cell with NiO/YSZ anode and Pt cathode reached a power density of $262\,mW\,cm^{-2}$ at 800°C. Rather than dip-coating, Liu et al. [94] drop coated a YSZ suspension on porous electrode support to forming a thin and dense dual SDC-YSZ electrolyte films. Zhang et al. [95] showed that slurry casting could also be used to deposit thin YSZ electrolyte film on a NiO/YSZ substrate disc assisted by a revolving rod to spread and compact the YSZ layer. However, the revolving or rotating-assisted slurry-casting technique would be difficult to scale up for the fabrication of large area planar cells.

Vacuum slip or slurry casting is used to deposit thin GDC electrolyte layer on LSCF/GDC composite cathode substrates [96]. In addition to the preparation of YSZ electrolyte layer, slurry coating or slurry spray is also used for the deposition of anode and cathode interlayers [97]. The results show that the inclusion of electrode interlayers significantly improved the performance of the anode-supported cells and the reasons for the improvement could be related to the introduction of a diffuse mixed conduction region associated with the interlayers. However, slurry or slip casting is commonly used to fabricate thick NiO/YSZ anode substrates or tubes for anode-supported thin electrolyte SOFCs.

5.4.1.2   Sol-Gel Method

In sol-gel methods, organometallic salts, such as metal alkoxides (e.g., zirconium propoxide and yttrium propoxide), are deposited on porous electrode substrates and hydrolyzed under controlled conditions, forming a colloidal sol and a condensation step with organic monomers to form a gel. The deposition by methods such as by spin coating or dip coating is followed by a drying and firing process, leading to the formation of a dense electrolyte film. The key feature in this sequence is sol-gel polymerization, which can be described by a two-step reaction: initiation via the hydrolysis of alkoxy ligands and polycondensation via an oxylation reaction. The particle concentration, viscosity, concentration, and stability of the sol-gel influence the deposition parameters and film quality and have to be controlled carefully. The nature of the porous substrate is also critical in the sol-gel processes. Large pores could lead to pore-induced defects. Thus the substrate should have a porosity that is both submicron and uniform. Dunn et al. [98] gave an overview and briefly discussed the sol-gel chemistry.

Mehta et al. [99] used sol-gel method to deposit a thin (100–300 nm) YSZ layer on yttrium-doped ceria (YDC) electrolyte to block the electron transfer. An ethyl alcohol solution of YSZ was made by dissolving 11 mol% yttrium isopropoxide and 89 mol% zirconium isopropoxide in anhydrous isopropanol. The solvent was slightly heated and a small amount of HNO_3 was added to promote dissolution and hydrolysis. The YSZ precursor solution was applied to a YDC electrolyte substrate by spin coating. The formation of the cubic YSZ phase was achieved at 600°C in air. A 0.25 M solution used in spin coating results in a ~100 nm thick YSZ layer after heat treatment. The open circuit voltage (OCV) of the two-layer YSZ–YDC electrolyte cell is increased by 150–200 mV as compared to an uncoated YDC electrolyte cell in the temperature range of 600°C–800°C in H_2/O_2. The compatibility of the two-layer YSZ–YDC electrolyte cell was also studied by Kim et al. [79] using a sol-gel spin-coating method. The YSZ film deposited by sol-gel spin-coating method showed a crack- and pinhole-free microstructure after sintered at 1400°C. A 2 µm thick YSZ film on YDC electrolyte can be obtained after six repetitive spin coatings. However, the maximum power density of the two-layer cell is comparable to a YSZ single layer cell with the same thickness at 1000°C.

Lee et al. [100] used sol-gel dip-coating method to deposit a lanthanum chromite–based perovskite coating on a ferritic stainless steel (SUS444) to reduce the oxidation rate of the metallic interconnect. Precursor solutions for $(La,Ca)CrO_3$ (LCCr) and LSCr coatings were prepared by adding nitric acid and ethylene glycol into an aqueous solution of lanthanum, strontium (or calcium), and chromium nitrates. Dried LCCr and LSCr gel films were heat treated at 400°C–800°C after dip coating on the SUS444 substrate. The SEM results show that the microstructure of LCCr film is denser than that of LSCr. The porous structure of the LSCr layer is attributed to the formation of the $SrCrO_4$ phase. The presence of LCCr and LSCr thin-film layers depress the oxidation of the SUS444 metallic substrate.

The sol-gel method possesses the advantages of precise composition control, simple processing procedure, and low processing temperatures. However, thin films based on the sol-gel process also have several demerits. For example, a large shrinkage during heat treatment and low density from inherent high organic content may cause local defects, resulting in serious gas leakage and crossover.

5.4.1.3 Screen-Printing Method

In the screen-printing process, a highly viscous paste consisting of a mixture of ceramic powder, organic binder, and plasticizer is forced through the open meshes of a screen using a squeegee. The screen-printed films are dried and sintered at high temperatures. Parameters such as grain size, grain form, slurry viscosity, and sintering temperature and time are important for the quality and densification of the screen-printed film. Slip casting and screen-printing usually involve large shrinkage associated with the removal of polymeric binders and plasticizers in subsequent sintering and heat treatment stages. This would deteriorate the quality of the thin films.

Zhang et al. [101] used screen-printing technique to fabricate $Sm_{0.2}Ce_{0.8}O_2$ (SDC, 15 µm) single layer and YSZ (5 µm) + SDC (15 µm) bilayer on Ni/YSZ cermet substrate, followed by co-firing. Co-firing at 1400°C led to the formation of Zr-rich micro-islands, indicating the Zr migration from the NiO/YSZ substrate during the co-firing process. Screen-printing was used to deposit $Sm_{0.2}Ce_{0.8}O_2$ (SDC) electrolyte thin film onto the green NiO/SDC anode substrates using a printing slurry consisted of SDC powder, methylcellulose, terpineol, and ethanol vehicle [102]. A dense SDC electrolyte layer with a thickness of ~30 µm was obtained after co-firing at 1350°C in air for 5 h. The cell with SSC cathode, Ni/GDC anode, and screen-printed SDC electrolyte achieved a power output of 397 mW cm^{-2} in H_2 and 304 mW cm^{-2} in methane at 600°C, an impressive performance at these low temperatures.

Joo and Choi [103] fabricated µ-SOFCs based on a porous and thin Ni substrate using the screen-printing technique. Figure 5.15 shows the fabrication process for µ-SOFCs using screen-printing techniques. The NiO ink was made of commercial NiO powder mixed with an organic solution of α-terpineol and ethyl cellulose in a weight ratio of 10:1. A phosphate ester–based surfactant and dihydroterpineol acetate were used as dispersants. The Ni film was fabricated by screen-printing a NiO film on a ceramic substrate and subsequently reducing the printed film at 700°C–750°C in hydrogen. After reduction, a freestanding and porous Ni film was obtained. A thin GDC electrolyte film (~3 µm) was deposited on Ni substrate by the PLD method at an oxygen partial pressure

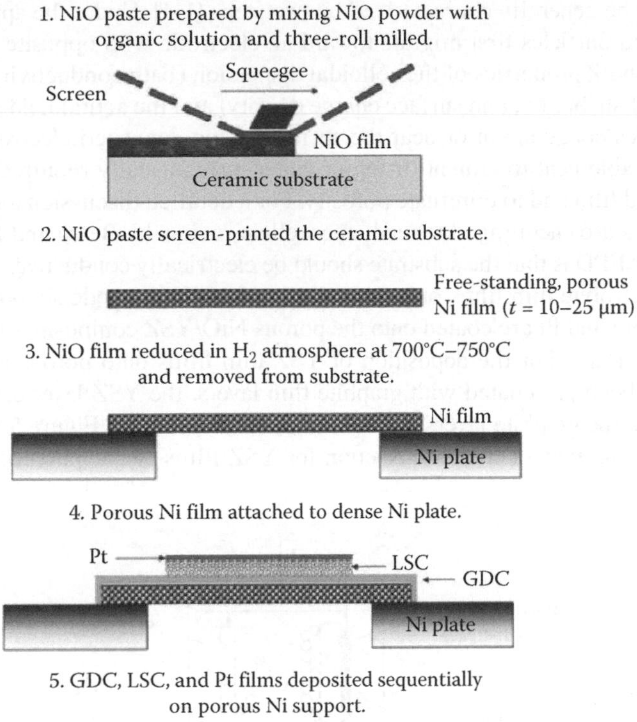
1. NiO paste prepared by mixing NiO powder with organic solution and three-roll milled.
Squeegee
Screen
NiO film
Ceramic substrate
2. NiO paste screen-printed the ceramic substrate.
Free-standing, porous Ni film (t = 10–25 μm)
3. NiO film reduced in H2 atmosphere at 700°C–750°C and removed from substrate.
Ni film
Ni plate
4. Porous Ni film attached to dense Ni plate.
Pt
LSC
GDC
Ni plate
5. GDC, LSC, and Pt films deposited sequentially on porous Ni support.

FIGURE 5.15 Fabrication process for μ-SOFC based on a screen-printed porous Ni substrate. (From Joo, J.H. and Choi, G.M., *J. Power Sources*, 182, 589, 2008. With permission.)

of ~30 mm torr, followed by the deposition of a porous LSCo cathode for 90 min at room temperature. The deposition at room temperature produces the porous structure of LSC as required for the cathode. The cell achieved a maximum power output of 26 mW cm^{-2} at 450°C in H_2/air. The advantage of the screen-printing is its simplicity and low cost particularly in comparison to the lithography and etching processes commonly used in the fabrication of μ-SOFCs.

5.4.1.4 Electrophoretic Deposition

Electrophoretic deposition (EPD) is one of the colloidal processes by which ceramic films are shaped directly onto substrates from an electrostatically stabilized colloidal suspension in a DC electrical field. A DC electrical field causes these charged particles to move forward and deposit on an electrode with opposite charge. The EPD process has been used for the fabrication of SOFC components including YSZ [104,105], and (La,Sr)(Ga,Mg)O_3 electrolyte films [106,107]. The EPD process is very simple and has the advantage of the uniformity of deposition and high deposition rates. In the case of the deposition of YSZ thin electrolyte layers, YSZ nanoparticles are dispersed in an organic suspension medium, such as ethanol, acetylacetone, or iodine-dissolved acetylacetone, instead of water, to avoid the detrimental effect of water electrolysis on the quality of deposited film. Positive charges are developed on the YSZ particles due to the presence of some residual water. A solid concentration of 10 g L^{-1} was found to be suitable for EPD using acetylacetone as solvent [108]. Iodine-dissolved acetylacetone is an effective solvent for EPD as it has a much lower resistance than ethanol [104,109]. Lower solution resistance reduces the applied voltage, depressing evolution of gases on the negative (i.e., deposited) electrode. The dispersion of YSZ in an iodine (I_2)-acetone solution is effective to form charged YSZ particles as the reaction between acetone and iodine produces protons that are absorbed by the YSZ particles; subsequently, the YSZ particles become positively charged by the addition of I_2 [110].

EPD process can be generally characterized by two steps [111]. Under the application of an electric field, the charged particles first migrate toward an electrode with opposite charge. The migration depends on the bulk properties of the colloidal dispersion (bath conductivity, viscosity, particle concentration, size distribution, and surface charge density) and the actual field strength in the bath. The charged particles coagulate at or near the surface of the deposited electrode, forming a solid deposit layer. A suitable heat treatment (firing or sintering) is usually required in order to further densify the deposited film and to eliminate porosity. For a detailed discussion and application of the EPD process, readers are encouraged to read an excellent review by Besra and Liu [112].

A prerequisite for EPD is that the substrate should be electrically conductive. Thus, for deposition of YSZ or GDC electrolyte thin films on nonconducting NiO/YSZ anode supports, thin conducting layers such as graphite and Pt are coated onto the porous NiO/YSZ composite substrates to facilitate conduction on the surface. For the deposition of YSZ thin films onto porous NiO/YSZ composite substrates that have been pre-coated with graphite thin layers, the YSZ layer can be deposited onto the substrates or onto the graphite layers on the substrates, as shown in Figure 5.16 [113]. Figure 5.17 shows the SEM micrographs of the cross section for YSZ films electrophoretically deposited onto

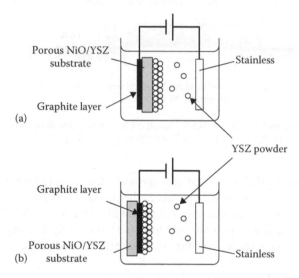

FIGURE 5.16 Experimental setup for (a) EPD of YSZ powders onto porous NiO/YSZ substrates whose reverse side is coated with conducting graphite layers and (b) EPD of YSZ powder onto conducting graphite layers on NiO/YSZ substrates after co-firing. (From Hosomi, T. et al., *J. Eur. Ceram. Soc.*, 27, 173, 2007. With permission.)

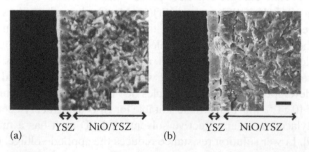

FIGURE 5.17 SEM micrographs of the cross-section EPD of YSZ thin films electrophoretically deposited onto porous NiO/YSZ substrates whose reverse side is coated with conducting graphite layers (a) and onto conducting graphite layers on NiO/YSZ substrates (b). (From Hosomi, T. et al., *J. Eur. Ceram. Soc.*, 27, 173, 2007. With permission.)

NiO/YSZ substrate surface [113]. For YSZ powders deposited onto porous NiO/YSZ substrates whose reverse side is coated with conducting graphite layers, the YSZ powders are subsequently transferred into dense and continuous films with thickness of 3–15 μm after co-firing with substrates (Figure 5.17a). For the YSZ films deposited onto graphite layer, an opening is observed at the interface after co-firing (Figure 5.17b), most likely due to the burning and decomposition of the graphite interlayer.

Besra et al. found that in the case of EPD of YSZ thin layers on nonconducting NiO/YSZ substrates with carbon sheet backing, the deposition of YSZ increases with increasing substrate porosity, indicating the presence of a continuous "conducting path" between the electrical contact and the particles through pores in the substrate [108]. The results indicate a minimum threshold porosity of the NiO/YSZ substrates, which is 52.5% and 58.5% porosity for 100 and 25 V applied potentials, respectively. Cherng et al. [114] showed that Ni/YSZ cermet presintered at 1200°C and reduced at 700°C behaves like a metal electrode and does not require the use of the additional electrical conducting backing layer.

Cherng et al. [114] studied the EPD of YSZ thin films using aqueous suspension. To prevent the colloids from agglomeration and sedimentation during EPD, negatively charged polyelectrolytes, such as ammonium polyacrylate (PAA-NH$_4$), are added as dispersant. PAA-NH$_4$ dissociates in water, forming negatively charged polyanions that would adsorb to YSZ particles to stabilize them electrostatically. An addition of ~0.1 wt% PAA-NH$_4$ is sufficient to stabilize the YSZ slurry. The aqueous EPD is characterized by low current (<3 mA cm^{-2}) and low voltage (<20 V) on conducting substrates. Will et al. [115] showed that through the adjustment of shrinkage and the shrinkage rate of the deposited zirconia layer on the presintered porous Ni/YSZ substrate, thin, dense layers without cracks can be prepared. The thickness of the deposited YSZ layer could be monitored from the total charge transfer during the EPD process. EPD method was also used to deposit a NiO/YSZ anode and a YSZ bilayer structure with a good interface bonding between the anode and electrolyte [116], and to deposit YSZ on a porous LSM cathode substrate [105,110].

The EPD can be used for mass production and has the advantages of short formation times, little restriction in the shape of deposition substrates, suitability for mass production, and a simple deposition apparatus. In addition, there is no requirement of binder burnout because the green coating contains little or no organics. However, it has been shown that five or more successive repetitions of the process are necessary to produce a gastight and dense YSZ layer and to achieve a good cell performance [104,105].

5.4.2 CERAMIC POWDER TECHNIQUES

5.4.2.1 Tape-Casting and Freeze-Tape-Casting Processes

Tape casting is a commercial processing technology that has been used extensively for the manufacturing of electronic and structural ceramics with thicknesses typically ranging from 25 to 1000 μm and is particularly suitable for making anode-supported planar SOFCs. Tape casting involves spreading of slurries of the ceramic powders and organic ingredients, such as the modifier, organic binders, dispersant (fish oil), and solvents (ethanol and toluene), onto a flat surface where solvents are allowed to evaporate. After drying, the resulting tape develops a leather-like consistency and can be stripped off from the casting surface; see Figure 5.18. To increase the porosity of the anode supports, graphite or other pore formers can be added.

Typically, tape-cast anode layer (1–2 mm thick) and tape-cast electrolyte layer (40–100 μm thick) are produced and these two tapes are laminated and after rolling or calendering, the green laminates are cut to size and sintered at 1300°C–1400°C in a specially designed kiln furnace to ensure the parts are all flat. The final anode-supported structures are produced with an anode support thicknesses of 500–1000 μm and electrolyte thicknesses of 15–40 μm. Tape-casting technique is also used to produce thick electrolyte of ~100 μm for electrolyte-supported cells. Figure 5.19 shows the typical structure of YSZ electrolyte-supported and Ni/YSZ cermet anode-supported thin YSZ cells. Mismatch in the densification of such bilayer structures will cause intolerable bending and/or

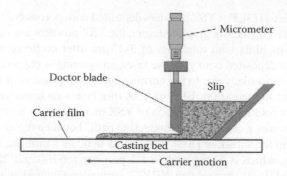

FIGURE 5.18   Typical tape-casting apparatus.

FIGURE 5.19   (a) YSZ electrolyte-supported cell and (b) Ni/YSZ anode-supported cells prepared by tape-casting technique.

cracking of the structure during co-firing [117]. Thus, match in the sintering profile of the laminated anode and electrolyte tapes is critical in producing flat anode-supported structures.

Tape casting is also used to prepare porous anode substrate for anode-supported thin electrolyte cells. Park et al. [118] reported the fabrication of anode-supported thin electrolyte SOFCs by using two tape-casting layers of YSZ, one containing a pore former as a porous YSZ anode matrix and one without pore former as the electrolyte layer. The porosity of the anode tape was controlled by the addition of pore formers. The addition of 30% pore former resulted in the formation of ~50% porosity for the YSZ anode substrate tape. The shape of pores in the porous YSZ matrix is related to the shape of the pore formers that are used. After the formation of the composite structure at 1550°C, copper and ceria were added to the porous YSZ layer by wet impregnation. However, the ability to engineer pore structures is limited to the manipulation of thermal fugitive particle orientation and stability during the slurry process.

A combination of the tape-casting process and freeze-casting process has resulted in a new freeze-tape-casting process that has been developed as a direct means of forming and controlling complex pore structures in green tapes. The freeze-tape-casting process not only allows the tailoring of continuously graded pores through the entire cross section but also slows for long-range alignment of acicular pores from the surface. The freeze-tape-casting process starts with the traditional tape-casting process, where an aqueous ceramic slip is cast onto a Mylar or Teflon carrier film via a doctor blade apparatus. A standard tape caster that has been modified only with a thermally isolated freezing bed to allow for unidirectional solidification of the slurry after casting is used. As

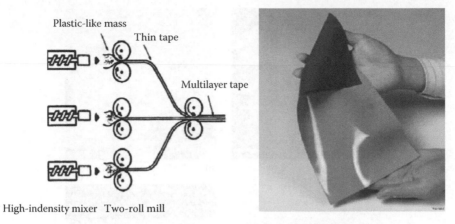

FIGURE 5.20 Tape-calendering process and green electrolyte/anode bilayer. (From Minh, N.Q., *Solid State Ionics*, 174, 271, 2004. With permission.)

with traditional tape casting, the slip contains sufficient organic binders that make the tape strong and flexible after solvent evaporation for handling and cutting. The freezing of the tape, typically solidified in several minutes, eliminates particles settling out of the suspension. After the solidification process, the tape is subsequently cut and freeze dried under a vacuum for quick solvent removal through sublimation, where the frozen liquid transforms from a solid to a gas without an intermediate liquid phase. Sofie [119] investigated in detail the fabrication of functionally graded and continuously aligned porous YSZ tapes (500–1000 μm) by a modified freeze-tape-casting process using aqueous and tertiary butyl alcohol (TBA) solvents. The result indicates that the freeze-tape-casting technology can be effective in fabricating fuel cell components.

5.4.2.2 Tape-Calendering Process

The tape-calendering process for making thin-film SOFCs is based on the progressive rolling of green (unfired) ceramic tapes to produce a thin electrolyte film (typically 0.5–10 μm) on an electrode support [120,121]. The tape-calendering process (using an anode as the support) and green electrolyte/anode bilayer are shown in Figure 5.20 [122]. In this fabrication process, electrolyte and anode powders are first mixed with organic binders to form ceramic masses. The masses, having a doughy consistency with many of the characteristics of a plastic, are rolled into tapes using a two-roll mill. Electrolyte and anode tapes of certain thickness ratios are laminated and rolled into a bilayer tape. This thin bilayer is then laminated with a thick anode tape, and the lamination is rolled again into a thin bilayer tape. The process can be repeated with different tape thickness ratios until a desired electrolyte film thickness is obtained. In general, the process requires only three rollings to achieve a bilayer with micrometer-thick electrolyte films. The bilayer is fired at elevated temperatures to remove the binders and sinter the ceramics. To form a thin-film SOFC single cell, a cathode layer is applied on the electrolyte surface of the sintered bilayer. The process is simple and scalable. Cell size as large as 500 cm² can be fabricated by this method.

5.4.2.3 Dry-Pressing Method

Xia and Liu developed a simple and cost-effective method to fabricate a thin GDC electrolyte film on NiO/GDC anode substrates by dry pressing [123–125]. In this method, the green NiO/GDC substrate was formed by the uniaxial pressing of NiO/GDC powder under 200 MPa. Highly porous or "foam" GDC powder, synthesized by a glycine nitrate combustion process, was carefully spread on top of the prepressed NiO/GDC substrate, and then compressed at 250 MPa, forming a bilayer structure. Co-firing of the bilayer structure at 1350°C for 5 h yields a dense GDC thin layer on a porous NiO/GDC electrode substrate. Care must be taken to ensure the uniform distribution of the

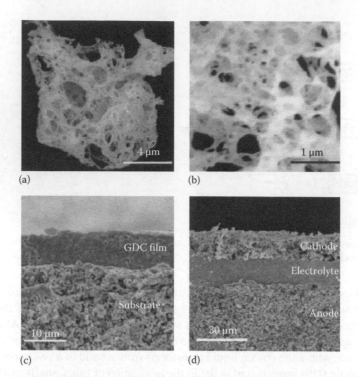

FIGURE 5.21 SEM micrographs of (a) a highly porous and foam-like GDC particle, (b) a portion of the particle shown in (a), (c) cross section of an 8 μm thick GDC film on a NiO/GDC substrate, and (d) a fuel cell consisting of a 15 μm thick GDC electrolyte, a Ni/GDC anode, and an $Sm_{0.5}Sr_{0.5}CoO_3$ cathode. (From Xia, C.R. and Liu, M.L., *J. Am. Ceram. Soc.*, 84, 1903, 2001. With permission.)

GDC powder on the substrate. Figure 5.21 are SEM micrographs of the as-synthesized GDC powder and the cross sections of a fuel cell consisting of thin GDC electrolyte layer prepared by dry pressing [123]. The GDC powder is highly porous; see Figure 5.21a and b. The relative density of the GDC foam powder is 0.84% (~0.06 g cm^{-3}), based on the theoretical density of GDC (7.12 g cm^{-3}). Low relative density indicates the low fill density of the powder. It is suggested that the extremely low fill density of the foam-like GDC is critical for the successful preparation of a thin, dense GDC electrolyte on a porous NiO/GDC substrate. The thickness of the GDC electrolyte is controlled by the amount of GDC powder.

The Gd-doped ceria electrolyte on anode substrates can also be formed in situ by solid-state reaction by dry pressing the stoichiometric Gd_2O_3 and CeO_2 oxides [126]. The XRD analysis indicates that a single solid solution of Gd-doped CeO_2 is formed after sintering at 1450°C. A maximum power density of 0.58 mW cm^{-2} at 600°C was obtained on a 10 μm thick GDC film and a LSCF/GDC composite cathode. The dry pressing technique is simple and cost-effective. However, the formation of a uniform and thin electrolyte film requires experience and skill particularly for the powder with high fill density. The scale-up of the dry pressing process may also be a problem due to the nature of the manual handing process.

5.5   FLAME-ASSISTED COLLOIDAL PROCESS

5.5.1   SPRAY PYROLYSIS AND FLAME-ASSISTED VAPOR DEPOSITION

Figure 5.22 shows schematically a typical spray pyrolysis setup. A sufficient force, e.g., using a stream of gas at high speed, applied to the surface of a metal salt precursor solution (usually aqueous or alcoholic) or colloidal suspension in the atomizer/nozzle causes the emission of droplets. The

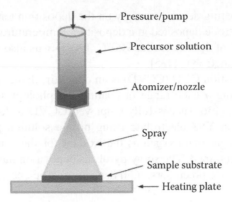

Pressure/pump

Precursor solution

Atomizer/nozzle

Spray

Sample substrate

Heating plate

FIGURE 5.22   A schematic diagram of a typical spray pyrolysis setup.

solution can also be forced through the nozzle with a syringe pump. Sprayed droplets reaching the substrate surface undergo pyrolytic decomposition. Newly deposited flat droplets, with thickness in the 10–20 nm range, pile up on the previously deposited droplets and undergo pyrolytic decomposition as well. This process continues until a film thickness of 100–500 nm is reached [127]. The degree of decomposition is determined by the relationship between the substrate temperature, the boiling temperature of the solvents, and the melting point of the salts used for the precursor. A compromise has to be found between sufficiently high deposition temperature to achieve complete decomposition if possible and the facility to deposit the droplets while still in a wet state on the substrate for a piling up of the droplets. The spray pyrolysis synthesis method has been used widely for the preparation of SOFC electrolyte thin films as well as porous electrode coatings. The spray pyrolysis method offers high film quality and low processing costs compared to other thin-film deposition techniques such as PLD and chemical or physical vapor deposition. The spray pyrolyzed thin films are usually amorphous after deposition [128], which can be converted to a nanocrystalline isotropic microstructure with grain boundaries perpendicular to the film surface [129].

Rupp et al. [130] performed a comprehensive characterization of a $Gd_{0.2}Ce_{0.8}O_{1.9-x}$ (GDC) thin film deposited by spray pyrolysis technique at 350°C and annealed at 1000°C. The film composition was $Gd_{0.23}Ce_{0.77}O_{1.9-x}$, slightly different from the ratio of the components in the precursor. The results also show that the pyrolysis process would lead to thin films that contain trace amounts of carbon and hydroxyl groups from the precursors. However, the residues of the spray pyrolysis film appear to have little effect on the electrical properties of the GDC films. The cracks formed during the thermal decomposition and heat treatment could be eliminated by optimizing the substrate temperature and repeating the film deposition and heating cycle [131].

Spray pyrolysis was also used to deposit porous and catalytic active NiO/SDC composite anode films at a substrate temperature 350°C and an annealing temperature 500°C [132]. Using a precursor sol consisting of a solution prepared from strontium acetate, lanthanum nitrate and manganese nitrate dissolved in propane-1,2-diol and LSM powder at a concentration of 0.2 mol L^{-1}, a thin porous LSM electrode with distribution of pores between 2 and 3 μm has been deposited by spray pyrolysis [133]. Porous coatings can also be obtained by powder spray without subsequent decomposition at the substrate (i.e., wet powder spraying). However, the adhesion and uniformity of the porous coating would be difficult to be controlled.

Flame assisted vapor deposition (FAVD) is a combination of spray pyrolysis and flame synthesis. In this method, an atomized solution is sprayed through a flame in an open atmosphere in which decomposition and combustion reactions occur, resulting in a stable film deposited on a heated substrate. This method requires simple apparatus and is performed at a relatively low temperature and at a high deposition rate. Choy et al. [134] prepared porous LSM cathodes by FAVD. Ethanol is added to the nitrite precursor solution to make the solution more inflammable. The morphology and porosity of the LSM coating depend strongly on the deposition temperature (i.e., the substrate temperature). An LSM electrode deposited at a deposition temperature of 710°C produced a rather high polarization resistance of 1.34 Ω cm² at 900°C. FAVD was also used to fabricate multilayer LSM/doped CeO_2 on YSZ electrolyte [135].

Similar to FAVD, combustion CVD (CCVD) is an open-air, flame-assisted chemical deposition process, capable of producing a wide range of coating morphologies from very dense to highly porous structures. Liu et al. [136] successfully employed CCVD to fabricate functionally graded LSM/LSC/GDC cathodes on YSZ electrolyte using nitrate solution precursors. In this method, methane is used as the fuel gas and oxygen as the oxidant for the combustion flame. Grain sizes as small as ~50 nm can be obtained. The spray pyrolysis deposition has the advantages of a simple setup, inexpensive and nontoxic precursors, high deposition efficiency, and direct deposition under ambient atmosphere.

5.5.2 Electrostatic Spray Deposition

Spray pyrolysis of aerosol under electrostatic field, electrostatic spray pyrolysis deposition or electrostatic spray deposition (ESD), has been developed and used to prepare thin electrolyte and electrode layers for SOFCs [137]. In the case of ESD, a high DC voltage (e.g., 6–10 kV) is applied between the nozzle (positive polarity) and the grounded substrate (negative polarity) of a normal spray pyrolysis device of Figure 5.22. The distance between the nozzle and the grounded substrate is in the range of 30–50 mm. ESD makes use of electrostatic charging to disperse the liquid. The advantage of electrostatic dispersion is that the unipolar (usually positive) charge helps to achieve very small drop sizes. The charge also prevents coalescence of drops, hence agglomeration of particles, during spray. Also, the electric field allows a high degree of control over the direction of flight and the distribution of the rate of deposition over the substrate. For fundamental aspects of the ESD or electrospraying techniques in general, readers should refer to a detailed review by Jaworek [138].

Nomura et al. [139] studied the effect of ESD process on the morphology and density of thin YSZ layers on Ni/YSZ anode substrates using a colloidal suspension of YSZ. The direct use of colloidal suspension would avoid the chemical reactions associated with the precursor solutions during the dispersion and deposition process. Operating parameters such as colloidal concentration, particle size of feed solution, solution medium, flow rate, nozzle tip shape, distance between nozzle and substrate, substrate temperature, and applied voltage were found to be important on the quality and morphology of the deposited YSZ films. For example, the shape of the nozzle tip strongly affects the type of spray. A thin-layer (~3 μm thickness) YSZ electrolyte was deposited on Ni/YSZ anode substrate, achieving an open circuit of 1.06 V at 800°C.

Fu et al. [137] used ESD to deposit porous LSCF films on stainless steel and glass substrates using metal nitrate precursor solutions. The as-deposited films are amorphous, but after calcining at 750°C, the deposited films crystallize to form the perovskite phase. Films obtained at the deposition temperature of 350°C are much more porous than those obtained at 150°C and 250°C, which is attributed due to the preferential landing of aerosol droplets and agglomeration of particles. Princivalle and Djurado [140] studied the effect of YSZ content and nozzle-to-substrate distance on the morphology of LSM/YSZ composite cathodes on YSZ substrate by ESD. Shown in Figure 5.23 are SEM micrographs of the LSM/YSZ composite films deposited at 350°C as a function of YSZ content in the composites [140]. The nozzle-to-substrate distance was 45 mm. When the YSZ content in the composites was lower than 50 wt%, the deposited composite film morphology was found reticulate and highly porous; see Figure 5.23a–c. In the case of 40% LSM/60% YSZ composite coatings, the ESD LSM/YSZ coating becomes much denser; see Figure 5.23d. The simultaneous boiling and drying of the solution with low YSZ content is suggested to be the origin of the formation of reticulate coatings. ESD was also used to deposit highly porous LSCF thin layers on SDC electrolyte before applying the screen-printed LSCF cathode [141]. The LSCF double layer cathode reduces the electrode polarization resistance by 50% as compared to single layered LSCF cathode.

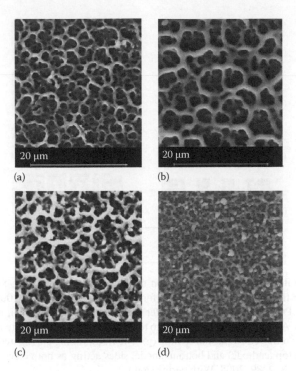

FIGURE 5.23   SEM micrographs of the LSM/YSZ composite films deposited at 350°C for YSZ content of (a) 21%, (b) 43%, (c) 50%, and (d) 60% in the composites. (From Princivalle, A. and Djurado, E., *Solid State Ionics*, 179, 1921, 2008. With permission.)

5.6   LITHOGRAPHY AND ETCHING TECHNIQUES FOR μ-SOFCs

Lithography in combination with an etching process based on substrates such as Ni or silicon wafers has been used in the design and fabrication of miniature or μ-SOFCs. Chen et al. [142] fabricated a micro thin-film SOFC based on thin-film deposition and microlithographic processes. The μ-SOFC is composed of a thin-film electrolyte deposited on a nickel foil substrate by PLD. The Ni foil substrate was then processed into a porous anode by photolithographic patterning and wet etching to develop pores for gas transport into the fuel cell. A $La_{0.5}Sr_{0.5}CoO_3$ thin-film cathode was then deposited on the electrolyte, and a porous NiO/YSZ cermet layer was added to the anode to improve the electrode performance. The μ-SOFC yielded a maximum output power density of $110\,mW\,cm^{-2}$ at 570°C. Huang et al. [37] fabricated a thin-film SOFC containing 50–150 nm thick YSZ or GDC electrolyte and 80 nm porous Pt cathode and anode, using sputtering, lithography, and etching techniques. The peak power densities are 200 and $400\,mW\,cm^{-2}$ at 350°C and 400°C, respectively. The high power densities achieved was attributed to the ultrathin electrolyte and the high charge-transfer reaction rates at the interfaces between the nanoporous electrodes (cathode and/or anode) and the nanocrystalline thin electrolyte. The unit cell area was $0.06\,mm^2$ and increasing the cell area may be difficult due to the inadequate mechanical strength of the freestanding electrolyte films.

Prinz's group [22] fabricated freestanding ultrathin corrugated YSZ electrolyte cells using a sequence of MEMS processing steps. The SOFC is based on a corrugated thin-film electrolyte, which is generated by a pattern transfer technique. Figure 5.24 shows the process flow for the fabrication of the corrugated thin film μ-SOFC. A 350 μm thick, (100) n-type silicon wafer is used to generate the template. One side of the silicon surface is patterned by standard photolithography and deep reactive ion etching (DRIE) to create cup-shaped trenches with smooth cup sidewall surface. A 100 nm thick low stress silicon nitride is deposited on both sides of the wafer with low pressure chemical vapor deposition (LPCVD). On the opposite side to the etched trenches, the silicon nitride

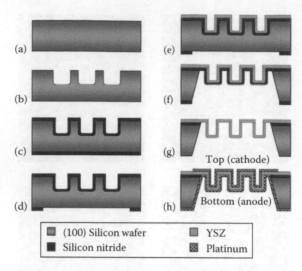

FIGURE 5.24 Process flow diagram for fabrication of the corrugated thin-film SOFC. The (100) silicon substrate is etched by DRIE to generate the template for pattern transfer (a, b). A 100 nm thick silicon nitride layer is deposited with LPCVD on both sides of wafer (c). The backside of silicon nitride is patterned with openings (d), followed by ALD deposition of YSZ onto template (e). Silicon template is etched in KOH (f) and silicon nitride etch stop is removed by plasma etching (g). The freestanding corrugated electrolyte is deposited with porous platinum on top (cathode) and bottom (anode) sides acting as both electrode and catalyst. (From Su, P.C. et al., *Nano Lett.*, 8, 2289, 2008. With permission.)

layer is patterned to generate square openings for subsequent silicon etching. ALD technique is applied to deposit the ultrathin YSZ electrolyte, forming a corrugated film (~70 nm thick). To release the YSZ membrane, the silicon substrate is immersed into a 30 wt% KOH solution for 5 h at 85°C. Pt anode and cathode are then sputtered on both sides of the freestanding YSZ membrane to form the cell. The maximum power density is 677 mW cm^{-2} at 400°C [22].

However, photolithography, in combination with the etching process, is complicated and often difficult since the etching process could damage the fuel cell materials. In addition, the resolution of wet etching is generally limited to ~10 μm. Dry etching would allow much finer patterns but is time-consuming for a multilayered Ni-based substrate.

5.7 CONCLUDING REMARKS

Many different techniques are available for thin-film deposition and the fabrication of SOFC components. They can be classified into three main groups: (1) the vapor-phase deposition, such as physical or chemical deposition, spray pyrolysis, plasma spray, etc.; (2) the liquid/colloidal-phase deposition such as EPD, sol-gel, slip-casting, and dip-coating methods; and (3) the particle deposition/consolidation methods such as tape casting, screen-printing, etc. Physical vapor deposition techniques such as PLD and sputtering are generally expensive techniques from the point of view of equipment and have low deposition rates of a few nanometer per minute. But the quality of the films can be precisely controlled. Similarly, vacuum deposition techniques such as CVD, EVD, MOCVD, etc., also suffer from relatively high equipment cost and complexity, particularly when compared to more conventional particle deposition/consolidation methods of thin-film techniques that include tape casting, EPD, screen-printing, and transfer-printing. Despite these challenges, physical deposition and vacuum methods offer a number of unique advantages. Very thin, fully dense films can be produced on either porous or dense substrates, and can be formed at temperatures much lower than those required in traditional ceramic processing, avoiding unwanted interfacial reactions. Physical deposition and vacuum methods are also well suited to the formation of interlayers, where small grain sizes and thin layer thicknesses are required. Among the most unique aspects of physical and vacuum deposition is the ability to produce unique structures that are not otherwise achievable [61]. On the other hand, VPS has the advantages of high deposition rate and the ability to be automated. However, there are two major concerns for the electrolyte deposited by VPS: anisotropic properties and significant variation in film thickness.

In contrast, the particle deposition/consolidation methods such as tape casting and screen-printing are popular, inexpensive, and are easily scaled up. The main issue is the thickness reduction. The minimum scale of the layers produced by state-of-the-art particle deposition/consolidation methods, such as tape casting, is several micrometers. On the other hand, slurry- or dip-coating techniques exhibit advantages of low capital cost and simplicity of equipment and can be applied to produce very thin and high-quality electrolyte films through the control of the surface microstructure of the substrate and the sol particle and particle size distribution [92]. Moreover, slurry or dip coating can be employed to fabricate thin electrolyte films not only for planar SOFCs, but also tubular SOFCs. The major concerns of the liquid- or colloidal-based deposition methods are the strain induced during drying or sintering, which results in some processing defects such as cracks, pores, or delamination. Thus, the most feasible way to overcoming these problems is by a multicoating process. However, the repeated processing approach may not be economically viable for the industrial fabrication and production of the SOFCs components. For the deposition processes involving colloidal solution or liquid slurry, care should also be taken to prevent the solution from filtration into the porous substrates.

The choice of an appropriate thin-film deposition technique for SOFC applications is strongly influenced by the material to be deposited, the desired film quality and microstructure, process complexity and scalability, areas, shape, and geometry of the substrates to be deposited, and the cost of the instrumentation and operation. The SOFC technology has demonstrated much higher energy efficiency with extremely low pollutant emission as compared to conventional energy technologies, but the cost of the current SOFC systems is still prohibitively high for wide commercial applications. One of the major cost items is the cost of SOFC stack fabrication. The cost of thin electrolyte films, electrode coatings, and protective coatings for metallic interconnect is an important factor in the commercial realization of SOFC technologies. Nevertheless, in cases such as μ-SOFCs the high film quality and ability to produce unusual configurations would outweigh other considerations for the selection of thin-film deposition techniques.

ABBREVIATIONS

AFL	Anode functional layer
ALD	Atomic layer deposition
APS	Atmosphere plasma spray
ASR	Area specific resistance
CCVD	Combustion chemical vapor deposition
CVD	Chemical vapor deposition
DC	Direct current
DMFCs	Direct methanol fuel cells
EB-PVD	Electron beam-physical vapor deposition
EPD	Electrophoretic deposition
ESD	Electrostatic spray deposition or electrostatic spray pyrolysis deposition
EVD	Electrochemical vapor deposition
FAVD	Flame-assisted vapor deposition
GDC	Gd-doped ceria or $(Gd, Ce)O_{2-\delta}$
IT-SOFC	Intermediate temperature solid oxide fuel cells
LAFAD	Large area filtered arc deposition
LCCr	Lanthanum calcium chromate or $(La, Ca)CrO_3$

LPCVD	Low pressure chemical vapor deposition
LSCF	Lanthanum strontium cobalt ferrite or $(La, Sr)(Co, Fe)O_3$
LSCo	Lanthanum strontium cobaltite or $(La, Sr)CoO_3$
LSCr	Lanthanum strontium chromite or $(La, Sr)CrO_3$
LSM	Lanthanum strontium manganite or $(La, Sr)MnO_3$
OCV	Open circuit voltage
PLD	Pulse laser deposition
RF	Radio frequency
SDC	Sm-doped ceria or $(Sm, Ce)O_{2-\delta}$
SOFCs	Solid oxide fuel cells
SSC	Samarium strontium cobaltite or $(Sm, Sr)CoO_3$
TPB	Three phase boundaries
VPS	Vacuum plasma spray
YDC	Yttria-doped ceria or $(Y, Ce)O_{2-\delta}$
YSZ	Yttria-stabilized zirconia or $(Y, Zr)O_2$
μ-SOFCs	Micro-solid oxide fuel cells

REFERENCES

1. Singhal S.C. 2002. Solid oxide fuel cells for stationary, mobile, and military applications. *Solid State Ionics* 152: 405–410.
2. Minh N.Q. 1993. Ceramic fuel-cells. *Journal of the American Ceramic Society* 76: 563–588.
3. Williams M.C., Strakey J.P., and Singhal S.C. 2004. US distributed generation fuel cell program. *Journal of Power Sources* 131: 79–85.
4. Jiang S.P. 2008. Development of lanthanum strontium manganite perovskite cathode materials of solid oxide fuel cells: A review. *Journal of Materials Science* 43: 6799–6833.
5. Yokokawa H., Sakai N., Horita T., Yamaji K., and Brito M.E. 2005. Electrolytes for solid-oxide fuel cells. *Materials Research Society Bulletin* 30: 591–595.
6. Jiang S.P. and Chan S.H. 2004. A review of anode materials development in solid oxide fuel cells. *Journal of Materials Science* 39: 4405–4439.
7. Gorte R.J., Vohs J.M., and McIntosh S. 2004. Recent developments on anodes for direct fuel utilization in SOFC. *Solid State Ionics* 175: 1–6.
8. Fergus J.W. 2005. Metallic interconnects for solid oxide fuel cells. *Materials Science and Engineering A: Structural Materials: Properties, Microstructure and Processing* 397: 271–283.
9. Will J., Mitterdorfer A., Kleinlogel C., Perednis D., and Gauckler L.J. 2000. Fabrication of thin electrolytes for second-generation solid oxide fuel cells. *Solid State Ionics* 131: 79–96.
10. Jiang S.P., Zhang S., and Zhen Y.D. 2005. Early interaction between Fe-Cr alloy metallic interconnect and Sr-doped LaMnO₃ cathodes of solid oxide fuel cells. *Journal of Materials Research* 20: 747–758.
11. Jiang S.P., Liu Z.C., and Tian Z.Q. 2006. Layer-by-layer self-assembly of composite polyelectrolyte-nafion membranes for direct methanol fuel cells. *Advanced Materials* 18: 1068–1072.
12. Wasmus S. and Kuver A. 1999. Methanol oxidation and direct methanol fuel cells: A selective review. *Journal of Electroanalytical Chemistry* 461: 14–31.
13. Shao Z.P., Haile S.M., Ahn J., Ronney P.D., Zhan Z.L., and Barnett S.A. 2005. A thermally self-sustained micro solid-oxide fuel-cell stack with high power density. *Nature* 435: 795–798.
14. Beckel D., Bieberle-Hutter A., Harvey A., Infortuna A., Muecke U.P., Prestat M., Rupp J.L.M., and Gauckler L.J. 2007. Thin films for micro solid oxide fuel cells. *Journal of Power Sources* 173: 325–345.
15. Aizawa M., Kobayashi C., Yamane H., and Hirai T. 1993. Preparation of ZRO_2-Y_2O_3 films by CVD using beta-diketone metal-chelates. *Nippon Seramikkusu Kyokai Gakujutsu Ronbunshi—Journal of the Ceramic Society of Japan* 101: 291–294.
16. Yamane H. and Hirai T. 1987. Preparation of ZRO_2-film by oxidation of $ZRCL_4$. *Journal of Materials Science Letters* 6: 1229–1230.
17. Yamane H. and Hirai T. 1989. Yttria stabilized zirconia transparent films prepared by chemical vapor-deposition. *Journal of Crystal Growth* 94: 880–884.
18. Chour K.W., Chen J., and Xu R. 1997. Metal-organic vapor deposition of YSZ electrolyte layers for solid oxide fuel cell applications. *Thin Solid Films* 304: 106–112.

19. Itoh H., Mori M., Mori N., and Abe T. 1994. Production cost estimation of solid oxide fuel-cells. *Journal of Power Sources* 49: 315–332.

20. Hausmann D.M., Kim E., Becker J., and Gordon R.G. 2002. Atomic layer deposition of hafnium and zirconium oxides using metal amide precursors. *Chemistry of Materials* 14: 4350–4358.

21. Shim J.H., Chao C.C., Huang H., and Prinz F.B. 2007. Atomic layer deposition of yttria-stabilized zirconia for solid oxide fuel cells. *Chemistry of Materials* 19: 3850–3854.

22. Su P.C., Chao C.C., Shim J.H., Fasching R., and Prinz F.B. 2008. Solid oxide fuel cell with corrugated thin film electrolyte. *Nano Letters* 8: 2289–2292.

23. Pal U.B. and Singhal S.C. 1990. Electrochemical vapor-deposition of yttria-stabilized zirconia films. *Journal of the Electrochemical Society* 137: 2937–2941.

24. Sasaki H., Otoshi S., Suzuki M., Sogi T., Kajimura A., Sugiura N., and Ippommatsu M. 1994. Fabrication of high-power density tubular type solid oxide fuel-cells. *Solid State Ionics* 72: 253–256.

25. Suzuki M. and Kajimura A. 1997. Geometrical analysis of SOFC anodes fabricated by electrochemical vapor deposition. *Denki Kagaku* 65: 859–864.

26. Ogumi Z., Ioroi T., Uchimoto Y., Takehara Z., Ogawa T., and Toyama K. 1995. Novel method for preparing nickel/YSZ cermet by a vapor-phase process. *Journal of the American Ceramic Society* 78: 593–598.

27. Kikuchi K., Tamazaki F., Okada K., and Mineshige A. 2006. Yttria-stabilized zirconia thin films deposited on NiO-$(Sm_2O_3)(0.1)$ $(CeO_2)(0.8)$ substrates by chemical vapor infiltration. *Journal of Power Sources* 162: 1053–1059.

28. Bae J.W., Park J.Y., Hwang S.W., Yeom G.Y., Kim K.D., Cho Y.A., Jeon J.S., and Choi D. 2000. Characterization of yttria-stabilized zirconia thin films prepared by radio frequency magnetron sputtering for a combustion control oxygen sensor. *Journal of the Electrochemical Society* 147: 2380–2384.

29. Srivastava P.K., Quach T., Duan Y.Y., Donelson R., Jiang S.P., Ciacchi F.T., and Badwal S.P.S. 1997. Electrode supported solid oxide fuel cells: Electrolyte films prepared by DC magnetron sputtering. *Solid State Ionics* 99: 311–319.

30. Wang L.S. and Barnett S.A. 1993. Sputter-deposited medium-temperature solid oxide fuel-cells with multilayer electrolytes. *Solid State Ionics* 61: 273–276.

31. Wanzenberg E., Tietz F., Panjan P., and Stover D. 2003. Influence of pre- and post-heat treatment of anode substrates on the properties of DC-sputtered YSZ electrolyte films. *Solid State Ionics* 159: 1–8.

32. Tsai T. and Barnett S.A. 1995. Bias sputter-deposition of dense yttria-stabilized zirconia films on porous substrates. *Journal of the Electrochemical Society* 142: 3084–3087.

33. Fedtke P., Wienecke M., Bunescu M.C., Barfels T., Deistung K., and Pietrzak M. 2004. Yttria-stabilized zirconia films deposited by plasma spraying and sputtering. *Journal of Solid State Electrochemistry* 8: 626–632.

34. Tao S.W. and Irvine J.T.S. 2002. Study on the structural and electrical properties of the double perovskite oxide $SrMn_{0.5}Nb_{0.5}O_{3-\delta}$. *Journal of Materials Chemistry* 12: 2356–2360.

35. Ai N., Lu Z., Chen K.F., Huang X.Q., Du X.B., and Su W.H. 2007. Effects of anode surface modification on the performance of low temperature SOFCs. *Journal of Power Sources* 171: 489–494.

36. Kim S.D., Lee J.J., Moon H., Hyun S.H., Moon J., Kim J., and Lee H.W. 2007. Effects of anode and electrolyte microstructures on performance of solid oxide fuel cells. *Journal of Power Sources* 169: 265–270.

37. Huang H., Nakamura M., Su P.C., Fasching R., Saito Y., and Prinz F.B. 2007. High-performance ultra-thin solid oxide fuel cells for low-temperature operation. *Journal of the Electrochemical Society* 154: B20–B24.

38. Uhlenbruck S., Jordan N., Sebold D., Buchkremer H.P., Haanappel V.A.C., and Stover D. 2007. Thin film coating technologies of $(Ce,Gd)O_{2-\delta}$ interlayers for application in ceramic high-temperature fuel cells. *Thin Solid Films* 515: 4053–4060.

39. Horita T., Yamaji K., Ishikawa M., Sakai N., Yokokawa H., Kawada T., and Kato T. 1998. Active sites imaging for oxygen reduction at the $La_{0.9}Sr_{0.1}MnO_{3-x}$/yttria-stabilized zirconia interface by secondary-ion mass spectrometry. *Journal of the Electrochemical Society* 145: 3196–3202.

40. Mizusaki J., Tagawa H., Saito T., Kamitani K., Yamamura T., Hirano K., Ehara S. et al. 1994. Preparation of nickel pattern electrodes on YSZ and their electrochemical properties in H_2-H_2O atmospheres. *Journal of the Electrochemical Society* 141: 2129–2134.

41. Lee C. and Bae J. 2008. Oxidation-resistant thin film coating on ferritic stainless steel by sputtering for solid oxide fuel cells. *Thin Solid Films* 516: 6432–6437.

42. Orlovskaya N., Coratolo A., Johnson C., and Gemmen R. 2004. Structural characterization of lanthanum chromite perovskite coating deposited by magnetron sputtering on an iron-based chromium-containing alloy as a promising interconnect material for SOFCs. *Journal of the American Ceramic Society* 87: 1981–1987.

43. Suzuki M., Sasaki H., and Kajimura A. 1997. Oxide ionic conductivity of doped lanthanum chromite thin film interconnectors. *Solid State Ionics* 96: 83–88.

44. Mengucci P., Barucca G., Caricato A.P., Di Cristoforo A., Leggieri G., Luches A., and Majnia G. 2005. Effects of annealing on the microstructure of yttria-stabilised zirconia thin films deposited by laser ablation. *Thin Solid Films* 478: 125–131.

45. Nair B.N., Suzuki T., Yoshino Y., Gopalakrishnan S., Sugawara T., Nakao S., and Taguchi H. 2005. An oriented nanoporous membrane prepared by pulsed laser deposition. *Advanced Materials* 17: 1136–1140.

46. Infortuna A., Harvey A.S., and Gauckler L.J. 2008. Microstructures of CGO and YSZ thin films by pulsed laser deposition. *Advanced Functional Materials* 18: 127–135.

47. Joo J.H. and Choi G.M. 2006. Electrical conductivity of YSZ film grown by pulsed laser deposition. *Solid State Ionics* 177: 1053–1057.

48. Chen L., Chen C.L., Huang D.X., Lin Y., Chen X., and Jacobson A.J. 2003. High temperature electrical conductivity of epitaxial Gd-doped CeO_2 thin films. *Solid State Ionics* 175:103–106.

49. Hobein B., Tietz F., Stover D., and Kreutz E.W. 2002. Pulsed laser deposition of yttria stabilized zirconia for solid oxide fuel cell applications. *Journal of Power Sources* 105: 239–242.

50. Meunier M., Izquierdo R., Hasnaoui L., Quenneville E., Ivanov D., Girard F., Morin F., Yelon A., and Paleologou M. 1998. Pulsed laser deposition of superionic ceramic thin films: Deposition and applications in electrochemistry. *Applied Surface Science* 129: 466–470.

51. Koep E., Jin C.M., Haluska M., Das R., Narayan R., Sandhage K., Snyder R., and Liu M.L. 2006. Microstructure and electrochemical properties of cathode materials for SOFCs prepared via pulsed laser deposition. *Journal of Power Sources* 161: 250–255.

52. Koep E., Compson C., Liu M.L., and Zhou Z.P. 2005. A photolithographic process for investigation of electrode reaction sites in solid oxide fuel cells. *Solid State Ionics* 176: 1–8.

53. de Larramendi I.R., Ortiz N., Lopez-Anton R., de Larramendi J.I.R., and Rojo T. 2007. Structure and impedance spectroscopy of $La_{0.6}Ca_{0.4}Fe_{0.8}Ni_{0.2}O_{3-\delta}$ thin films grown by pulsed laser deposition. *Journal of Power Sources* 171: 747–753.

54. Chen X., Wu N.J., and Ignatiev A. 1999. Structure and conducting properties of $La_{1-x}Sr_xCoO_{3-\delta}$ films. *Journal of the European Ceramic Society* 19: 819–822.

55. Endo A., Ihara M., Komiyama H., and Yamada K. 1996. Cathodic reaction mechanism for dense Sr-doped lanthanum manganite electrodes. *Solid State Ionics* 86–8: 1191–1195.

56. Mikkelsen L., Pryds N., and Hendriksen P.V. 2007. Preparation of $La_{0.8}Sr_{0.2}Cr_{0.97}V_{0.03}O_{3-\delta}$ films for solid oxide fuel cell application. *Thin Solid Films* 515: 6537–6540.

57. Gannon P., Deibert M., White P., Smith R., Chen H., Priyantha W., Lucas J., and Gorokhousky V. 2008. Advanced PVD protective coatings for SOFC interconnects. *Symposium on Materials in Clean Power Systems II Held at the 2007 TMS Annual Conference and Exposition*, pp. 3991–4000.

58. Jung H.Y., Hong K.S., Kim H., Park J.K., Son J.W., Kim J., Lee H.W., and Lee J.H. 2006. Characterization of thin-film YSZ deposited via EB-PVD technique in anode-supported SOFCs. *Journal of the Electrochemical Society* 153: A961–A966.

59. Gibson I.R., Dransfield G.P., and Irvine J.T.S. 1998. Influence of yttria concentration upon electrical properties and susceptibility to ageing of yttria-stabilised zirconias. *Journal of the European Ceramic Society* 18: 661–667.

60. Laukaitis G. and Dudonis J. 2008. Microstructure of gadolinium doped ceria oxide thin films formed by electron beam deposition. *Journal of Alloys and Compounds* 459: 320–327.

61. Pederson L.R., Singh P., and Zhou X.D. 2006. Application of vacuum deposition methods to solid oxide fuel cells. *Vacuum* 80: 1066–1083.

62. Kulkarni A.A., Sampath S., Goland A., Herman H., Allen A.J., Ilavsky J., Gong W.Q., and Gopalan S. 2003 Plasma spray coatings for producing next-generation supported membranes. *Symposium on Synthetic Clean Fuels from Natural Gas and Coal-Bed Methane*, pp. 241–249.

63. Tai L.W. and Lessing P.A. 1991. Plasma spraying of porous-electrodes for a planar solid oxide fuel-cell. *Journal of the American Ceramic Society* 74: 501–504.

64. Li C.J., Li C.X., and Wang M. 2005. Effect of spray parameters on the electrical conductivity of plasma-sprayed $La_{1-x}Sr_xMnO_3$ coating for the cathode of SOFCs. *Surface and Coatings Technology* 198: 278–282.

65. Lim D.P., Lim D.S., Oh J.S., and Lyo I.W. 2005. Influence of post-treatments on the contact resistance of plasma-sprayed $La_{0.8}Sr_{0.2}MnO_3$ coating on SOFC metallic interconnector. *Surface and Coatings Technology* 200: 1248–1251.

66. Rambert S., McEvoy A.J., and Barthel K. 1999. Composite ceramic fuel cell fabricated by vacuum plasma spraying. *Journal of the European Ceramic Society* 19: 921–923.

67. Van herle J., McEvoy A.J., and Thampi K.R. 1994. Conductivity measurements of various yttria-stabilized zirconia samples. *Journal of Materials Science* 29: 3691–3701.

68. Ma X.Q., Zhang H., Dai J., Roth J., Hui R., Xiao T.D., and Reisner D.E. 2005. Intermediate temperature solid oxide fuel cell based on fully integrated plasma-sprayed components. *Journal of Thermal Spray Technology* 14: 61–66.

69. Fauchais P., Rat V., Delbos U., Coudert J.F., Chartier T., and Bianchi L. 2005. Understanding of suspension DC plasma spraying of finely structured coatings for SOFC. *IEEE Transactions on Plasma Science* 33: 920–930.

70. Zheng R., Zhou X.M., Wang S.R., Wen T.L., and Ding C.X. 2005. A study of Ni+8YSZ/8YSZ/ $La_{0.6}Sr_{0.4}CoO_{3-\delta}$ ITSOFC fabricated by atmospheric plasma spraying. *Journal of Power Sources* 140: 217–225.

71. White B.D., Kesler, O., and Rose, L. 2007. Electrochemical characterization of air plasma sprayed LSM/ YSZ composite cathodes on metallic interconnects. *ECS Transactions* 7: 1107–1114.

72. Vassen R., Kassner H., Stuke A., Hauler F., Hathiramani D., and Stover D. 2008. Advanced thermal spray technologies for applications in energy systems. *Surface and Coatings Technology* 202: 4432–4437.

73. Lang M., Henne R., Schaper S., and Schiller G. 2001. Development and characterization of vacuum plasma sprayed thin film solid oxide fuel cells. *Journal of Thermal Spray Technology* 10: 618–625.

74. Schiller G., Henne R., Lang M., Ruckdaschel R., and Schaper S. 2000. Development of vacuum plasma sprayed thin-film SOFC for reduced operating temperature. *Fuel Cells Bulletin* 3: 7–12.

75. Meier L.P., Urech L., and Gauckler L.J. 2004. Tape casting of nanocrystalline ceria gadolinia powder. *Journal of the European Ceramic Society* 24: 3753–3758.

76. Song J.H., Park S.I., Lee J.H., and Kim H.S. 2008. Fabrication characteristics of an anode-supported thin-film electrolyte fabricated by the tape casting method for IT-SOFC. *Journal of Materials Processing Technology* 198: 414–418.

77. Kim S.D., Hyun S.H., Moon J., Kim J.H., and Song R.H. 2005. Fabrication and characterization of anode-supported electrolyte thin films for intermediate temperature solid oxide fuel cells. *Journal of Power Sources* 139: 67–72.

78. Cai Z., Lan T.N., Wang S., and Dokiya M. 2002. Supported Zr(Sc)O-2 SOFCs for reduced temperature prepared by slurry coating and co-firing. *Solid State Ionics* 152: 583–590.

79. Kim S.G., Yoon S.P., Nam S.W., Hyun S.H., and Hong S.A. 2002. Fabrication and characterization of a YSZ/YDC composite electrolyte by a sol-gel coating method. *Journal of Power Sources* 110: 222–228.

80. Xu X.Y., Xia C.R., Huang S.G., and Peng D.K. 2005. YSZ thin films deposited by spin-coating for IT-SOFCs. *Ceramics International* 31: 1061–1064.

81. Zhang Y.H., Huang X.Q., Lu Z., Ge X.D., Xu J.H., Xin X.S., Sha X.Q., and Su W.H. 2006. Effect of starting powder on screen-printed YSZ films used as electrolyte in SOFCs. *Solid State Ionics* 177: 281–287.

82. Chu W.F. 1992. Thin-film and thick-film solid ionic devices. *Solid State Ionics* 52: 243–248.

83. Chen K.F., Lu Z., Ai N., Huang X.Q., Zhang Y.H., Ge X.D., Xin X.S., Chen X.J., and Su W.H. 2007. Fabrication and performance of anode-supported YSZ films by slurry spin coating. *Solid State Ionics* 177: 3455–3460.

84. Chen Y.Y. and Wei W.C.J. 2006. Processing and characterization of ultra-thin yttria-stabilized zirconia (YSZ) electrolytic films for SOFC. *Solid State Ionics* 177: 351–357.

85. Wang Z.C., Weng W.J., Chen K., Shen G., Du P.Y., and Han G.R. 2008. Preparation and performance of nanostructured porous thin cathode for low-temperature solid oxide fuel cells by spin-coating method. *Journal of Power Sources* 175: 430–435.

86. Wang J.M., Lu Z., Huang X.Q., Chen K.F., Ai N., Hu J.Y., and Su W.H. 2007. YSZ films fabricated by a spin smoothing technique and its application in solid oxide fuel cell. *Journal of Power Sources* 163: 957–959.

87. Matus Y.B., De Jonghe L.C., Jacobson C.P., and Visco S.J. 2005. Metal-supported solid oxide fuel cell membranes for rapid thermal cycling. *Solid State Ionics* 176: 443–449.

88. Tucker M.C., Lau G.Y., Jacobson C.P., DeJonghe L.C., and Visco S.J. 2007. Performance of metal-supported SOFCs with infiltrated electrodes. *Journal of Power Sources* 171: 477–482.

89. Wang C.H., Worrell W.L., Park S., Vohs J.M., and Gorte R.J. 2001. Fabrication and performance of thin-film YSZ solid oxide fuel cells. *Journal of the Electrochemical Society* 148: A864–A868.

90. Zhang L., He H.Q., Kwek W.R., Ma J., Tang E.H., and Jiang S.P. 2009. Fabrication and characterization of anode-supported tubular solid-oxide fuel cells by slip casting and dip coating techniques. *Journal of the American Ceramic Society* 92: 302–310.

91. Yamaguchi T., Suzuki T., Shimizu S., Fujishiro Y., and Awano M. 2007. Examination of wet coating and co-sintering technologies for micro-SOFCs fabrication. *Journal of Membrane Science* 300: 45–50.

92. Van Gestel T., Sebold D., Meulenberg W.A., and Buchkremer H.P. 2008. Development of thin-film nano-structured electrolyte layers for application in anode-supported solid oxide fuel cells. *Solid State Ionics* 179: 428–437.

93. Wang Z.H., Sun K.N., Shen S.Y., Zhang N.Q., Qiao J.S., and Xu P. 2008. Preparation of YSZ thin films for intermediate temperature solid oxide fuel cells by dip-coating method. *Journal of Membrane Science* 320: 500–504.

94. Liu M.F., Dong D.H., Zhao F., Gao J.F., Ding D., Liu X.Q., and Meng G.Y. 2008. High-performance cathode-supported SOFCs prepared by a single-step co-firing process. *Journal of Power Sources* 182: 585–588.

95. Zhang Y.H., Liu J., Huang X.Q., Lu Z., and Su W.H. 2008. Performance evaluation of thin membranes solid oxide fuel cell prepared by pressure-assisted slurry-casting. *International Journal of Hydrogen Energy* 33: 775–780.

96. Serra J.M., Vert V.B., Buchler O., Meulenberg W.A., and Buchkremer H.P. 2008. IT-SOFC supported on mixed oxygen ionic-electronic conducting composites. *Chemistry of Materials* 20: 3867–3875.

97. Reitz T.L. and Xiao H.M. 2006. Characterization of electrolyte-electrode interlayers in thin film solid oxide fuel cells. *Journal of Power Sources* 161: 437–443.

98. Dunn B., Farrington G.C., and Katz B. 1993. Sol-gel approaches for solid electrolytes and electrode materials. *Ninth International Conference on Solid State Ionics*, the Hague, the Netherlands, pp. 3–10.

99. Mehta K., Xu R., and Virkar A.V. 1998. Two-layer fuel cell electrolyte structure by sol-gel processing. *Journal of Sol-Gel Science and Technology* 11: 203–207.

100. Lee E.A., Lee S., Hwang H.J., and Moon J.W. 2006. Sol-gel derived $(La_{0.8}M_{0.2})CrO_3$ (M = Ca, Sr) coating layer on stainless-steel substrate for use as a separator in intermediate-temperature solid oxide fuel cell. *Journal of Power Sources* 157: 709–713.

101. Zhang X.G., Robertson M., Deces-Petit C., Xie Y.S., Hui R., Yick S., Styles E., Roller J., Kesler O., Maric R., and Ghosh D. 2006. NiO-YSZ cermets supported low temperature solid oxide fuel cells. *Journal of Power Sources* 161: 301–307.

102. Xia C.R., Chen F.L., and Liu M.L. 2001. Reduced-temperature solid oxide fuel cells fabricated by screen printing. *Electrochemical and Solid State Letters* 4: A52–A54.

103. Joo J.H. and Choi G.M. 2008. Simple fabrication of micro-solid oxide fuel cell supported on metal substrate. *Journal of Power Sources* 182: 589–593.

104. Ishihara T., Sato K., and Takita Y. 1996. Electrophoretic deposition of Y_2O_3-stabilized ZrO_2 electrolyte films in solid oxide fuel cells. *Journal of the American Ceramic Society* 79: 913–919.

105. Ishihara T., Shimose K., Kudo T., Nishiguchi H., Akbay T., and Takita Y. 2000. Preparation of yttria-stabilized zirconia thin films on strontium-doped $LaMnO_3$ cathode substrates via electrophoretic deposition for solid oxide fuel cells. *Journal of the American Ceramic Society* 83: 1921–1927.

106. Mathews T., Rabu N., Sellar J.R., and Muddle B.C. 2000. Fabrication of $La_{1-x}Sr_xGa_{1-y}Mg_yO_{3-(x+y)/2}$ thin films by electrophoretic deposition and its conductivity measurement. *Solid State Ionics* 128: 111–115.

107. Matsuda M., Ohara O., Murata K., Ohara S., Fukui T., and Miyake M. 2003. Electrophoretic fabrication and cell performance of dense Sr- and Mg-doped $LaGaO_3$-based electrolyte films. *Electrochemical and Solid State Letters* 6: A140–A143.

108. Besra L., Compson C., and Liu M.L. 2007. Electrophoretic deposition on non-conducting substrates: The case of YSZ film on NiO-YSZ composite substrates for solid oxide fuel cell application. *Journal of Power Sources* 173: 130–136.

109. Chen F.L. and Liu M.L. 2001. Preparation of yttria-stabilized zirconia (YSZ) films on $La_{0.85}Sr_{0.15}MnO_3$ (LSM) and LSM-YSZ substrates using an electrophoretic deposition (EPD) process. *Journal of the European Ceramic Society* 21: 127–134.

110. Peng Z.Y. and Liu M.L. 2001. Preparation of dense platinum-yttria stabilized zirconia and yttria stabilized zirconia films on porous $La_{0.9}Sr_{0.1}MnO_3$ (LSM) substrates. *Journal of the American Ceramic Society* 84: 283–288.

111. Sarkar P., De D., and Rho H. 2004. Synthesis and microstructural manipulation of ceramics by electrophoretic deposition. *Journal of Materials Science* 39: 819–823.

112. Besra L. and Liu M. 2007. A review on fundamentals and applications of electrophoretic deposition (EPD). *Progress in Materials Science* 52: 1–61.

113. Hosomi T., Matsuda M., and Miyake M. 2007. Electrophoretic deposition for fabrication of YSZ electrolyte film on non-conducting porous NiO-YSZ composite substrate for intermediate temperature SOFC. *Journal of the European Ceramic Society* 27: 173–178.

114. Cherng J.S., Sau J.R., and Chung C.C. 2008. Aqueous electrophoretic deposition of YSZ electrolyte layers for solid oxide fuel cells. *Journal of Solid State Electrochemistry* 12: 925–933.

115. Will J., Hruschka M.K.M., Gubler L., and Gauckler L.J. 2001. Electrophoretic deposition of zirconia on porous anodic substrates. *Journal of the American Ceramic Society* 84: 328–332.

116. Besra L., Zha S.W., and Liu M.L. 2006. Preparation of NiO-YSZ/YSZ bi-layers for solid oxide fuel cells by electrophoretic deposition. *Journal of Power Sources* 160: 207–214.

117. Jean J.H., Chang C.R., and Chen Z.C. 1997. Effect of densification mismatch on camber development during cofiring of nickel-based multilayer ceramic capacitors. *Journal of the American Ceramic Society* 80: 2401–2406.

118. Park S., Gorte R.J., and Vohs J.M. 2001. Tape cast solid oxide fuel cells for the direct oxidation of hydrocarbons. *Journal of the Electrochemical Society* 148: A443–A447.

119. Sofie S.W. 2007. Fabrication of functionally graded and aligned porosity in thin ceramic substrates with the novel freeze-tape-casting process. *Journal of the American Ceramic Society* 90: 2024–2031.

120. Minh N.Q. 1988. Development of monolithic solid oxide fuel-cells for aerospace applications. *Journal of the Electrochemical Society* 135: C344–C344.

121. Singh P. and Minh N.Q. 2004. Solid oxide fuel cells: Technology status. *International Journal of Applied Ceramic Technology* 1: 5–15.

122. Minh N.Q. 2004. Solid oxide fuel cell technology-features and applications. *Solid State Ionics* 174: 271–277.

123. Xia C.R. and Liu M.L. 2001. A simple and cost-effective approach to fabrication of dense ceramic membranes on porous substrates. *Journal of the American Ceramic Society* 84: 1903–1905.

124. Xia C.R. and Liu M.L. 2002. Microstructures, conductivities, and electrochemical properties of $Ce_{0.9}Gd_{0.1}O_2$ and GDC-Ni anodes for low-temperature SOFCs. *Solid State Ionics* 152: 423–430.

125. Xia C.R. and Liu M.L. 2001. Low-temperature SOFCs based on $Gd_{0.1}Ce_{0.9}O_{1.95}$ fabricated by dry pressing. *Solid State Ionics* 144: 249–255.

126. Leng Y.J., Chan S.H., Jiang S.P., and Khor K.A. 2004. Low-temperature SOFC with thin film GDC electrolyte prepared in situ by solid-state reaction. *Solid State Ionics* 170: 9–15.

127. Perednis D. and Gauckler L.J. 2005. Thin film deposition using spray pyrolysis. *Journal of Electroceramics* 14: 103–111.

128. Perednis D. and Gauckler L.J. 2004. Solid oxide fuel cells with electrolytes prepared via spray pyrolysis. *Solid State Ionics* 166: 229–239.

129. Rupp J.L.M., Infortuna A., and Gauckler L.J. 2006. Microstrain and self-limited grain growth in nanocrystalline ceria ceramics. *Acta Materialia* 54: 1721–1730.

130. Rupp J.L.M., Drobek T., Rossi A., and Gauckler L.J. 2007. Chemical analysis of spray pyrolysis gadolinia-doped ceria electrolyte thin films for solid oxide fuel cells. *Chemistry of Materials* 19: 1134–1142.

131. Setoguchi T., Sawano M., Eguchi K., and Arai H. 1990. Application of the stabilized zirconia thin-film prepared by spray pyrolysis method to SOFC. *Solid State Ionics* 40–41: 502–505.

132. Patil B.B., Ganesan V., and Pawar S.H. 2008. Studies on spray deposited NiO-SDC composite films for solid oxide fuel cells. *Journal of Alloys and Compounds* 460: 680–687.

133. Charpentier P., Fragnaud P., Schleich D.M., and Gehain E. 2000. Preparation of thin film SOFCs working at reduced temperature. *Solid State Ionics* 135: 373–380.

134. Choy K.L., Charojrochkul S., and Steele B.C.H. 1997. Fabrication of cathode for solid oxide fuel cells using flame assisted vapour deposition technique. *Solid State Ionics* 96: 49–54.

135. Charojrochkul S., Choy K.L., and Steele B.C.H. 1999. Cathode electrolyte systems for solid oxide fuel cells fabricated using flame assisted vapour deposition technique. *Solid State Ionics* 121: 107–113.

136. Liu Y., Compson C., and Liu M.L. 2004. Nanostructured and functionally graded cathodes for intermediate temperature solid oxide fuel cells. *Journal of Power Sources* 138: 194–198.

137. Fu C.Y., Chang C.L., Hsu C.S., and Hwang B.H. 2005. Electrostatic spray deposition of $La_{0.8}Sr_{0.2}Co_{0.2}Fe_{0.8}O_3$ films. *Materials Chemistry and Physics* 91: 28–35.

138. Jaworek A. 2007. Electrospray droplet sources for thin film deposition. *Journal of Materials Science* 42: 266–297.

139. Nomura H., Parekh S., Selman J.R., and Al-Hallaj S. 2005. Fabrication of YSZ electrolyte using electrostatic spray deposition (ESD): I - a comprehensive parametric study. *Journal of Applied Electrochemistry* 35: 61–67.

140. Princivalle A. and Djurado E. 2008. Nanostructured LSM/YSZ composite cathodes for IT-SOFC: A comprehensive microstructural study by electrostatic spray deposition. *Solid State Ionics* 179: 1921–1928.

141. Hsu C.S., Hwang B.H., Xie Y., and Zhang X. 2008. Enhancement of solid oxide fuel cell performance by $La_{0.6}Sr_{0.4}Co_{0.2}Fe_{0.8}O_{3-\delta}$ double-layer cathode. *Journal of the Electrochemical Society* 155: B1240-B1243.

142. Chen X., Wu N.J., Smith L., and Ignatiev A. 2004. Thin-film heterostructure solid oxide fuel cells. *Applied Physics Letters* 84: 2700–2702.

6 Nanoscale Organic Molecular Thin Films for Information Memory Applications

J.C. Li, D.C. Ba, and Y.L. Song

CONTENTS

6.1   INTRODUCTION

Silicon-based semiconductor industry has been very successful in the past decades, and has led us to the information age. According to Moore's law, the semiconductor device density in a chip has to be doubled every one-and-half year [1]. To date, the semiconductor devices have been miniaturized to sub-50 nm dimensions, where several factors limit the bit density increase, including lithography resolution, electromigration, and capacitance. With the onslaught of the digital age, the demands of ultrahigh data density have fueled the need for information memory media with sub-10 nm scale that falls to the dimensions of a single molecule [2,3]. However, there are some crucial effects emerging from such low-dimension devices: (1) when too few atoms are used in a wire, the nanostructure moves in response to currents; (2) when too few electrons are used in a nanoscale device, the fluctuation in their number becomes significant in the device performance; and (3) the electron noise in a nanoscale world may come from thermodynamic fluctuations, defect scattering, and finite-size statistics.

To develop ultrahigh density information storage systems, some other critical factors have to be taken into account too: density, cost, writing–erasing time, reversibility, power consumption, and durability. Now, researchers are also trying to continue the current trend of device miniaturization to fulfill the requirement of Moore's law. Such an effort calls for the development of new techniques, materials, and theories for nanoscale information storage. An important technique of scanning tunneling microscopy (STM) was invented in the 1980s. The development of various scanning probe microscope (SPM) techniques has brought with it the possibilities to construct and further characterize molecular junctions with different conformation, substituent, and molecule number within the junction. From then on, it has been seen in the past 20 years that there have been rapid and significant advances of both basic and applied research in the field of molecular electronics. As a bottom-up approach in molecular electronics, we have to exploit the possibility of using organic functional molecules as active information storage media, which includes molecular design, synthesis, active media growth, device fabrication and structure–property characterizations [4]. This has opened up a new research area in the field of molecular electronics.

Since Avriam and Ratner proposed the donor-sigma-acceptor model of single-molecular diodes in the 1970s, molecular materials including molecular clusters and thin layers have been intensively studied owing to their unique optical/electrical properties and potential applications in the field of new generation of information technology. The molecular switching performance, especially optoelectronic switching, of these nanostructures are the very important functional features for the usage of information memories. To understand the switching mechanisms, one has to construct a molecular device, usually in the form of metal-organic-metal (MOM) junction. The current–voltage (I–V) characteristics of the molecular junction can then be investigated under various controlled conditions (e.g., temperature, gas and/or solvents, optical illumination, etc.). The correlation between these researching areas is schematically shown in Figure 6.1.

In this chapter, we have reviewed functional molecules for application in the field of information storage, i.e., with molecular switching performance, from the aspects of chemical conformation/substituent, thin film growth, and structure–property relationships. Two kinds of molecular materials are discussed in detail, which include redox dendrimeric materials and self-assembly monolayer (SAM). In addition, the fabrication and growth of polymer, Langmuir–Blodgett (LB), and charge transfer salt-based thin films/devices are briefly discussed too. For easy reference of this handbook chapter, most of the data is presented in the form of tables and figures.

The organization of this chapter is as follows. First, we give a brief presentation of the techniques for organic active layer growth, device fabrication, and molecular junction switching mechanism. More emphasis has been given to the fabrication of self-assembled monolayers and junctions. Second, we discuss the redox dendrimeric materials–based molecular memories highlighting the structure–property relationship and ultrahigh density charge storage. Then, as the focus of this chapter, we describe the experimental studies on self-assembled monolayer-based molecular memories from

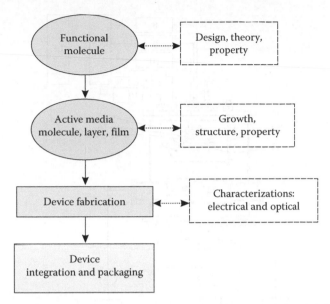

FIGURE 6.1 Schematic correlation between the main researching contents of molecular electronics.

the aspects of molecular structure–property relationship, device fabrication and characterization. Finally, we briefly introduce the usage of polymers, LB layers, and charge transfer salts as functional media for molecular memories.

6.2 ACTIVE LAYER GROWTH, DEVICE FABRICATION, AND SWITCHING MECHANISM

This section generally introduces the techniques and knowledge about thin-film deposition, growth of monolayer, fabrication and characterization of molecular junctions. The switching mechanism for usual molecular electronic devices is also briefly presented, which will be further discussed in later sections.

6.2.1 ACTIVE LAYER GROWTH

There are several types of molecular active media for potential application in information memory devices, which include thin films, self-assembled monolayers, LB layers, molecular patterns, and single molecules. In this section, the techniques for active layer fabrication are briefly presented in the sequence of the frequency used and significance.

6.2.1.1 Vacuum Thermal Vapor Deposition

Vacuum thermal evaporation is a traditional method used widely for the fabrication of various thin films, including metal electrodes and organic layers. It has the advantages of clean and controlled environment (excluding the factors of water and oxygen), known thickness, multisources and in situ fabrication and characterization, etc. For example, Figure 6.2 shows a schematic vacuum system that we use for the in situ fabrication and characterization of molecular electronic devices. The chamber is pumped by a combined vacuum station consisting of a turbo pump and a mechanical pump. There are two separate evaporating stations within the chamber and each one has two evaporation sources. The substrate is sheltered during the exchange of the sources when a layer deposition is finished. The substrate can be cooled down to liquid nitrogen temperature during the metal electrode deposition so as to protect the organic layer. The film thickness, substrate temperature,

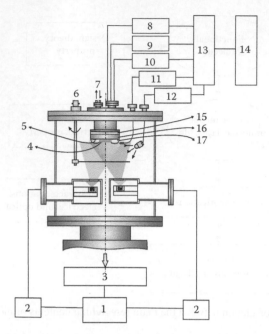

FIGURE 6.2 Schematic vacuum thermal vapor deposition system with in situ vary-temperature current–voltage characterizations. Here, 1, power; 2, evaporation crucible heating controllers; 3, to ultrahigh vacuum pumps; 4, sample; 5, temperature sensor; 6, shutter; 7, liquid nitrogen feedthrough; 8, current/voltage measurement; 9, temperature detector; 10, substrate heating controller; 11, film thickness monitor; 12, LED light; 13, data processing board; 14, computer; 15, thermal insulator; 16, substrate holder; and 17, electrical bridging connector.

and optical illumination will be continually monitored and controlled during the film fabrication and device characterization processes by using a computer through a data board. The application of such systems is further discussed in the section on self-assembled monolayers.

6.2.1.2 Self-Assembly Monolayer

Self-assembly monolayer method is a promising approach for the mass production of organic electronic devices with nanoscale dimension, high reproducibility, and low-cost advantages, which is preferable in the next generation of ultrahigh density information memory applications [5–8]. There are usually two kinds of approaches to make self-assembled monolayers in the field of molecular electronics. One is based on thiolated molecular materials that can be easily self-assembled onto fresh deposited gold surface. The other is to make monolayers on silicon surface with or without thermal dioxide layer through an organosilane chemical process. We first introduce some necessary knowledge about organic solvents that are used in the preparation of molecular monolayers. Then we present the procedures for the growth of two kinds of self-assembled monolayers, i.e., both thiolated molecule-based and organosilane-based monolayers. The common solvents include methanol, ethanol (EtOH), propanol, butanol, acetone (ACE), tetrahydrofuran (THF), acetic acid (HAc), and chloroform. Their formula and physical properties are given in Table 6.1. *Caution! The handling of hydrofluoric acid (HF) and sulfuric acid needs to be done very careful with full personal protection and in well-ventilated chemical hood.*

6.2.1.2.1 Thiolated Material-Based Monolayers

Table 6.2 lists out the schematic procedures to make *n*-alkanethiol SAMs onto gold substrate surface. The key points are substrate clean (if not freshly deposited) and concentration of target

TABLE 6.1
Common Organic Solvents Used for Molecular Monolayer and/or Substrate Treatments

Solvent (Abbr.)	Chemical Formula	Boling Point (°C)	Solubility in 25°C Water	Toxic, Corrosive, and Dangerous Scale
Methanol	CH_3OH	65	Completely	High
Ethanol (EtOH)	CH_3CH_2OH	78	Completely	Low
Propanol	$CH_3CH_2CH_2OH$	97	Completely	Medium
Butanol	$CH_3CH_2CH_2CH_2OH$	118	$0.08\,g/cm^3$	High
Acetone (ACE)	CH_3COCH_3	58	Completely	Low
Acetic acid	CH_3COOH	118	Infinite	High
Chloroform[a]	$CHCl_3$	61.7	Not	Very high
Tetrahydrofuran (THF)	$OCH_2CH_2CH_2$	65.4	Not	Low
Sulfuric acid[a]	H_2SO_4	338	Completely	Very high
Hydrogen peroxide[a]	H_2O_2	152.1	Completely	Very high
Hydrofluoric acid[a]	HF	112.2	$1.14\,g/cm^3$	Extremely hazardous

[a] These solvents need to be handled carefully with full protections from any direct contact with skin, eye, or inhalation.

TABLE 6.2
A Typical Procedure to Make Thiolated Molecular SAM (e.g., 1-Decanethiol, $C_{10}SH$) onto Au Surface

Entry	Action	Note
1	Prepare 1 mM target molecular solution; usually pure ethanol is used as the solvent	Make sure the gold substrate and container have been previously cleaned as the process given before
2	Immerse the Au substrate into the molecular solution and deoxygenate by using nitrogen gas	Keep it in dark for 18–24 h to allow the SAMs growth
3	Take out the sample and rinse with the sequential flow of pure ethanol, DI water, and pure ethanol, respectively	About 60–90 s for each steps
4	Dry with nitrogen gun	60 s
5	Use the sample immediately	Otherwise, keep it in pure ethanol

molecular solution. To make high-quality monolayer, the gold film is always deposited just before the growth of SAMs. One millimole molecular solution is usually an appropriate concentration for the fabrication of n-alkanethiol SAMs. Figure 6.3 shows two schematic molecular junctions made from alkanethiol and alkanedithiol SAM between gold electrodes. It is obvious that the top molecule/electrode interfaces for the two kinds of SAMs are different, which will inevitably affect the device performance.

The experimental conditions and growth procedure for conjugated molecular SAMs are different due to their relatively high rigidity and non-flexible natures comparing to that of the alkanethiol counterparts. In this case, it will be a good idea to make conjugated molecular SAMs with densely packing characteristics out of solution mixed with a few percent of alkanethiols. This has been named as monolayer matrix technique that will be further discussed in the later corresponding section of this chapter.

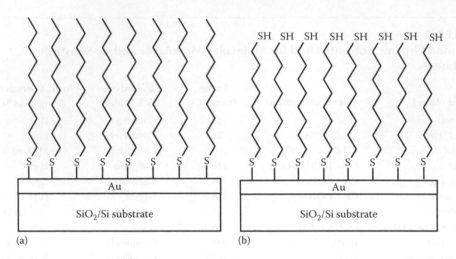

FIGURE 6.3 Schematic SAM of (a) 1-decanethiol and (b) 1-nonanedithiol sandwiched between Au electrodes.

6.2.1.2.2 *Organosilane-Based Molecular Monolayers*

To make silane monolayer, the silicon wafer needs to be carefully cleaned and pretreated with flat dry thermal dioxide surface. The wafer Radio Corporation of America (RCA) clean is a standard set of wafer cleaning steps which needs to be performed before high-temperature processing steps (e.g., oxidation) of silicon wafers in semiconductor manufacturing. Werner Kern developed this basic procedure in 1965 while working for the RCA. It involves the following steps: (1) removal of the organic contaminants (i.e., organic clean); (2) removal of thin oxide layer (i.e., oxide strip); (3) removal of ionic contamination (i.e., ionic clean). The first step is performed with a 1:1:5 solution of $NH_4OH + H_2O_2 + DI \ H_2O$ at 75°C–80°C for 15 min. This treatment results in the formation of about 1 nm thin silicon dioxide layer on the silicon surface, along with a certain degree of metallic contamination (especially iron) that shall be removed in subsequent steps. The second step is a short immersion of the wafer in a 1:50 solution of HF:DI H_2O at 25°C for no more than 10 s, in order to remove the thin oxide layer and some fraction of ionic contaminants. The third step is performed with a 1:1:6 solution of HCl:H_2O_2:DI H_2O at 75°C–80°C for about 10 min. This procedure can effectively remove the remaining traces of metallic ionic contaminants from the silicon wafer surface. The experimental details of RCA clean are summarized in Table 6.3.

After wafer RCA clean, the substrate is now ready for organosilane monolayer growth. Table 6.4 describes a regular experimental procedure to make trichlorosilane monolayer onto Si(100) substrate surface [9–12]. As the thickness for such a monolayer is just a few nanometers, the assembly side of the substrate should be pre-polished and atomic flat. One millimolar molecular solution usually yields a good monolayer result with growth time of 12–24 h under ambient conditions. To avoid overlayers, the sample has to be intensively rinsed with the same solvent as used for the solution and then isopropyl alcohol and di-ionized water. This process is schematically shown in Figure 6.4.

TABLE 6.3
Procedure for Silicon Wafer RCA Clean

Step	Solution and Ratio	Temperature (°C)	Time	Purpose
1	$NH_4OH:H_2O_2:DI \ H_2O = 1:1:5$	75–80	~15 min	Organic clean
2	HF:DI $H_2O = 1:50$	25	~10 s	Remove oxide
3	$HCl + H_2O_2 + DI \ H_2O = 1:1:6$	75–80	~10 min	Clean metallic ionic

TABLE 6.4

Experimental Procedure to Make Organosilane Monolayer (e.g., Trichlorosilane) on Silicon Dioxide Substrate

Step	Action	Note
1	Silicon wafer RCA clean	Refer to the RCA clean table
2	Growth of dry thermal dioxides	800°C, 30 min in N_2 gas to grow several nm silicon dioxide layer
3	Immerse the treated substrate immediately into target molecular solution and deoxygenate with nitrogen gas flow	Dip in 1 mM molecular solution for 12–24 h
4	Rinse the sample thoroughly to avoid overlayers	Using solvents chloroform, isopropyl alcohol, and DI water, respectively (few minutes for each step)
5	Dry with nitrogen gun	1 min
6	Annealing in vacuum	150°C for 10–15 min to improve the monolayers quality and get rid of water and solvent
7	A substituted treatment is to bake the sample under nitrogen environment	75°C–80°C for 10 min, keep the sample covered (e.g., with aluminum foil) during this process

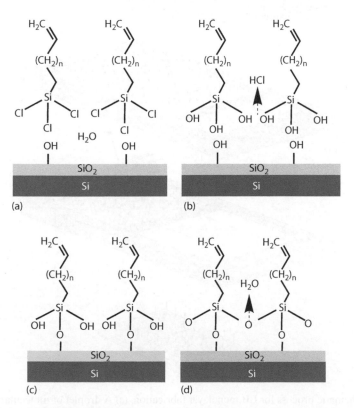

FIGURE 6.4 Schematic self-assembly process for trichlorosilane monolayer formation on silicon dioxide surface. A very thin water layer is a necessary factor in this chemical process. (Adapted from Collet, J. et al., *Microelectron. Eng.*, 36, 119, 1997.)

6.2.1.3 Langmuir–Blodgett Monolayers

To make a Langmuir–Blodgett monolayer, minute droplets of pre-prepared target molecular solution are carefully cast on the pure water surface in the LB trough at intervals of tens of seconds. As shown in Figure 6.5, the hydrophobic chain-ended molecules remained on the surface of water after evaporation of the solvent. The molecules are then compressed by moving the barriers at some speed (e.g., 3–6 mm/min). The surface pressure isotherm has to be recorded during the compression. The target molecules are transferred to the substrate by retracting the substrate, which is pre-immersed in pure water before the molecular solution is spread in the trough. The retracting speed is kept at a speed (e.g., 0.3–0.7 mm/min). The temperature of the pure water in the trough is normally kept at about 20°C–25°C.

6.2.1.4 Vacuum Spray Deposition

Vacuum spray deposition is a new technique for the fabrication of high-molecular-weight polymer thin films with quality better than that of the other approaches. As shown in Figure 6.6, to make a spray film, the polymer predissolved in a suitable solvent is first fed into a glass-bottle with two openings. The concentration of the dye can be varied according to experimental requirements. The

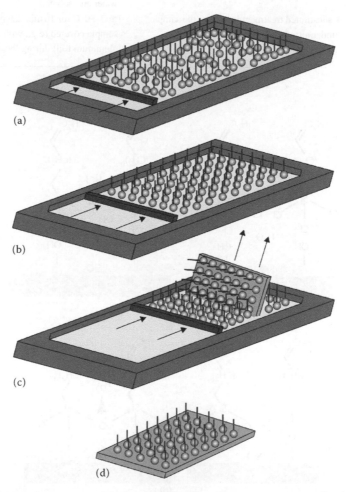

FIGURE 6.5 Schematic process for LB monolayer fabrication. (a) A droplet of molecular solution is spread on the water surface in the trough. (b) The molecules are compressed by slowly moving the barrier bar. (c) The pre-dipped substrate is pulled out of the water with the molecules transferred onto its surface. (d) The as-fabricated sample with a single LB layer.

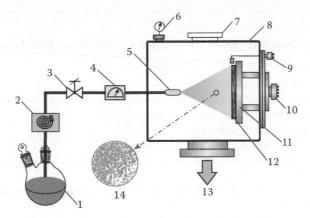

FIGURE 6.6 Schematic diagram of vacuum spray deposition system. 1, Polymer solution; 2, compress pump; 3, valve; 4, flow meter; 5, nozzle; 6, pressure gauge; 7, observation window; 8, vacuum chamber; 9, film thickness monitor; 10, substrate heating electrical feedthrough; 11, substrate holder; 12, sample; 13, to high vacuum pump system; and 14, zoom of the spray.

chamber is vacuumed by using a rotary pump through a liquid nitrogen trap, whose pressure is monitored with a vacuum gauge. A valve separates the spray nozzle specially designed with a small mouth (10–20 μm in diameter) from the high performance liquid chromatography pump in order to control the spray process. A flow counter is used to monitor the spray rate of the polymer solution through the nozzle. The substrate is heated to 137°C–157°C before the spray process. When the polymer solution is introduced into the deposition room through the nozzle, it will be aspirated into extremely fine aerosol consisting of numerous fine droplets with diameters less than 100 nm. The solvent can be evaporated very fast due to the factors of vacuum and high substrate temperature. The polymer is thus adsorbed onto the substrate surface forming a smooth thin film.

6.2.1.5 Spin Coating and Dip Coating
Spin coating and dip coating are some of the other fabrication methods involving the usage of solution. The quality and thickness of the molecular thin film are mainly controlled through the spinning/dipping speed and molecular concentration. In addition, the baking (or annealing) of the sample under a vacuum condition is also a necessary procedure to get rid of the residue solvent. Like that of vacuum spray, these two approaches are suitable for thin film fabrication from high molecular weight organic materials with high melting points, which tend to be thermally decomposed in traditional vacuum vapor deposition process. Unlike that of the vacuum spray method, they are an easy and cheap way with no sophisticated growth systems needed.

6.2.2 DEVICE FABRICATION
Generally, two terminal molecular junctions are the common type of devices used in the investigation of various molecular electronic systems. A molecular junction consists of a molecular active layer connected with two electrodes that can be made from metal or semiconductor materials. To make a molecular device, first we need to prepare a clean substrate before proceeding with experimental procedures such as electrode deposition, active layer growth, junction fabrication, and characterization.

6.2.2.1 Preparations of Substrate and Electrode
The regular conductive substrates used in the researches of molecular electronics include at least doped silicon wafer, highly oriented pyrolitic graphite (HOPG), indium tin oxide (ITO), and metal

TABLE 6.5
Silicon Wafer Clean Procedure for Thermal Oxidation or Thin Film Deposition

Step	Action	Time (s)
1	Place the wafer in a sample holder	N/A
2	Place the holder in a 400 mL beaker with acetone and then set it in ultrasonic bath	180
3	Place the holder in a 400 mL beaker of isopropyl alcohol	N/A
4	Set the beaker in the ultrasonic bath	180
5	Rinse with DI water and blow dry with nitrogen	N/A
6	Clean with *Piranha etch* (*Caution! Keep away from this explosive and corrosive solution*)	N/A
7	Cover wafer and take to HF acid bench (*Caution! Wear safety apron, gloves, and face shield*)	N/A
8	Place sample holder in 400 mL plastic beaker filled with diluted 10:1 solution DI water:HF acid (*Caution! Extremely hazardous, keep away from any skin contact*)	60
9	Place sample holder in DI water cascade and rinse at each stage	120
10	Dry with nitrogen gun and cover wafer	N/A
11	Immediately follow with the next experimental procedure	N/A

film (such as Au, Ag, Al, and Cu) deposited onto mica, glass, or silicon dioxide. In this section, we simply introduce how to prepare various substrates for molecular junction fabrication. Table 6.5 gives another wafer cleaning procedure besides the standard RCA method, where a step involved with HF acid is used to etch the native oxide layer off the silicon wafer.

Mica and HOPG substrates are normally prepared through fresh cleaving just before the relevant experiment like metal electrode deposition. Sometimes glass slide is used, which can be cleaned using the following procedures: (1) ultrasonically wash with detergent plus deionized (DI) water for 3 min; (2) rinse with DI water flow; (3) clean with *Piranha etch*, i.e., dipping into hot H_2SO_4 and DI H_2O_2 mixed solution (volume ratio 4:1) for about 10 min followed by rinsing with water, acetone, and ethanol, respectively.

ITO glass substrates can be cleaned sequentially in ultrasonic baths of acetone and 2-propanol, before oxygen plasma etching under a pressure of 100 Torr at the power of 100 W for about 4–6 min.

6.2.2.2 Molecular Materials

To realize organic molecule-based information storage, chemists have designed various unique molecular materials with specific opto/electro functional groups and units. The first kind of prime molecules is charge transfer materials with electron donor–acceptor groups, which can be bridged with each other by either a σ or a π linker. In some cases, co-deposition is used to fabricate complex thin film out of two different molecular materials with electron donor and acceptor groups, respectively. Since there are many intensive review articles and books on charge transfer materials and devices, it is not included in this chapter.

Dendrimeric molecules with core-shell structure are the second kind of important active materials for application as information memory media. The redox gradient nature of dendrimers enables them preferable for single-molecule charge storage application [13]. Self-assembled monolayers based on thiolated and organosilane molecular materials are the third kind of molecular materials with fantastic properties, which make them as an ideal active media for nanoscale information memory device [14]. With the use of SAM techniques, scientist can easily make high-quality monolayers and even nanoscale patterns on various metal and semiconductor surfaces [6]. Other molecular materials with potential for information memory applications are polymers and LB layers, which will be discussed in the later section.

The molecular structure including functional groups and conformation can greatly affect the device performance and application. Figure 6.7 shows the building units used in molecular design and construction, while Table 6.6 lists out the functional substituents usually encountered in molecular electronics research. This knowledge will be a necessary preparation for later sections when we discuss the molecular and/or device structure–property relationships.

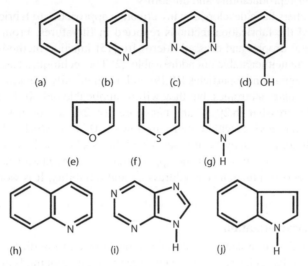

(a)   (b)   (c)   (d) OH

(e)   (f)   (g) H

(h)   (i)   (j)

FIGURE 6.7 Several ring compounds containing nitrogen, oxygen, and sulfur incorporated in heterocyclic system, which are used as building units in the design of molecular electronic materials. The molecules are named as (a) benzene, (b) pyridine, (c) pyrimidine, (d) phenol, (e) furan, (f) thiophene, (g) pyrrole, (h) quinoline, (i) purine, and (j) indole.

TABLE 6.6
Functional Substituents Usually Encountered in Molecular Electronic Materials, Especially in Molecules for Monolayer Devices

Entry	Substituent (Abbr.)	Formula
1	Acid (Ac)	$-COOH$
2	Alcohol	$-OH$
3	Amine	$-NH_2$
4	Ethyne	$-C_2H_2$
5	Ethene	$-C_2H_4$
6	Ethyl (Et)	$-C_2H_5$
7	Methoxy	$-OCH_3$
8	Methylene	$=CH_2$
9	Methyl (Me)	$-CH_3$
10	Nitro	$-NO_2$
11	Nitrile	$-CN$
12	Thiol	$-SH$
13	Thiolacetyl (AcS)	$-SCOOH$
14	Trichlorosilane	$-Si(Cl)_3$
15	Triethoxysilane	$-Si(OC_2H_5)_3$
16	Trimethylsilane	$-Si(CH_3)_3$
17	Vinyl	$-C_2H_3$

6.2.2.3 Approaches for Molecular Junction Fabrication

Single-molecule-scale memory devices and their integrated system are the long-term objects for the intense research in the field of molecular electronics. Experimental and theoretical efforts by numerous scientists and engineers confirmed that it is very promising to realize molecular-scale information memories. One of the biggest obstacles is how to fabricate addressable single-molecular junctions with a higher reproducibility and durability.

To date, the researchers have developed many effective approaches to fabricate MOM junctions. Table 6.7 summarized the fabrication methods reported in literatures. From this table, we learn that (1) the liquid metal droplet and crossed microwire [15] junction methods are simple, but the microscale junctions are neither stable nor addressable. (2) The techniques based on SPM [16], electrodeposited [17], and against nanoparticles [18,19] and mechanically breaking nanowires are powerful in studying molecular electronics, but they will be applicable only if the addressing problems can be solved. (3) The crossbar [6,20,21] and nanopore [22–24] approaches are applicable in the investigation of molecular devices with dimensions down to 0.1 μm, which is limited by the resolution of e-beam and photolithography. (4) It is suggested that SAM-based crossed nanowire junction be the most promising approach to meet the application requirements for nanoscale information memory from the aspects of junction size, addressing and reliability. It is worth more researching attention.

6.2.2.4 Device Characterizations

The electronic information of molecular material can be deduced from direct device electrical performance like current–voltage characteristics. There are two regular methods to measure a molecular

TABLE 6.7
Summary and Comparison of Molecular Junction Fabrication Approaches

Fabrication Approach			Junction Dimension	Addressable	Temperature Variation	Nanoscale
Liquid metal droplet	Hg		50–200 μm	Not	Not	Not
	GaIn					
SPM tip	STM	STM tip	1–5 nm	Not	Not	Yes
		Nanoparticle coupled STM	1–5 nm	Yes	Not	Yes
	CAFM	CAFM tip	5–50 nm	Not	Not	Not
		Nanoparticle coupled CAFM	1–5 nm	Yes	Not	Yes
Crossbar	Magnetically controlled crosswire		0.5–5 μm	Not	Not	Not
	Lithography crossbar Stamp-printing crossbar		2–50 μm	Yes	Yes	Not
	Crossed nanowire		1–5 nm	Yes	Yes	Yes
Etched hole	Etched hole plus nanotubes		2–10 μm	Yes	Yes	Not
	Nanopore		30–50 nm	Yes	Yes	Not
Against nanowire	Electromigrated		1–70 nm	Yes	Yes	Unknown
	Electrodeposited		1–50 nm	Yes	Yes	Unknown
	Mechanically broken		1–50 nm	Not	Not	Unknown
Against nanoparticle	Nanoparticle bridged		1–5 nm	Yes	Yes	Yes

Source: Adapted from Li, J.C. and Y.L. Song, *High Density Data Storage: Principle, Technology, and Materials*, World Scientific Publishing Co., Singapore, 2009.

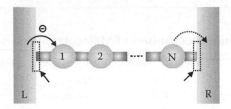

FIGURE 6.8 Schematic structure of a junction with one molecular wire contacted with left (L) and right (R) electrodes. The left and right electrode–molecule interfaces can drastically affect the junction's charge transport and performance. The molecular functional units are labeled from 1 to N.

junction: (1) apply a current and measure voltage, or (2) apply a voltage then measure the current. If the molecular junction has a higher conductance, the first technique generally yields a good result. On the other hand, if the device's resistance is very high, the second method is a better choice.

As shown in the schematic diagram of Figure 6.8, the charge transport performance of a molecular junction depends not on the organic molecule itself but on the molecule/electrode interface(s). The influencing molecule-specific factors include molecular length, functional group/ligand, conformation and electronic structure, etc. The work function values of metal electrodes are also an important element. For regular electrode materials, the work functions of GaIn, Al, Cr, Cu, ITO, Au, and Ni are 4.15, 4.28, 4.5, 4.65, 4.95, 5.1, and 5.15, respectively.

Although there are some significant research results reported, there is still only a poor understanding about the dependence of junction switching performances on molecule-specific properties. To characterize the device performance of a molecular junction, the following factors need to be carefully examined:

1. The substrate effect, i.e., how the substrate property affect the junction performance.
2. The solvent-induced effect, if the junction is investigated under solution.
3. Active layer thickness and morphology effects.
4. Effects of oxygen and/or water in the measuring environments.
5. Electrode/molecule interfacial effects including contacting nature (ohmic or non-ohmic), charge transfer, and electronic energy structure (Table 6.8 lists out the work function values for some metals used in molecular electronics).
6. Influence of optical, temperature, molecular self-heating, and molecule-specific properties (such as length, ligand, and conformation).
7. Finally, the device contamination, variability, and reproducibility also need to be examined.

6.2.3 Switching Mechanisms

To effectively control and enhance the molecular switching behaviors at room temperature, it is required to understand the device dominant switching mechanism from the viewpoint of structure–property relationship. The investigation of the effect of chemical structure/functional groups

TABLE 6.8
Work Function of Metals Used in Molecular Electronics

Electrode	GaIn	Al	Cr	Cu	ITO	Au	Ni
Work function (eV)	4.15	4.28	4.5	4.65	4.95	5.1	5.15

Note: The values are just for clean metal electrodes; it may drastically change upon contact with organic layers.

TABLE 6.9
Comparison of Switching Mechanisms of Molecular Memory Systems

Switching Mechanism	Molecule Intrinsic	Reversible	Transition Time	Stability
Metal filament formation	Not	Not	Long	Bad
Molecule–electrode interface	Not	Sometimes	Short	OK
Charge trapping/detrapping	Not	Yes	Short	OK
Conformation change	Yes	Yes	Short	Good
Charge transfer	Yes	Yes	Shorter	Good
Redox effect	Yes	Yes	Shorter	Better

on the device switching performance can yield a great deal of information about both the charge transport and switching mechanisms. The electrical and/or optical switching mechanism of a molecular junction may result from many factors that at least include (1) metal filament formation, (2) molecule–electrode interface effect, (3) charge transfer effect, (4) charge trapping/detrapping or doping/dedoping effect, (5) redox effect, and (6) molecular conformation change. Table 6.9 presents the comparison result for different molecular device systems operated under these mechanisms. It is indicated that the molecular systems with charge transfer and redox switching performance are promising for nanoscale information memory applications.

6.3 REDOX-MOLECULE-BASED MEMORY

Recently, Blackstock et al. reported the growth of pinhole-free thin films for a series of polyarylamine materials with redox gradient properties [25–27]. Except for their good hole transport characteristics [28–31], one unique aspect of the compounds in this family is their spatial relationship of the redox active moieties, using easy-oxidize group as the molecular "core" and harder-oxidize group as the "shell." The other special feature of this kind of material is their application for multilevel redox state information memory [32–34]. We investigated the fabrication of organic diodes and conduction properties of these materials in order to evaluate the correlation between the molecule-specific properties and the device performance [7,13,14]. We also introduce a simple method to directly measure the rectifying barrier between the molecular thin films and a removable GaIn liquid cathode. This simple approach can not only prevent the damage to the organic active layers, but also has the advantage of reproducibly measuring devices with active layer as thin as 50 nm thickness.

6.3.1 CHARGE CARRIER INJECTION AND TRANSPORT

Carrier injection into a traditional inorganic semiconductor is usually described by using either Fowler–Nordheim tunneling or Richardson–Schottky thermionic emission theories. The average mean free path of carriers in organic semiconductor is of the order of the molecular scale, which is far less than that of the inorganic one. Therefore, the charge carrier injection and transport in organic solids are more difficult than that in regular semiconductors. The carriers in organic solids have to overcome random energy barriers caused by disorder like polycrystal boundaries. However, we can roughly use the normal semiconductor theories to analyze the experimental results of organic thin-film devices. As we discussed in the later section of this chapter, the following theoretical models and equations are usually used in analyzing the I–V curves of molecular junctions with an active layer of either organic thin film or molecular monolayer.

Richardson–Schottky thermionic emission is based on the lowering of the image charge potential by the external electrical field. In this model, the current density, J, is given as a function of the external field, F, in the form of

$$J = A * T^2 \exp\left(-\frac{\phi_B - \beta\sqrt{F}}{k_B T}\right) \tag{6.1}$$

with the Richardson constant $A* = 4\pi q m * k_B^2/h^3$, constant $\beta = \sqrt{q^3/4\pi\varepsilon\varepsilon_0}$, and the zero-field injection barrier ϕ_B. While q is elementary charge, $m*$ effective electron mass, k_B Boltzmann's constant, h Planck's constant, T temperature, ε relative dielectric constant, and ε_0 vacuum permittivity.

Fowler–Nordheim tunneling mechanism ignores Coulombic effects and considers mere tunneling through a triangular barrier into continuum states. In this model, the current density is given by

$$J = \frac{A* q^2 F^2}{\phi_B \alpha^2 k_B^2} \exp\left(-\frac{2\alpha\phi_B^{3/2}}{3qF}\right) \tag{6.2}$$

with constant $\alpha = \left(4\pi\sqrt{2m*}\right)/h$. Unlike that of the Richardson–Schottky theory, there is no temperature dependence of the current density in this model. For the charge carrier transport in a thin-film-based molecular junction, the conductance can be simply regarded as either space charge limited or injection limited. The readers can refer to some good review articles for further details.

6.3.2 Redox Materials, Thin Film, and Property

Figure 6.9 shows the molecular structures of 16 molecular compounds studied for redox dendrimeric family. Molecules **a** and **j–o** are designed in the lab of Prof. Blackstock, while the synthesis of **k**, **l**, and **n** has been previously reported [25–27]. Based on the molecular structure, the materials could be intentionally divided into three groups (1st group **a–c**, 2nd group **d–i**, and 3rd group **j–p**, respectively). Note that the moderate size and dendritic structures are the common characters for the molecules of the third group. As discussed later, this classification would be helpful in understanding the correspondence between the molecular structure and the device performance.

Rectification was observed in the organic diodes made from each molecular material. Figure 6.10 shows the representative *I–V* curves for six materials. Inset shows the schematic device structure and the measuring system used. Forward bias corresponds to holes emitted from the Ag anode and collected by the GaIn cathode. Negligible leakage current was measured below the turn-on voltage. Above the turn-on voltage, the current shows nonlinear behavior. A preliminary analysis of these characteristics can be described by the space charge limiting conduction model.

Typical tapping-mode AFM images of several representative molecular thin films deposited on Ag surface are presented in Figure 6.11. We found that the film morphology and device performance are closely correlated with the molecular chemical structures. Briefly, the small molecular materials, with symmetry structures, tend to form rough film (molecules **a–c**) or smooth films either with lots of pinholes (molecules **e**, **f**, and **i**) or with loose-packed film morphology (molecules **d**, **g**, and **h**). Consequently, their devices show very poor performance as indicated by the low good device rates. In contrast, the films deposited from the materials with moderate molecular size and dendrimeric structures, usually show both smooth and close-packed characters. As a result, most of these devices exhibit a very high good device rate. These results indicate that the morphology effect induced by the molecular structure plays an important role in the single-carrier organic diodes [13]. The data is summarized in Table 6.10.

It is interesting to compare our results with those of the multilayer devices reported in literatures. For example, molecule **p** exhibits high durability in multilayer devices, which has a good device rate in our diodes. Contrarily, the molecules (**b**, **c**, **g**, and **i**), with poor good device rate in our case, show very low durability in the multilayer devices. Therefore, our results are well consonant with that of

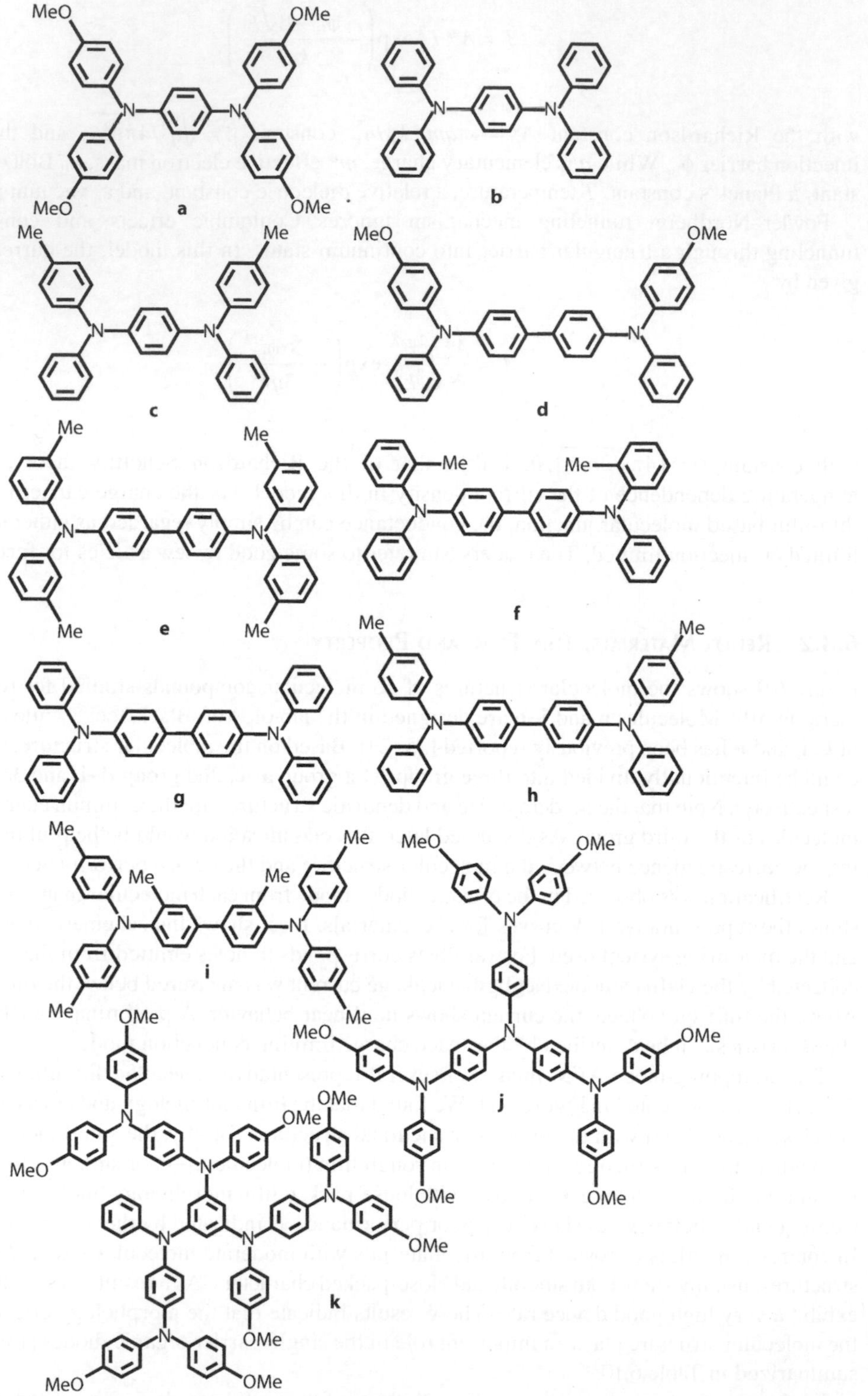

FIGURE 6.9 Chemical structure of a series of redox arylamine molecular materials. The first group includes molecules **a**, **b**, and **c**, the second group are molecules **d–i**, while the left redox arylamine molecules are classified as the third group. Here, molecule **i** is widely known as TPD, a hole transport material used in organic LEDs.

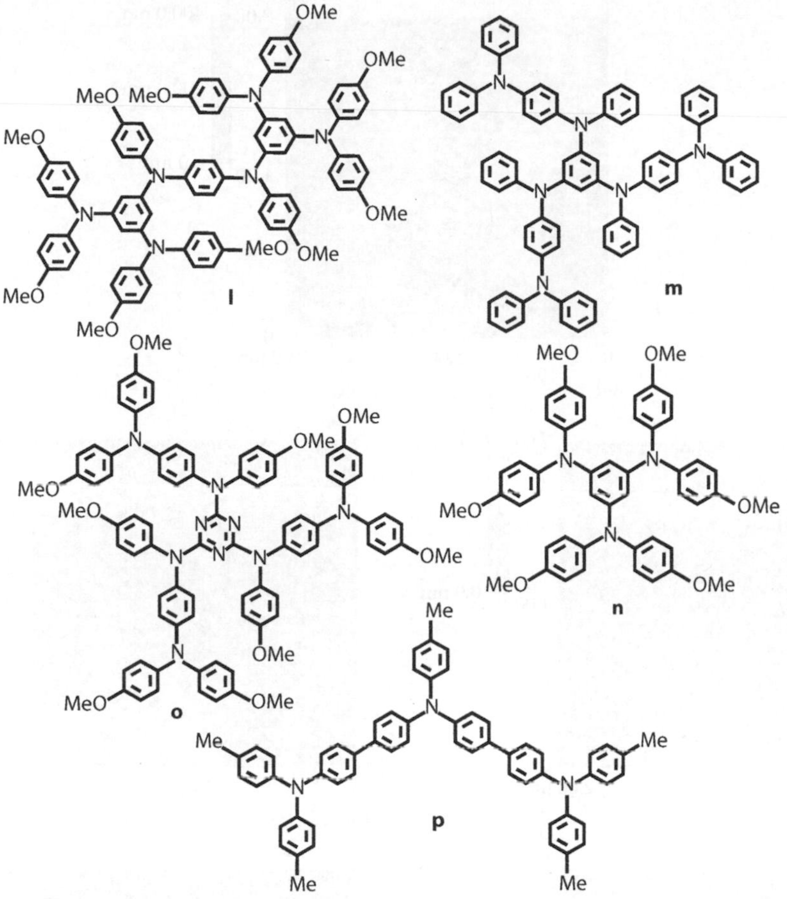

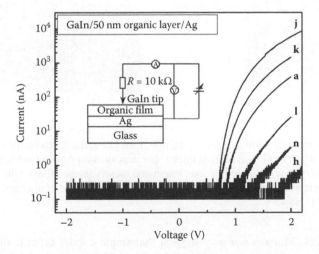

FIGURE 6.9 (continued)

FIGURE 6.10 Representative *I–V* curves for six typical materials. It is clear that the turn-on voltage of different molecular thin films are drastically different although the active layer was kept in the same thickness. Inset shows the schematic device structure and measuring system used. (From Li, J.C. et al., *Chem. Mater.*, 16, 4711, 2004.)

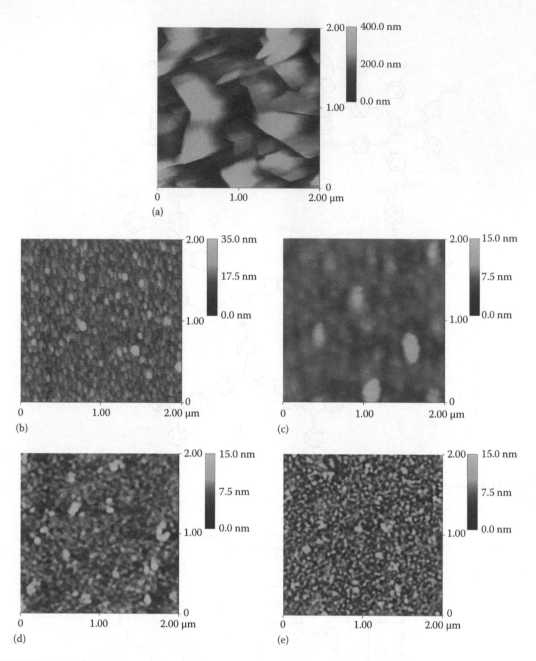

FIGURE 6.11 Tapping mode AFM images for the thin films of Figure 6.9 molecules **a**, **h**, **n**, **k**, and **l** (from top to down), respectively. The films are 50 nm thick vapor deposited on Ag electrodes. Low molecular weight materials with symmetry structures tend to form rough and loosely packed films, while moderate molecular weight and dendrimeric materials usually show smooth and close-packed characters. (From Li, J.C. et al., *Chem. Mater.*, 16, 4711, 2004.)

the multilayer devices. Moreover, it may suggest that single-carrier organic diodes might provide a simple way to screen molecular materials for the study of practical multilayer devices. Several large compounds, with higher dendrimeric degree and bigger molecular size than that of molecule **k**, have been intentionally synthesized and measured. No reasonable results could be obtained due to material thermal decomposition during the thermal evaporation process.

TABLE 6.10

Comparison of the Molecular First Oxidation Potential, Thin Film Morphology, and Electrical Performance of GaIn/50nm Redox Film/Ag Junctions

Molecule	First Oxidation Potential (V vs. SCE)	Turn-On Voltage (V_t)	Mean Roughness (nm)	Good Junction Percent (%)
a	0.46	0.79 ± 0.09	75	Poor (<10)
b	0.56	N/A	80	Very poor
c	0.60	N/A	35	Very poor
d	0.76	N/A	1.2	Very poor
e	0.80	N/A	2.1	Very poor
f	0.81	N/A	3.9	Very poor
g	0.83	N/A	1.1	Very poor
h	0.85	1.80 ± 0.20	1.8	Poor (<25)
i	0.9	N/A	1.0	Very poor
j	0.32	0.67 ± 0.09	1.7	80
k	0.45	0.71 ± 0.06	1.0	95
l	0.49	1.03 ± 0.11	1.4	85
m	0.64	1.52 ± 0.17	1.9	30
n	0.67	1.36 ± 0.09	1.1	85
o	0.71	1.53 ± 0.06	1.2	70
p	0.74	1.14 ± 0.06	1.8	90

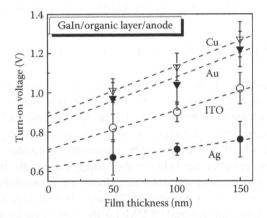

FIGURE 6.12   Effect of anode metal work function on the device turn-on voltage for molecule **k**.

The influence of anode work function on the device electrical performance was carefully investigated. As shown in Figure 6.12, the electrical response of molecule **k**, exhibiting the best performance in these molecules, was characterized with using Ag, Au, Cu, and ITO anodes, respectively. No apparent relationship between the anodes and the good device rate is observed. Independent of anode metals, the thickness dependence of the device turn-on voltage shows linear behavior too. The ordinate intercepts of the turn-on voltage at zero film thickness just agree with the work function difference between the GaIn cathode and the corresponding anodes. This result confirms that there are no evident interfacial effects in our single-carrier organic devices.

The junction turn-on voltage (V_t) can be described as $V_t = \Delta W + kd$. Here, ΔW is the work function difference between the anode and cathode; k is a constant related to the molecular properties; and d

is the thickness of the organic layer. In this formula, the first term represents the built-in potential, the second part is the voltage reduction consumed in the organic layer. When both electrodes and organic layer thickness are fixed, turn-on voltage could be tuned by changing the value of k. This result clearly demonstrates how the molecular properties could play a unique role in tuning the device turn-on voltage. On the contrary, when both the electrodes and molecular material are selected, turn-on voltage can only be lowered by reducing the layer thickness d. However, this will result in a high leakage current and low quantum efficiency.

6.3.3 INTERFACES BETWEEN REDOX MOLECULAR FILM AND ELECTRODE

The interfaces between metals and organic molecular thin films play a crucial role in the electrical and/or optical performance of molecular electronics, which include organic light emitting diodes (LEDs), thin-film transistors, solar cells, sensors, and information memory devices. The metal/organic interfacial properties may be affected by various factors such as (1) molecular structure and functional groups, (2) metal work function, (3) approaches of thin film and metal electrode deposition, (4) interface morphology, and (5) charge transfer and buffer layers.

We previously studied the interface between arylamine thin films and various metal electrodes [14]. As shown in Figure 6.13, it was found that there exist a charge transfer (CT) interface between metal electrode and arylamine molecular thin film when the molecule has an electron acceptor group like –CN. Moreover, the CT interface can be enhanced if the metal surface is premodified with a SAM of alkanethiol molecules, suggesting there exists a new CT interface between metal and SAM. The CT interface can be controlled by inserting a buffer layer between the metal and electron acceptor film. As a buffer layer material, the molecule should have a very weak interaction with the metal electrode. The interfacial interactions between metal electrodes and organic materials with various functional groups usually follow the next, from weak to strong, sequence [35–45]:

$$-OCH_3 < -OH < -CH_3 < -Br < -CF_3 < -Cl < -CN < -F \qquad (6.3)$$

Kahn et al. reviewed the interface between metal electrode and organic thin films for a series of conjugated molecular materials with similar properties [46]. The chemical structure of the molecules is shown in Figure 6.14, which are well-studied materials used in organic LEDs as charge transporting materials. The ionization energy (IE) for these materials are schematically shown in Figure 6.15 to make a clear comparison. The IE was measured by ultraviolet photoemission spectroscopy (UPS), which is defined as the energy difference between the leading edge of the highest occupied molecular orbital (HOMO) and the vacuum level obtained from the photoemission cutoff. Figure 6.16 shows the comparison among metal work function, IE, and electron affinity (EA) (i.e., HOMO and LUMO) positions of the molecular materials. LUMO is the molecular lowest unoccupied molecule orbital. The "energy zero" is defined as the vacuum level. The IE and EA are determined by UPS and inverse photoemission spectroscopy (IPES). These data are very useful in studying the organic/metal interfaces such as charge transfer, dipole, work function changes, and energy barrier, etc. [38–45].

6.3.4 MOLECULAR SWITCHING AND MEMORY

Switchable redox molecular materials have drawn much researching attention owing to their special core-shell structure and multimode redox properties. Electrical switching and memory behavior was recently observed in the experiment of redox dendrimer sandwiched junctions with the structure of Ag/barrier layer/active layer/barrier layer/Ag [7]. Here, the barrier layers are molecular thin films that were vacuum vapor deposited from the molecule **i**, while the active

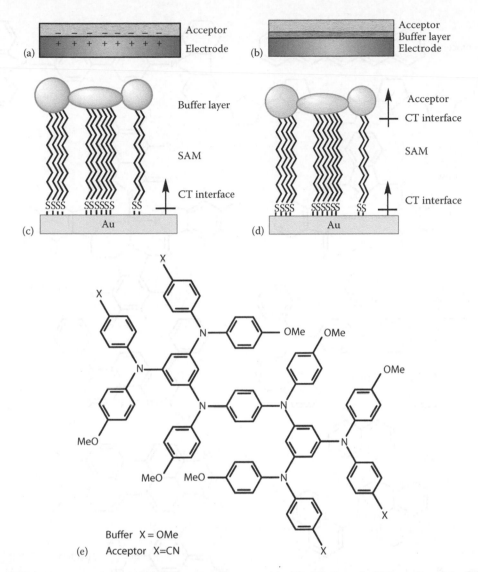

FIGURE 6.13 Schematic drawings show the charge transfer (CT) interfaces between redox molecular thin films with electron acceptor group (–CN) on substrates: (a) without buffer layer, (b) with buffer layer, (c) buffer layer on SAM, and (d) acceptor film on SAM. Buffer layer means the molecular thin films made from molecules without electron acceptor group. The molecular structure of the two kinds of materials is also shown.

layer was deposited from molecule **1** (see Figure 6.9 for molecular structures). The molecular junctions were fabricated through a shadow mask with a dimension of $1 \, mm^2$. Control experiments confirmed that the switching mechanism is due to charge trapping in the redox molecules sandwiched by more insulated barriers. The switching threshold voltage (0.3–0.6 V) and on/off ratio (as high as 1000) depend on both the thickness and the ratio of the active and barrier layers. This work demonstrates the possibility of using solid-state redox thin films as information memory media.

As shown in Figure 6.17b, Wakayama et al. [47] observed optical switching in molecular multiplayer device with the structure of Au/porphyrin-based molecules/Si (100), where the organic molecules are inserted into silicon oxide matrix layer. Their experiment was conducted under ultrahigh vacuum conditions. The molecules of porphyrin derivative, tetrakis-3,5, di-butylphenyl-porphyrin,

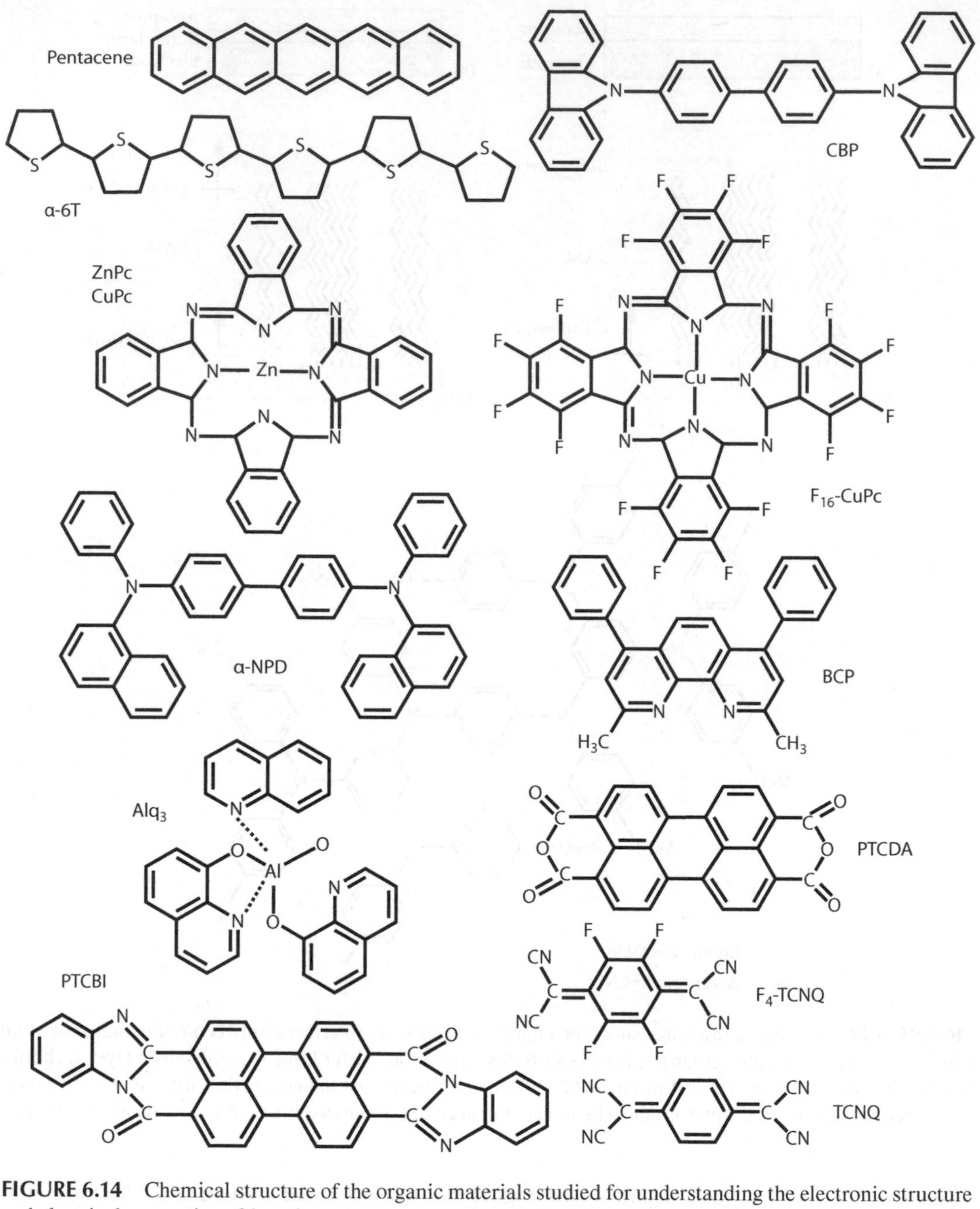

FIGURE 6.14 Chemical structure of the organic materials studied for understanding the electronic structure and electrical properties of interfaces between metals and pi-conjugated molecular thin films. (From Kahn, A. et al., *J. Polym. Sci. B: Polym. Phys.*, 41, 2529, 2003.)

were deposited from an effusion cell and were sandwiched between two silicon dioxide layers (with total thickness of 5 nm). The current–voltage characteristics were measured at a low temperature of 5 K. The light wavelength is 430 nm with intensity of $13\,\mu W/mm^2$ irradiated from a Xe lamp. The top gold electrode was deposited through shadow mask with a thickness of 12 nm and a transparency of 60%. The results indicated that the device current–voltage curves behaved obviously like a Coulomb staircase originating from single-electron tunneling. The Coulomb staircase can be

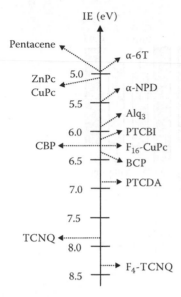

FIGURE 6.15 Comparison of the molecular ionization energy values of organic materials measured by UPS. Here, IE is defined as the energy difference between the leading edge of the HOMO and the vacuum level obtained from the photoemission cutoff. The chemical structure of the molecules is given in Figure 6.14. (From Kahn, A. et al., *J. Polym. Sci. B: Polym. Phys.*, 41, 2529, 2003.)

reversibly switched on/off by a threshold voltage of 300 mV under optical illumination light/dark conditions. It was suggested that the possible switching mechanism is photoinduced charge carriers trapping/detrapping at energy levels at the Si–SiO$_2$ interface. Such interfacial change will influence the junction tunneling parameters and thus result in optical switching. Liu et al. also reported their studies on the charge trapping/detrapping of photoconductive zinc porphyrin film [48].

In a recent work [49], Chotsuwan and Blackstock reported on the observation of charge switching of small domains of polyarylamine isolated in polymethylmethacrylate (PMMA) through co-spin coating on SiO$_2$/Si substrate (see Figure 6.17c). The experiment was conducted under ambient conditions by using ambient conducting atomic force microscopy/Kelvin probe microscopy writing–reading technique. The thickness of the molecule mixed with PMMA is about 3 nm, while the thermal oxide layer of silicon is 25 nm. The device structure is Au-tip/molecules in PMMA/SiO$_2$-Si that is very similar to that of Figure 6.17b. It was shown that the thin layer could be charged with a threshold voltage of 6 V. This value is much larger than that of the molecular first oxidation potential measured from cyclic voltammetry. The higher the amine content of the mixed film is, the easier the film can be charged. Increasing the thickness of silicon dioxide layer can decrease the surface potential decay rate, suggesting the reduction of the molecular charge by electrons from the silicon substrate. Moreover, they observed stepped discharge behavior, indicating that multimode redox switching happened in the experimental process. However, this method has the disadvantages of (1) unknown distribution of the redox molecules in the PMMA and (2) difficulty to realize sub-10 nm scale information storage.

A molecular approach was described for information storage application that involves the usage of porphyrin derivative monolayer self-assembled on Au microelectrode as the memory storage element [50]. As shown in Figure 6.17d, the experiments were conducted in dried, distilled CH$_2$Cl$_2$ containing 0.1 M Bu$_4$NPF$_6$ on a two-electrode potentialstat with a 5 MHz bandwidth. A set of four zinc prophyrins were examined, with each molecule bearing three mesityl groups and one S-acetylthio-derivatized linker with the structure of 1-[AcS-(CH$_2$)$_n$]-4-phenylene ($n = 0$, 1, 2, or 3). It was shown that multimode information storage could be realized through multiple oxidation states (neutral, monocation, and dication) of the porphyrin molecules. The charge retention time is in the scale of hundreds of seconds and the redox process can be cycled thousands of times under ambient conditions. However, this approach involves the use of electrolyte and organic solvent, which is unfavorable for the practical applications of molecular-scale information storage.

To avoid the problems mentioned above, a promising STM tip-based information storage approach based on redox molecular monolayer with multilevel memory states has been proposed [51,52]. This method can easily reach a resolution of sub-5 nm scale. Experiments on similar molecular materials had been extensively used [53–55]. Moreover, we can easily add some special groups or ligands on the target molecules such as azobenzene [56–58], spiropyran/spirooxazine [59], and cyclohexadiene [60]. These functional groups will enable the molecules with optoelectronic switching properties that is very crucial for application in devices such as photodetector, photovoltatics, and optically gated single-molecule transistors.

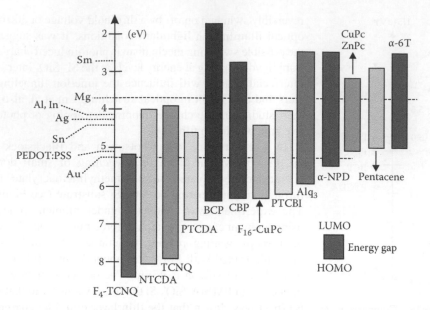

FIGURE 6.16 Comparison between metal work function, IE, and electron affinity (EA) (i.e., HOMO and LUMO) positions of various molecular materials. The zero is defined as the vacuum level. The IE and EA are determined by UPS and inverse photoemission spectroscopy. (From Kahn, A. et al., *J. Polym. Sci. B: Polym. Phys.*, 41, 2529, 2003.)

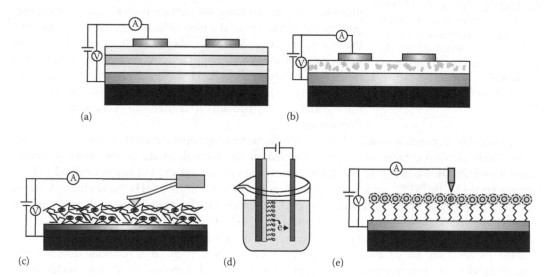

FIGURE 6.17 Schematic experimental setups for charge storage or electrical switching on redox dendrimeric thin films or monolayers. (a) Molecular junction of Ag/TPD/4AAPD/TPD/Ag [7], (b) Au/molecule in SiO_2/Au [47], (c) Au-tip/molecules in PMMA/SiO_2/Si [49], (d) Au/SAM/electrolyte [50], and (e) STM tip/molecule/Au [53–55].

6.4 SELF-ASSEMBLY MONOLAYER-BASED MEMORY

In general, to realize molecular-scale data storage, we need to thoroughly understand the structure–property relationships of the molecule, the active layer, and the device. The goal of this section is to address the following questions: (1) What kinds of SAMs molecular materials are suitable for nanoscale memory applications? (2) How such monolayers can be patterned? (3) What is the structure–property relationship? (4) How can their information storage characteristics be greatly enhanced to function at room temperature?

6.4.1   SELF-ASSEMBLY MONOLAYER AND PROPERTY

For self-assembled monolayer-based single-molecular junctions, the charge transport properties and thus device performance may be drastically modified by factors of molecular structure or ligand [61], light illumination, thermal effect, and stress [62]. In fact, the electron tunneling in a single-molecule junction can be drastically varied even after a very small change of the molecule from the aspects of charge transfer, vibration, rotation, defects, and/or conformation [63]. These effects must be taken into account in the investigation of molecular electronics, especially in the design and study of nanoscale memory devices [64].

6.4.1.1   Charge Transport in Molecular Monolayer Junctions

To discuss the charge transport mechanisms of monolayer junctions, Simmon's model is the simplest equation used to analyze the electron tunneling behavior through a rectangular barrier in the metal–insulator–metal junctions [65]. The equation of Simmon's model is

$$I = \frac{Ce}{4\pi^2 \hbar d^2} \left\{ \left(\phi - \frac{eV}{2} \right) \exp\left[-\frac{2(2m)^{1/2}}{\hbar} \alpha \left(\phi - \frac{eV}{2} \right)^{1/2} d \right] - \left(\phi + \frac{eV}{2} \right) \exp\left[-\frac{2(2m)^{1/2}}{\hbar} \alpha \left(\phi + \frac{eV}{2} \right)^{1/2} d \right] \right\}$$

(6.4)

where

- e is the charge of an electron
- $\hbar$ is Planck's constant divided by 2π
- d is the tunneling distance (or wire length in the case of molecular junctions)
- ϕ is the barrier height
- V is the bias voltage applied between the electrodes
- m is the mass of an electron
- C is the proportionality constant
- α is a unitless adjustable parameter used in fitting

At lower bias voltage [66], the above equation can be approximated to $I = I_0 \exp(-\beta d)$ with conductance decay constant, β, described as

$$\beta = \frac{2(2m)^{1/2}}{\hbar} \alpha(\phi)^{1/2}$$

(6.5)

This simplification has been widely used in characterizing molecular monolayers with the contacting atomic force microscopy (CAFM) and the scanning tunneling microscopy (STM) tip-based molecular junction techniques. The decay constant, β, can serve as an indicator about the electrical properties of the molecules in the MOM junction. Keep in mind that, unlike that of the molecular junction in CAFM approach, there exists a small air gap between the STM tip and the molecular monolayer in the STM tip-based method [67]. Although the current–voltage characteristics of Simmon's model is independent of temperature, the situation may be extremely complicated in practical junctions, especially in high bias voltage regions [68].

Datta et al. calculated the current through the molecular junction of a STM tip/α,α′-xylyl dithiol molecule/Au substrate by using the standard scattering theory of charge transport [69]. After comparing with the experimental results, Datta et al. pointed out that we need to develop a more complete theory that includes other details such as the correct self-consistent potential profile or the structure in the density of states in the contacts. Later, Gonzalez et al. modeled the influence of bridge electronic defects (site substitutions) and weak links (local weak bonds) on the potential profile of the molecular junction [70]. The potential is determined self-consistently by solving Poisson and Schrödinger equations simultaneously. Their calculated results show some asymmetry and rectification effects as that observed in real chemically modified molecular junctions. Lehmann et al. theoretically investigated the incoherent charge transport through molecular wire junctions with the presence of Coulomb interactions [71]. The current for spinless electrons is determined in the limit of strong Coulomb repulsion. It is indicated that the voltage profile along the molecular wire crucially influences the dependence of the current on the wire length. Blocking effect is found upon the inclusion of the spin degree, which depends both on the interaction strength and on the number of wires contributing to the current.

These works further remind us again that the charge transport of molecular junctions is very complicated due to many factors, such as metal/molecule contacts, substitutions, chemical conformations, and external effects. These will be further discussed in the next experimental sections.

6.4.1.2 Thiolated Self-Assembly Monolayers

Salomon et al. reviewed the electron transport measurements on organic molecules self-assembled gold surface [72]. Table 6.11 summarizes the comparison results of sigma-bonded alkanethiol molecules investigated by using different junction approaches. Obviously, the current density values are drastically different for the same molecular structure with various investigation methods and/or under different measuring conditions. The SPM tip-based results are usually higher orders of magnitude than that of the others. It is indicated that the less molecules within the junction, the higher of the current through per-molecule at the given bias voltage. This can be ascribed to the exclusion of the intermolecular interactions in these measurements because of the nanoscale size and the limited number of molecules within the junctions. This table highlights the importance of the chemically well-defined contacts and controlled numbers of molecule of the molecular junction.

The charge transport through pi-bonded conjugated molecular wires were also compared and summarized in Table 6.12 [72]. The chemical structure of the molecules are shown in Figure 6.18. The absolute current per molecule of a given length at a given bias—higher currents were obtained experimentally for conjugated pi-bonded molecules than for saturated ones with comparable length. Such observations agreed with that of the theoretically calculated results. From this table, it is also suggested that the electrode–molecule interface, the molecular conformation, and the molecular energy gap between HOMO and LUMO all play very critical roles for the charge transport of the molecular junctions.

The fundamental investigations of the correlation between molecular structure/conformation and the electronic conductance of single molecules are crucial for a deep understanding of the charge transport process, device operating mechanisms, and further development of molecular electronics. Molecules based on oligothiophenes, phyenylacetylenes, and phenylvinylnes have drawn much research interest owing to their good charge transport and chemical properties than that of other thiolated materials. Moth-Poulsen et al. [73] systematically studied a series of single thiol end-capped oligo-phenylenevinylenes (OPVs) molecules by using an STM plus n-alkanethiol matrix technique. According to the theoretical model described in a previous section, they calculated the decay constant, β, values for those molecules. Their results revealed that a big change in the electronic transparency of the various OPV derivatives due to the insertion of a methylene spacer group or nitro substituent. However, changes in the conjugation path through the central benzene ring from para to meta substitution does not affect the molecular electronic transmission too much. The experimental data is summarized in Table 6.13. Figure 6.19 shows the corresponding molecular structures.

Furthermore, other studies confirmed that the molecular conformation variation may affect the conductance of molecular junction to a large extent. For example, Venkataraman et al. [77] fabricated single-molecular junction by using the approach of breaking Au point contacts in the target

TABLE 6.11

Comparison of Current per Molecule (I) through Saturated Sigma-Bonded Single Molecules and Monomolecular Layers Sandwiched between Two Electrodes Measured by a Variety of Different Experimental Approaches at Two Different Voltages (0.5 and 0.2 V)

Entry	Junction	I at 0.5V (pA)	I at 0.2V (pA)	Gap (eV)	Area (No. of Mol.)	Length (Å)	Method (Force)
1	Au/Vacuum/Au	N/A	N/A	N/A	$0.2\,nm^2$	16	Simmon's model
2	Au-S-C8/Au	30	12	~7	$25\,nm^2$ (~100)	11	CAFM (2nN)
3	Au-S-C8/Au	0.035	0.013	~7	$10\,nm^2$ (~40)	11	CAFM (6nN in solvent)
4	Au-S-C8-S-Au	1400	310	~7	Single	12	CAFM (6nN in solvent)
5	Au-S-C8-S-Au	1300	520	~7	$25\,nm^2$ (~100)	12	CAFM (2nN)
6	Au-S-C8-S-Au	15000	2500	~7	Single	12	Pico-STM
7	Au-S-C10/Au	0.007	0.0002	~7	$10\,nm^2$ (~40)	14	CAFM (6nN in solvent)
8	Au-S-C10/Au	5.0	2.0	~7	$25\,nm^2$ (~100)	14	CAFM (2nN)
9	Hg-S-C10/p-Si	N/A	6	~7	$0.002\,cm^2$ (~1012)	14	Hg droplet
10	n-Si-C10/Hg	N/A	2	~7	N/A	13	Hg droplet
11	Au-S-C10-S-Au	800	200	~7	Single	14	CAFM (6nN in solvent)
12	Au-S-C10-S-Au	1200	370	~7	Single	14	Pico-STM
13	Au-S-C12/Au	0.2	0.1	7	$1600\,nm^2$ (~6400)	16	Thermal deposition (nanopore)
14	Au-S-C12/Au	0.5	0.2	7	$25\,nm^2$ (~100)	16	CAFM (2nN)
15	Hg-S-C12/p-Si	N/A	0.6	~7	$0.002\,cm^2$ (~1012)	16	Hg droplet
16	n-Si-C12/Hg	N/A	0.5	~7	N/A	15	Hg droplet
17	Au-S-C12/(Pt/Ir)	0.55	N/A	~7	$0.25\,nm^2$	16	STS
18	Au-S-C12-S-Au	40	10	~7	$250\,nm^2$ (~1000)	17	Crossed wires

Source: Adapted from Salomon, B.A. et al., *Adv. Mater.*, 15, 1881, 2003.

Note: In the table, "gap" represents the molecular HOMO–LUMO energy gap and "length" is the calculated molecular length.

TABLE 6.12

Comparison of Currents through Conjugated pi-Bonded Single Molecules and Monolayers Sandwiched between Two Electrodes Measured by a Variety of Different Experimental Approaches at Two Different Voltages

Entry	Junction	Molecule	I at 0.5V (pA)	I at 0.2V (pA)	Gap (eV)	Area (No. of Mol.)	Length (Å)	Method (Force)
1	Au-S-phenyl-S-Au	a	1300	300	~5	Single	6	Mechanical break
2	Au-S-phenyl-S/(Pt/Ir)	a	110	30	~5	0.25 nm² single	6	STS
3	Au-S-C-phenyl-C-S/(Pt/Ir)	b	560	130	~5	0.25 nm² single	8	STS
4	Au-S-biphenyl-S-Au	c	36	4	~5	Single	11	CAFM (12 nN)
5	Au-S-biphenyl/Ti	d	300	7	~5	700 nm² (2800)	10	Thermal deposition (nanopore)
6	Au-S-molecules/Au	e	N/A	500	~5	25 nm² (~100)	11	CAFM (2 nN)
7	Au-S-terphenyl/Au	f	N/A	130	~5	25 nm² (~100)	15	CAFM (2 nN)
8	Au-S-OPV-S-Au	g	1000	500	3.1	250 nm² (~1000)	19	Crossed wires
9	Au-S-OPE-S-Au	h	500	200	3.5	250 nm² (~1000)	20	Crossed wires
10	Au-S-OPE-S-Au	h	10	4	3.5	Single	20	CAFM (6 nN in solvent)
11	Au-S-molecule-S-Au	i	14×104	4×104	3.5	Single	20	Mechanical break
12	Au-S-molecule-S-Au	i	3×104	2500	3.3	Single	20	Electromigrated
13	Au-S-caroteno dithiol-S-Au	j	100	40	2.4	Single	32	CAFM (6 nN in solvent)

Source: Adapted from Salomon, B.A. et al., *Adv. Mater.*, 15, 1881, 2003.

Note: In this table, "gap" represents the molecular HOMO–LUMO energy gap, and "length" is the calculated molecular length. The molecular structure is shown in Figure 6.19.

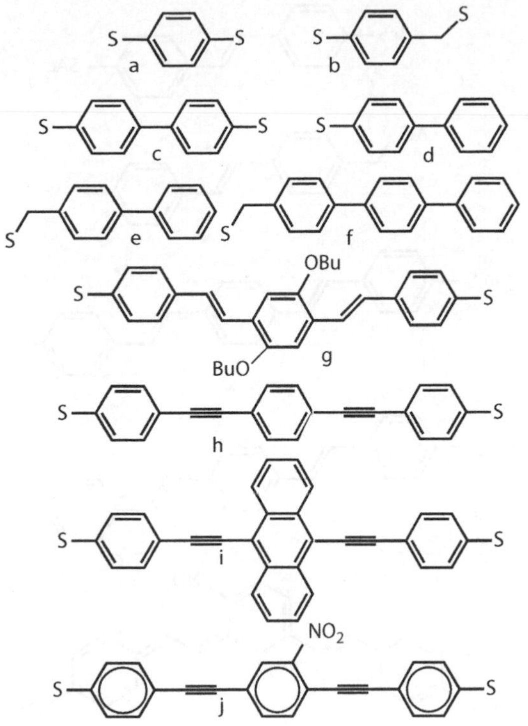

FIGURE 6.18 Chemical structure of the molecules listed in Table 6.12.

TABLE 6.13
Comparison of Molecular Length, Apparent Height, and Decay Constant, β, Values for a Series of Molecular Materials Experimentally Measured by Using Scanning Probe Microscopy Techniques

Entry	Molecule	Molecule Length (Å)	Apparent Height (Å)	β (Å)	Approach
1	Oligophenylene	N/A	N/A	0.42 ± 0.07	CAFM (~2 nN)
2	a	13.9	4.0 ± 0.7	0.53 ± 0.12	STM
3	b	14.2	3.5 ± 1.0	0.65 ± 0.16	STM
4	c	19.5	7.5 ± 1.1	0.63 ± 0.13	STM
5	d	19.4	6.1 ± 2.0	0.78 ± 0.23	STM
6	e	16.5	4.1 ± 1.6	0.80 ± 0.22	STM
7	f	15.9	4.2 ± 0.9	0.72 ± 0.13	STM
8	g	13.7	2.0 ± 0.9	0.84 ± 0.15	STM
9	h	19.3	4.70 ± 1.01	0.94 ± 0.12	STM
10	i	19.3	4.30 ± 0.54	0.99 ± 0.06	STM
11	j	14	0.0	1.2	STM
12	k	12	N/A	1.1	CAFM (2 nN)
13	n-alkanedithiols	N/A	N/A	0.53 ± 0.03	CAFM (2 nN)
14	n-alkanethiols	N/A	N/A	0.94 ± 0.6	CAFM (2 nN)
15	n-C$_4$SMe	N/A	N/A	0.86 ± 0.03	STM
16	n-C$_4$NH$_2$	N/A	N/A	0.88 ± 0.02	STM
17	n-C$_4$PMe$_2$	N/A	N/A	0.97 ± 0.02	STM

Source: Adapted from Moth-Poulsen, K. et al., *Nano Lett.*, 5, 783, 2005.
Note: Molecular structures are given in Figure 6.19 [37,73–76].

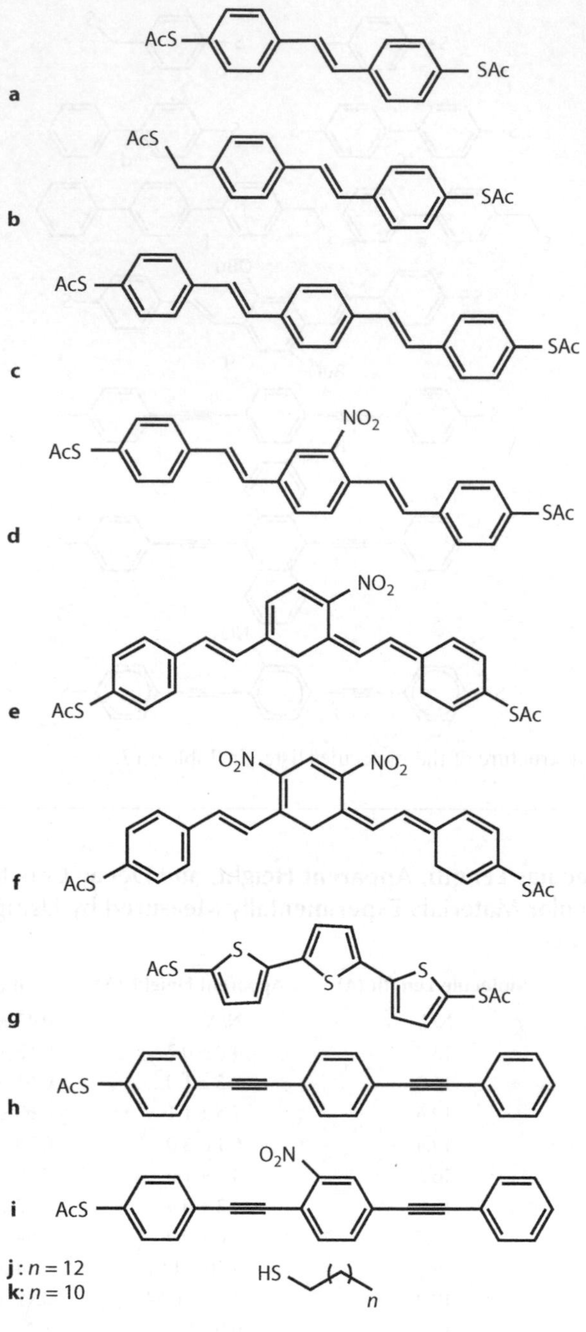

FIGURE 6.19   Chemical structures for the molecules studied by using STM shown in Table 6.13.

molecular solution. They systematically investigated a series of pi-conjugated biphenyl molecular systems (see Figure 6.20). As shown in Table 6.14, it was found that the molecular twist angle, altered by different aromatic ring substituents, can intensively affect the molecular conductance. The larger the twist angle is, the lower the molecular conductance.

Recently, Venkataraman et al. reported their experimental observations about the influence of molecular chemical substituents on its single-molecule junction conductance [35]. They use the same method as mentioned above to measure the low bias conductance of a series of substituted benzene

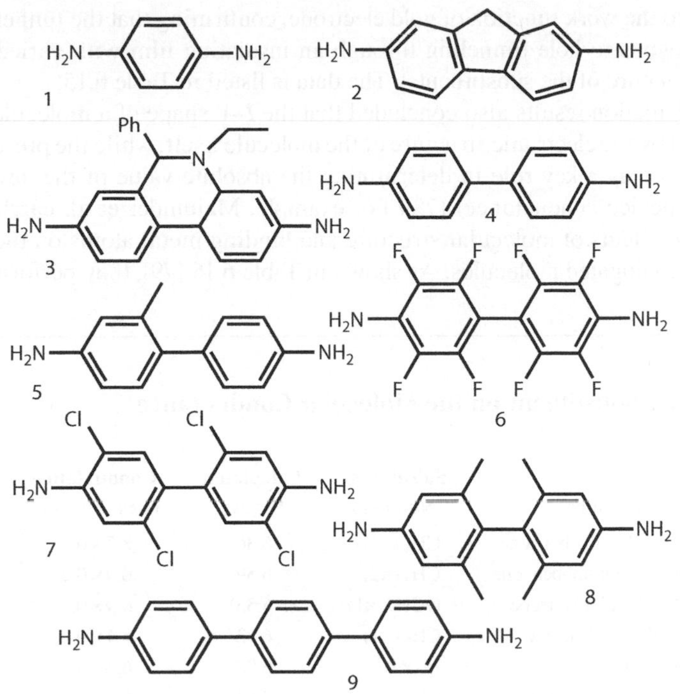

FIGURE 6.20 Chemical structure of the molecules listed in Table 6.14.

TABLE 6.14
Effect of Molecular Conformation on the Electronic Conductance
(Unit in G_0)

Molecule	Measured (G_0)	Calculated (G_0)	Peak Width	Twist Angle (°)
1	6.4×10^{-3}	6.4×10^{-3}	0.4	N/A
2	1.54×10^{-3}	2.1×10^{-3}	0.8	0
3	1.37×10^{-3}	2.2×10^{-3}	0.8	17
4	1.16×10^{-3}	1.6×10^{-3}	0.9	34
5	6.5×10^{-4}	1.2×10^{-3}	1.3	48
6	4.9×10^{-4}	7.1×10^{-4}	0.6	52
7	3.7×10^{-4}	5.8×10^{-4}	0.9	62
8	7.6×10^{-5}	6.4×10^{-5}	N/A	88
9	1.8×10^{-4}	3.5×10^{-4}	2.1	N/A

Source: Adapted from Venkataraman, L. et al., *Nature*, 442, 904, 2006.

Note: The single-molecular junction is fabricated by using the method of breaking Au point contacts in the corresponding molecular solution. Molecular structures are given in Figure 6.20.

diamine molecules in the molecular solution. In this molecular junction, the electron transport follows the mechanism of nonresonant tunneling, with the molecular conductance depending on the alignment of the electrode Fermi level to the closet molecular energy level. The results indicated that electron-donating substituents attempt to drive the highest occupied molecular orbital up. This will in turn result in higher molecular junction conductance. On the other hand, the electron-accepting functional groups pose the opposite effect. Their experiments revealed that the diamine molecules'

HOMO is closest to the work function of gold electrode, confirming that the tunneling through these molecules is analogous to hole tunneling through an insulating film with varied potential barrier (depending on the nature of the substituents). The data is listed in Table 6.15.

Theoretical calculation results also concluded that the $I–V$ shape of a molecular junction can be largely determined by the electronic structure of the molecule itself, while the presence of electrode–molecule interfaces play a key role in determining the absolute value of the device current (or in another word the device conductance) [78]. For example, Majumder et al. carried out theoretical calculations on the effects of molecular structure and binding metal atoms on the electronic structure of a series of conjugated molecules. As shown in Table 6.16 [79], they performed first-principle

TABLE 6.15
Effect of Molecular Substituent on the Molecular Conductance

Entry	Molecule Name	Substituent (Numbers)	Calculated IP (eV)	Conductance Peak (10^{-3} G_0)	Calculated Relative Conductance (10^{-3} G_0)
1	Tetramethyl-1,4-diaminobenzene	CH_3 (×4)	6.36	8.2×0.2	7.6
2	2,5-dimethyl-1,4-diaminobenzene	CH_3 (×2)	6.59	6.9×0.2	6.7
3	2-methoxy-1,4-diaminobenzene	OCH_3 (×1)	6.55	6.9×0.2	7.1
4	2-methyl-1,4-diaminobenzene	CH_3 (×1)	6.72	6.4×0.6	6.5
5	1,4-diaminobenzene	H (×4)	6.83	6.4×0.2	6.4
6	2,5-dichloro-1,4 diaminobenzene	Cl (×2)	7.14	6.1×0.2	6.0
7	2-bromo-1,4-diaminobenzene	Br (×1)	7.02	6.1×0.6	6.1
8	Trifluoromethyl-1,4-diaminobenzene	CF_3 (×1)	7.22	6.1×0.2	6.2
9	2-chloro-1,4-diaminobenzene	Cl (×1)	7.00	6.0×0.4	6.2
10	2-cyano-1,4-diaminobenzene	CN (×1)	7.30	6.0×0.3	5.9
11	2-fluoro-1,4-diaminobenzene	F (×1)	7.03	5.8×0.4	6.3
12	Tetrafluoro-1,4-diaminobenzene	F (×4)	7.56	5.5×0.3	5.2

Source: Adapted from Venkataraman, L. et al., *Nano. Lett.*, 7, 502, 2007.

Note: The ionization potential (IP) and relative conductance are also calculated and given out.

TABLE 6.16
First-Principle Molecular Electronic Structure Calculated Using Hartree–Fock (HF), Density Functional Theory (DFT), and HF-DFT Hybrid Methods

Molecule	HOMO (eV)	LUMO (eV)	Energy Gap (eV)
(a)	−6.00	−1.48	4.52
(b)	−5.97	−1.65	4.14
(c)	−6.13	−2.34	3.79
(d) M=H	−6.25	−2.88	−3.37
(e) M=Au	−6.00	−3.729	−2.27
(f) M=Ag	−5.79	−3.23	−2.56
(g) M=Cu	−5.81	−2.79	−2.84

Source: Adapted from Majumder, C. et al., *Jpn. J. Appl. Phys.*, 41, 2770, 2002.

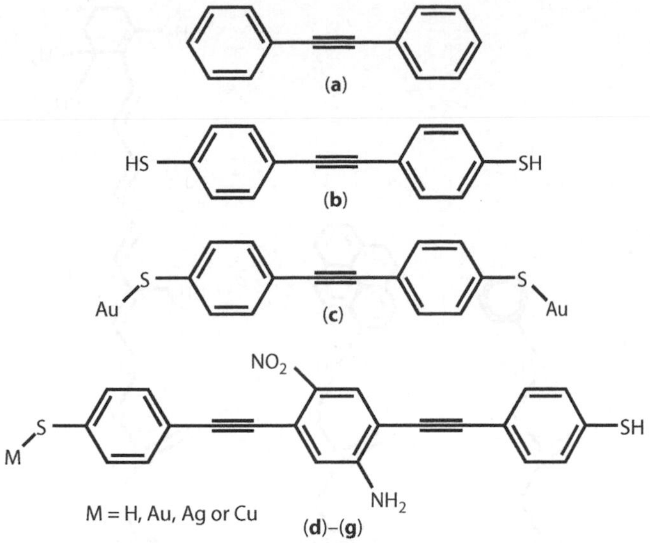

FIGURE 6.21   Molecular structures of a series of conjugated molecules calculated, as given in Table 6.16.

electronic structure calculations using Hartree–Fock (HF), density functional theory (DFT), and HF-DFT hybrid methods for a series of molecular wires (shown in Figure 6.21). The calculated energy gaps for molecules (**a** to **g**) are 4.52, 4.14, 3.79, 3.37, 2.27, 2.56, and 2.84, respectively. The results indicated that about 2.2 eV bias voltage is required to achieve electron transport through the target molecule. The connection with metal atoms (Au, Ag, and Cu) will lead to an obvious reduction in the molecular energy gap compared to that of the free molecules.

6.4.1.3   Organosilane-Based Self-Assembly Monolayers

In Section 6.2, we introduced that organosilane-based SAM is a very useful active media in molecular electronics. Collet et al. investigated the formation and modification of a series of organosilane monolayers by measuring their electrical properties in junction form of Al/Silane SAM/Si [9]. They first fabricated a vinyl-terminated trichlorosilane monolayer on an n-type Si (100) following the procedures descried in Section 6.2. Once the monolayer is grafted, they can obtain $-CH_2OH$- or $-COOH$-terminated monolayers through the chemical reactions of hydroboration, hydrolysis, or oxidation. Furthermore, they can attach highly conjugated moieties onto these SAMs by using esterification reactions between the $-COOH$ end-groups and different alcohols (such as benzyl alcohol, retinal, and pyrene methanol). The resulted SAM is schematically shown in Figure 6.22. The authors performed physical characterizations by measuring the monolayer embedded in an Al/SAM/Si junction. Their results are summarized in Table 6.17. It is indicated that the good insulating properties of the SAM are not altered by the chemical modification process, while the permittivity of the monolayer can be modified via adding different dipolar moieties. From the table, we can see that the silane monolayers with phenyl and pyrene ending groups have a lower barrier height and higher current density at the same bias voltage than that of the others. This is due to the conjugation effect of the monolayers. Again, the experimental results remind us that the molecular chemical structure and functional groups have a large effect on the device physical performances.

The silane monolayers can be further used in the patterning or directing placement of various nano-materials like nanoparticles and nanowires. Ma et al. reported their investigations on the electrostatic funneling for precise nanoparticle placement [11]. They use the current complementary metal-oxide-semiconductor transistor (CMOS) fabrication technology to define a pattern of structures on the substrate, which can be then selectively modified by using different SAMs (see Figure 6.23a and b). The $-NH_3$-terminated silane molecules will be formed on silicon dioxide

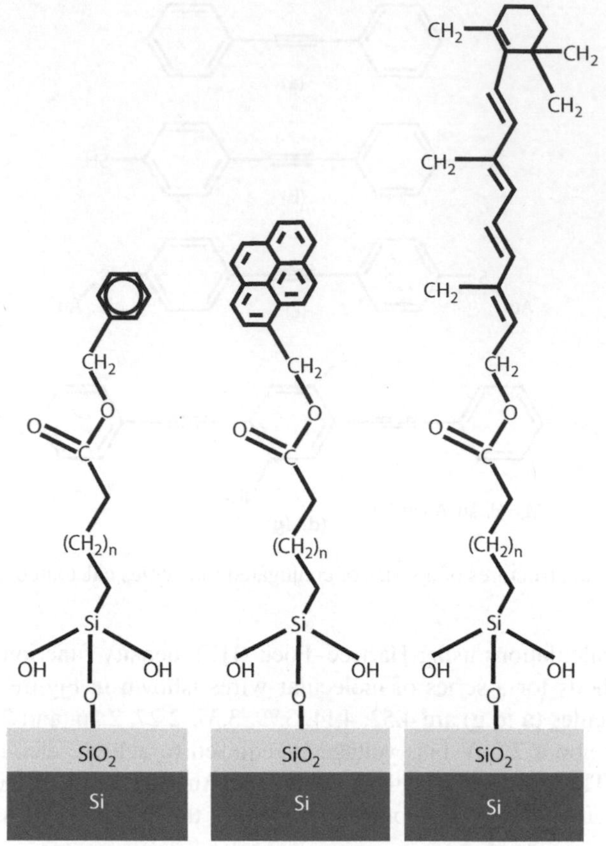

FIGURE 6.22   Organosilane monolayer can be used as surface anchor to direct the growth of a variety of conjugated parts on the substrate (from left to right: phenyl, pyrene, and retinal ended monolayer). This technique will be useful in the bottom-up approach of molecular memory system. (Adapted from Collet, J. et al., *Microelectron. Eng.*, 36, 119, 1997.)

TABLE 6.17
Summary of Electrical Properties of Organosilane Monolayers with Different Ending Groups with Junction Structure of Al/Silane SAM/Si

Ending Group	Current at 1V (A/cm²)	Barrier Height (eV)	Monolayer Capacitance (µF/cm²)	Ellipsometry n.d (nm)	Monolayer Permittivity	Monolayer Thickness (nm)
–CH$_3$	10^{-8}	4.3–4.5	1.07	2.75	2.2	1.84 (1.86)
–CH=CH$_2$	3×10^{-8}	4.1–4.2	0.78	2.81	1.82	2.1 (2.1)
–COOH	8×10^{-8}	4.1–4.2	0.95	3.79	2.55	2.4 (2.1)
–CH$_2$OH	2×10^{-8}	4.1–4.2	N/A	N/A	N/A	N/A
–COOCH$_2$–Phenyl	5×10^{-8}	3.6–3.8	0.85	5.03	2.85	2.98 (2.73)
–COO–Retinol	2×10^{-8}	N/A	0.75	6.12	3	3.53 (3.48)
–COOCH$_2$–Pyrene	10^{-7}	3.9	0.89	4.92	2.9	2.9 (2.98)

Source: Adapted from Collet, J. et al., *Microelectron. Eng.*, 36, 119, 1997.

Note: The substrate is n-type Si (100). The values in parenthesis are calculated ones, while n is the optical indice of the monolayer.

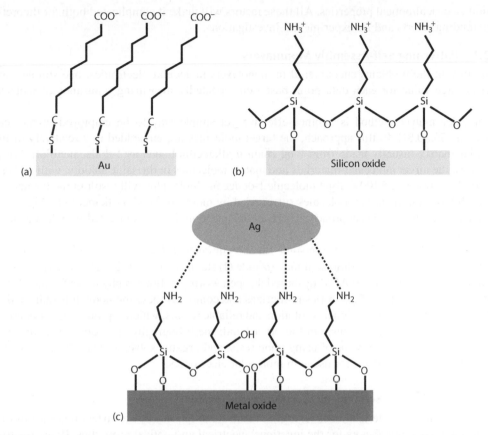

FIGURE 6.23 Organosilane monolayers can be used to modify the substrate surface and direct the assembly of metal nanoparticles. (a) $-COO^-$-terminated SAM is selectively formed on gold surfaces and (b) $-NH_3^+$-terminated molecules are selectively formed onto silicon dioxide surfaces. (From Ma, L.C. et al., *Nano. Lett.*, 7, 439, 2006.) (c) Silver nanoparticle is selectively attached to the $-NH_2$-terminated aminosilane surfaces. (From Morrill, A.R. et al., *Chem. Phys. Lett.*, 473, 116, 2009.)

surfaces, while the $-COO^-$-terminated molecules will be selectively grown on gold surfaces. In this way, they can obtain positively and negatively charged patterns on the substrate in an aqueous solution. When such a substrate is immersed into a colloidal solution containing negatively charged Au nanoparticles, the nanoparticles will be directed into the positively charged areas. Using a similar approach, Morrill et al. recently realized the selective placement of silver nanoparticles on surfaces covered with 3-aminopropyltriethoxysilane SAM. The mechanism arises from the silver's ability to donate electrons to the nitrogen's antibonding orbital via π back bonding (see Figure 6.23c). These knowledge and techniques can be further used in the fabrication of molecular junctions with specially designed functional active units and metal–molecule interfaces that will enable us the molecular-level ability to engineer the molecular device.

6.4.2 MOLECULAR SWITCHING AND MEMORY

One big obstacle in studying the device performance of SAMs is the uncertainty of the molecule numbers within the molecular junction area. Generally, it is not known exactly how many molecules contribute to the device's conduction. In addition, we are not very sure if there are any intermolecular effects between the neighboring molecules. For molecular-scale information memory, random fluctuations resulted from intermolecular interactions may have a strong influence on the junction electrical/optical properties. All these factors will make it complicated both for theoretical understanding efforts and for experimental investigations.

6.4.2.1 Patterning Self-Assembly Monolayers

To circumvent the problems encountered in monolayers molecular electronics, one simple way is to get average value for each data point based on repeatedly measuring thousands of molecular junctions.

Molecular matrix method is another effective yet simple way to be employed in molecular electronics [75,80,81]. In this approach, the target molecules are embedded in a relatively insulate molecular matrix (usually made from long chain *n*-alkanethiol such as 1-dodecanethiol). We just co-dissolve the target molecular materials and matrix molecules in the same solution with an appropriate molar ratio (e.g., 5:100 = target molecule:1-dodecanethiol). This will result in the formation of a mixed SAM with the target molecules surrounded by more insulated *n*-alkanethiols. Figure 6.24 presents an example for this approach, where a conjugated molecule is inserted into dodecanethiol monolayer.

As shown in Figure 6.25, there are some other techniques to fabricate SAM's patterns on various substrates at different conditions. It may include (1) stamp-printing [82], (2) selective growth on pre-patterned substrate [6], (3) AFM tip-based dip-pen writing or lithography [83–85], and (4) STM tip-based lithography [86]. Table 6.18 summarizes and compares these methods. It is indicated that SPM-based methods provide higher resolution and reliable results, which is promising for nanoscale information memory applications. On the other hand, the lithography approaches usually have a poor resolution and involve steps of using photoresist and organic solvents. The lithography methods are obviously unfavorable for usage in molecular electronics.

6.4.2.2 Switching of Self-Assembly Monolayer Devices

Low-temperature characterization under high vacuum conditions is a powerful technique to eliminate most of the unfavorable factors for the junctions' electrical and optical properties. Referring to the ultrahigh vacuum system given in Section 6.2, various control experiments can be easily conducted

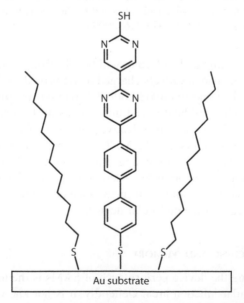

FIGURE 6.24 Schematic structure showing a conjugated dithiol molecular wire inserted within 1-dodecanethiol matrix.

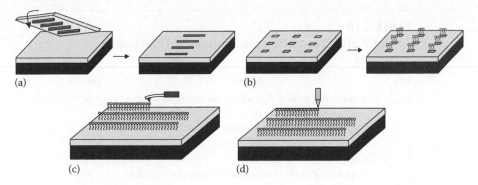

FIGURE 6.25 Schematic process for patterning SAM on substrate: (a) Stamp-printing, (b) selective growth on pre-patterned substrate (e.g., lithography predefined Au regions), (c) AFM tip-based dip-pen writing, and (d) STM tip-based lithography.

TABLE 6.18
Comparison of the Fabrication Approaches for SAM Patterning

Approach	Process	Resolution (nm)	Advantages	Disadvantage	References
Photolithography	Stamp-printing	5000	Low cost and mass production	Poor resolution, contamination from stamp	[82,87]
	Selective growth on pre-patterned substrate	2000	Low cost	Low resolution	[6]
E-beam lithography	Stamp-printing	200	Medium resolution and cost	High cost, contamination from stamp	N/A
	Selective growth on pre-patterned substrate	50	Good resolution and compatibility with semiconductor industry	High cost, complicated process	N/A
Atomic force microscopy	Dip-pen writing	100	Good resolution	Reproducible	[84,88]
	CAFM tip-based lithography	50	High resolution	Middle cost, reliable	[83,85,89,90]
STM	STM tip-based lithography	2	Highest resolution	High cost, low reliability	[86]

with fine variation of temperature, oxygen, water, solvent, or optical illumination. Since the molecules are in their low energetically state, the measurements at low temperatures will offer a good chance to explore the molecular-specific properties and their effect on the device characteristics.

In a previous work [6], we reported the observation of high photo-responsivity with intrinsic amplification for a novel sandwich structure. The devices are made from molecular monolayers softly sandwiched between two thin gold electrodes. The bottom electrodes were fabricated by using photolithography method, while the top electrode bars were printed from elastic stamps made from a polymer. The experimental parameters and procedures for making polymer stamps are given in Tables 6.19 and 6.20. The experiments are conducted under conditions of high vacuum and low temperature. As shown in Figure 6.26, the junctions with a set of molecular wires, with different length

TABLE 6.19
Silicon Wafer Preparation for Photoresist Lithography Patterning

Step	Action	Time (s)
1	Nitrogen gun to blow clean the holder and close lid after loading the sample	N/A
2	Ultrasound in 50 mL acetone bath	20
3	Ultrasound in 50 mL methanol bath	20
4	DI rinse under running water	60
5	Remove from beaker bottom and nitrogen gun blow dry	N/A

TABLE 6.20
Experimental Procedure for S1813 Photoresist Photolithography

Step	Action	Time (s)
1	Prepare silicon wafer according to RCA cleaning procedure	N/A
2	Center clean wafer on spinner, set to 4000 rpm	45
3	With clean pipette cover entire wafer with S1813 photoresist	N/A
4	Spin wafer at 4000 rpm	45
5	Soft-bake at 90°C (keep covered with petri dish but allow ventilation)	180
6	Cooling down (keep covered with petri dish)	60
7	Blow off any junk that may be on mask	N/A
8	Check the wavelength and power of the UV light	N/A
9	Pattern exposure	13
10	Bake at 110°C (keep covered with petri dish)	150
11	Cooling down	60
12	Develop in 1:1 of photoresist developer: DI water mixed solution	75
13	Rinse the sample with DI water flow and dry with nitrogen gun	N/A

and ending groups and conjugation degrees, are comparably investigated through current–voltage measurements at low temperatures. The device showed reversible optoelectronic switching with on/off ratio of 3 orders of magnitude at 95 K. The switching phenomenon is independent of both optical wavelength and molecular structure, while it strongly depends on the temperature. However, the switching on/off ratio is dependent on the molecular structure to some extent: The higher the conjugation of the molecular wires is, the higher the on/off ratio. The switching ratio for molecular junctions with dithiol molecules is relatively lower than that of the molecules with a single thiolated group. The results are very intriguing, potentially providing a novel method to obtain extremely sensitive and fast, UV to IR, photodetectors.

Temperature dependence results are shown in Figure 6.27. It is indicated that the Au–molecule interfaces in the crossbar junctions have distinct chemical nature and different electron tunneling parameters. The charge transport in the junction can be mainly attributed to Richardson–Schottky thermionic emission process under light condition, while Fowler–Nordheim tunneling plays a dominant role in dark (refer to previous section for these models). An explanation for the optical switching is that the injected light induces some physical changes within the crossbar junction. It may be the tunneling parameter at the top wire–lead interface and/or the transport mechanism through the wire. Two evidences support this hypothesis. First, under light irradiation, the temperature dependence of the junction is much weaker than that in dark, especially for the junctions with conjugated wires. This behavior is a clear indication of a change in the conduction mechanism, which may

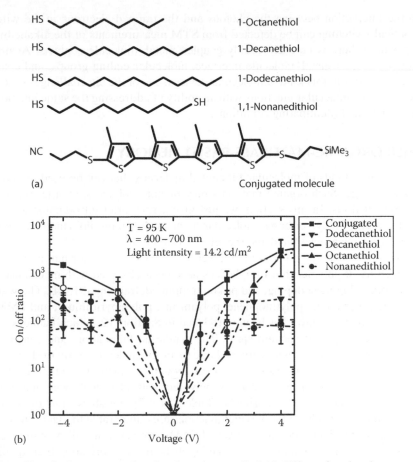

(a)

(b)

FIGURE 6.26 Chemical structure of molecular wires studied (a). Effect of molecular structure on the current switching ratio between light and dark conditions (b). It is clear that the junction with the conjugated molecules can yield a higher switching ratio than that without. (From Li, J.C., *Chem. Phys. Lett.*, 473, 189, 2009.)

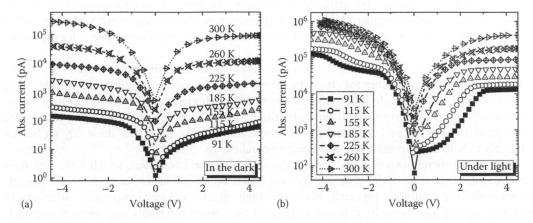

FIGURE 6.27 *I–V* temperature dependence of a 1-decanethiol junction (a) in the dark and (b) under illumination of florescent light. (From Li, J.C., *Chem. Phys. Lett.*, 473, 189, 2009.)

result from the interaction between the photons and the trapped electrons in the wires. Second, no similar optical switching can be detected from STM measurements of the alkanethiols at room temperature, where there is no such weakly coupled metal–molecule contacts. Again, our work shows the importance of metal–molecule interface, molecular ending groups, and conformations. For further applications of this kind of molecular devices, one of the big challenges is how to realize single-molecule-scale molecular junctions with an ability of addressing the same junction at various temperature and optical illuminating conditions.

6.5 OTHER ORGANIC-MATERIALS-BASED MEMORY

There are some other kinds of molecular functional materials that can be used as active layers in nanoscale data storage. Such organic materials may include polymers, charge transfer salts, and Langmuir–Blodgett films. In this section, we just give a very brief introduction about the recent advances in the area of polymer-based molecular memory. For more information, the readers can refer to the good review articles cited and the references therein.

Functional polymers are the other kind of molecular materials for information memory applications. The switching of polymer thin-film devices have been observed with mechanisms of charge trapping/detrapping, doping/dedoping, and proton trapping/detrapping [91–94]. The widely studied polymers include polypyrrole, polythiophene, polyaniline, polybipyridinium, and PMMA [95,96]. The growth of polymer thin layer has been introduced in Section 6.2, which includes spin coating, dip coating, drop casting, and vacuum spray. The synthesis, fabrication, and characterization of polymer memory materials and devices are too large to review in detail within the boundaries of this chapter. Here, our purpose is to give some examples about this widely studied material. Figure 6.28 presents the molecular structure of a group of polymers with electrical switching properties and their memory performances are summarized in Table 6.21 [97]. It is shown that the switching time, on/off ratio, and threshold voltage for polymer junctions are comparable to that of the devices made from small molecular weight redox materials and SAMs. However, it will be a great challenge to fabricate nanoscale polymer junctions due to the big difficulty in fabricating high-quality polymer monolayers.

6.6 SUMMARY AND OUTLOOKS

The investigation of organic-molecule-based information memories has been developing rapidly and many significant results are being reported everyday. However, there are far too many one-time "wanders" in the field of molecular electronics, which do not stand for the test of time because they are based on some "exciting" but artificial observations. So there has to be quantitative measures of the responsivity including thermal/dynamic responses, temperature dependence, molecule–electrode interfaces, junction size effect, molecular structural factors, and theory models for the device performance.

It is very applicable to realize ultrahigh density single-molecule-scale data storage in the future. We believe that two approaches may have the most potential for promising applications as building blocks in practical nanoscale information storage. One is single-molecule-based memory device with sub-10 nm characteristics built on molecular SAMs. Multimode information storage is the other powerful way to make breakthrough in nanoscale data storage. All of these rely on further advances in both experimental and theoretical researches.

Most of the current experiments were conducted at ambient conditions, which were very poor for molecular electronic device characterizations. It may be a good idea to test the target molecule under controlled conditions, especially under ultrahigh vacuum and low temperature situations. In this way, we can then carefully examine and extract the correct information of molecular-specific properties from experimental results, which in turn provides a solid foundation for the development and application of molecular-scale memories. Obviously, the researchers first have to develop

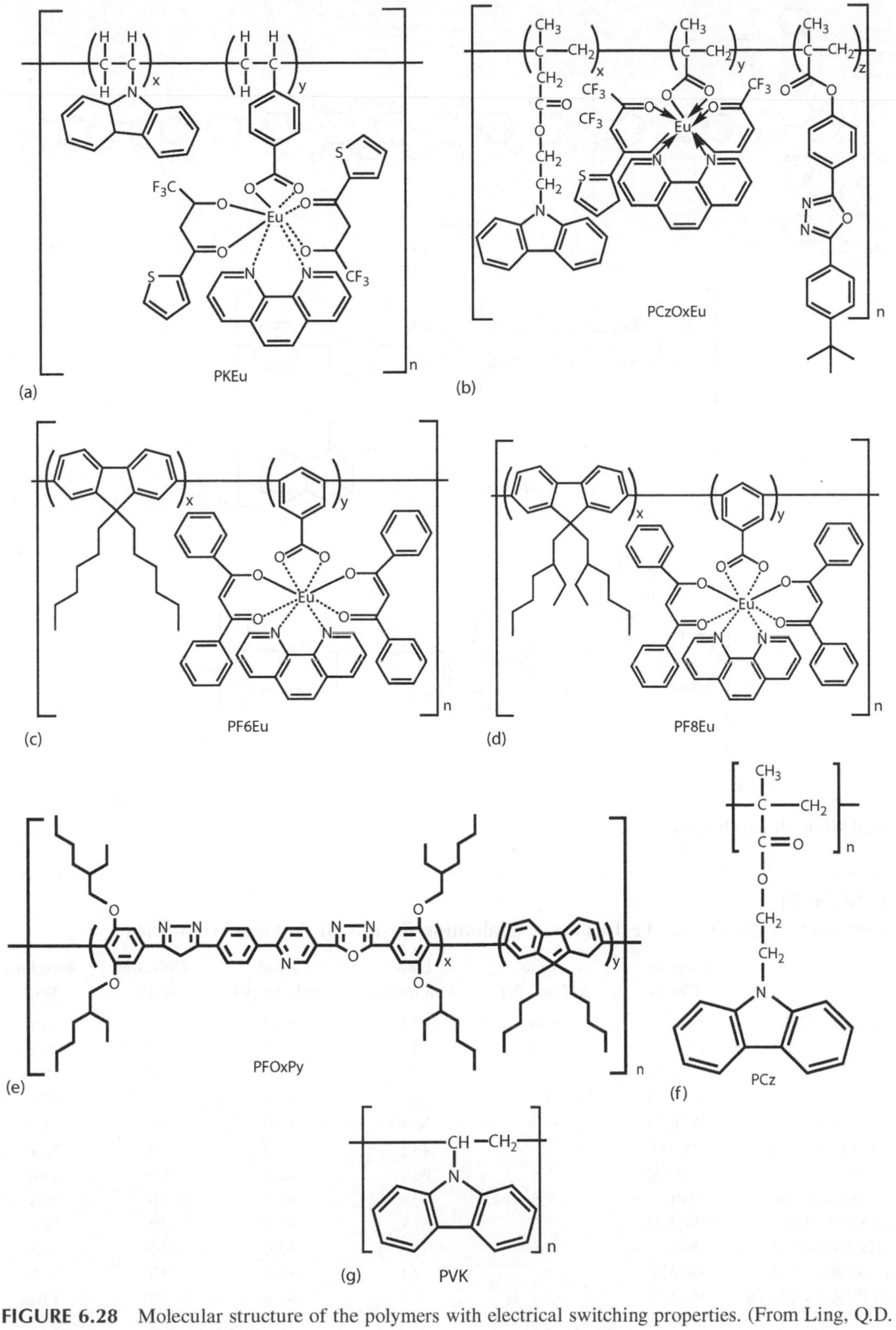

FIGURE 6.28 Molecular structure of the polymers with electrical switching properties. (From Ling, Q.D. et al., *Polymer*, 48, 5182, 2007.)

(*continued*)

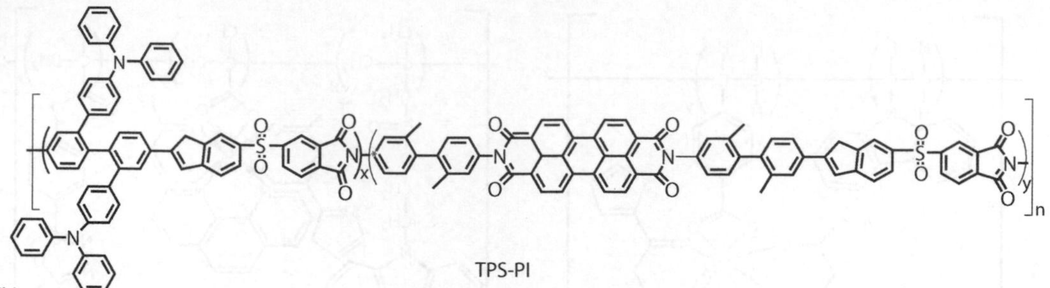

(h) TPS-PI

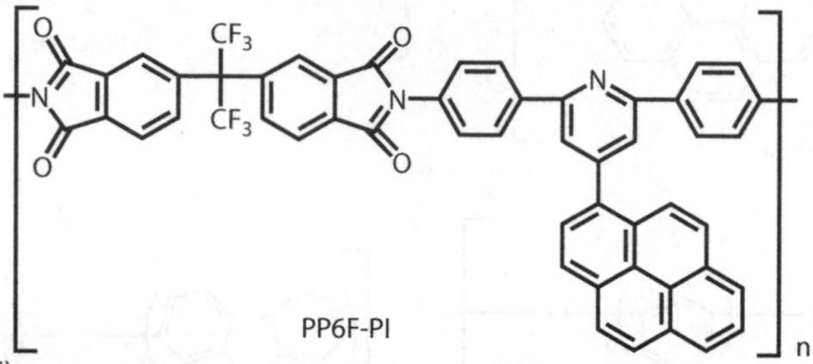

(i) PP6F-PI

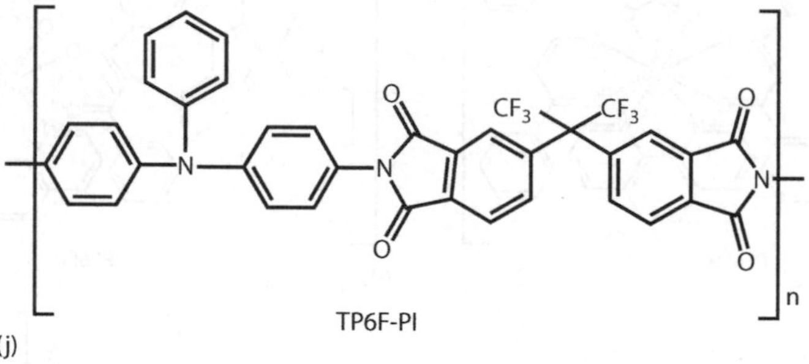

(j) TP6F-PI

FIGURE 6.28 (continued)

TABLE 6.21

Summary of the Device Performance Evaluation Results of Polymer Memories

Devices	Memory Effects	Write Voltage (V)	Erase Voltage (V)	Read Voltage (V)	ON/OFF Ratio	Switching Time
AL/PKEu/ITO	Flash	−2.0	+4.0	−1.0	10^4	20 μs
ITO/PCzOxEu/Al	Flash	−2.8	+4.4	+1.0	10^5	1.5 μs
Al/PF6Eu/ITO	WORM	+3.0	N/A	+1.0	10^7	N/A
Al/PF6Eu/n-Si	WORM	+2.2	N/A	+1.0	10^4	N/A
Al/PF8Eu/ITO	WORM	+3.0	N/A	+1.0	10^6	~1 μs
ITO/PFOxPy/Al	DRAM	−2.8	+3.5	−1.0	10^6	N/A
ITO/PCz/Al	WORM	−1.8	N/A	−1.0	10^6	1 ms
ITO/PVK-C$_{60}$/Al	Flash	−2.8	+3.0	+1.0	10^5	N/A
ITO/TPS-PI/Al	WORM	−5.7	N/A	−1.0	10^5	N/A
ITO/PP6F-PI/Al	Flash	~4.5	−5.0	+2.0	10^6	N/A
ITO/TP6F-PI/Al	DRAM	+3.2	−2.1	+1.0	10^5	N/A
Al/PVK-AuNP/TaN	Flash	+3.0	−1.7	+1.0	10^5	1.0 μs

Source: Adapted from Ling, Q.D. et al., *Polymer*, 48, 5182, 2007.

some simple but effective approaches to fabricate reliable, single-molecule-scale and addressable metal–molecule–metal junctions with well-defined electrode–molecule contacts. The combined approach of crossed nanowires plus nano-patterning of SAMs may be an effective way to volume-fabricate molecular junction arrays for nanoscale information memory applications.

On the other hand, the contacting interface between metal electrodes and molecular thin films are not thoroughly understood yet, more hard and systematic researching works are needed. In addition, there is lack of deep theoretical calculations, simulations, and predications on the experimental results and advances. It calls for interdisciplinary collaboration between scientists of physics, chemistry, engineering, electronics, and materials, etc.

ACKNOWLEDGMENTS

JCL acknowledges the support from Profs. G. J. Szulczewski and S. C. Blackstock, Dr. K.-Y. Kim (University of Alabama), Profs. Luping Yu and H. M. Jaeger (University of Chicago), Prof. Z. Q. Xue (Peking University, China), and Prof. P. Mulvaney (University of Melbourne, Australia). Graduate students Han Xiaobo, Han Na, and Han Yu were of great help in preparing the schematic drawings. Financial support comes from NEU young scholar program and National University Basic Research Fund (N090403001).

ABBREVIATIONS

ACE	Acetone
AcS	Thiolacetyl
AFM	Atomic force microscopy
Ag	Silver
Au	Gold
CAFM	Contacting atomic force microscopy
CMOS	Complementary metal-oxide-semiconductor transistor
CT	Charge transfer
Cu	Copper
DFT	Density functional theory
DI	Deionized
EA	Electron affinity
Et	Ethyl
EtOH	Ethanol
GaIn	Gallium indium
h	Hour
HAc	Acetic acid
HF	Hydrofluoric acid or Hartree–Fock
Hg	Mercury
HOMO	Highest occupied molecular orbital
HOPG	Highly oriented pyrolitic graphite
H_2O_2	Hydrogen peroxide
H_2SO_4	Sulfuric acid
IE	Ionization energy
IP	Ionization potential
IPES	Inverse photoemission spectroscopy
ITO	Indium tin oxide
$I–V$	Current–voltage
LB	Langmuir–Blodgett

LED   Light emitting diode
LUMO   Lowest unoccupied molecular orbital
Me   Methyl
mL   Milliliter
mM   Millimole
MOM   Metal-organic-metal
NN   Nano-Newton
OPV   Oligophenylenevinylene
PMMA   Polymethylmethacrylate
RCA   Radio Corporation of America
Redox   Reduction oxidation
Rpm   Rotation per minute
SAM   Self-assembly monolayer
SPM   Scanning probe microscopy
STM   Scanning tunneling microscopy
STS   Scanning tunneling spectroscopy
THF   Tetrahydrofuran
UPS   Ultraviolet photoemission spectroscopy
V   Voltage
V_t Turn-on voltage
W   Watt

SYMBOLS

C Proportionality constant
d Tunneling distance (or wire length in the case of molecular junctions)
e Charge of an electron
F Electrical field
h Planck's constant
$\hbar$ Planck's constant divided by 2π
I Current
J Current density
k_B Boltzmann's constant
m Mass of an electron
m^* Effective electron mass
q Elementary charge
T Temperature
V Bias voltage applied between the electrodes
α A unitless adjustable parameter used in fitting
β Decay constant
ε Relative dielectric constant
ε_0 Vacuum permittivity
ϕ Barrier height
ϕ_B Zero-field injection barrier

REFERENCES

1. Lundstom, M. 2003. Moore's law forever? *Appl. Phys.* 299:210–211.
2. Yasutomi, S., T. Morita, Y. Lmanishi et al. 2004. A molecular photodiode system that can switch photo-current direction. *Science* 304:1943–1948.
3. Lahann, J., S. Mitragotri, T. Tran et al. 2003. A reversibly switching surface. *Science* 299:371–374.

4. Smits, E. C. P., S. G. J. Mathijssen, and P. A. V. Hal. 2008. Bottom-up organic integrated circuits. *Nature* 455:956–959.
5. Li, J. C. and Y. L. Song. 2009. Nanoscale data storage. In *High Density Data Storage: Principle, Technology, and Materials*, Y. L. Song and D. B. Zhu (Eds.), pp. 193–260, World Scientific Publishing Co., Singapore.
6. Li, J. C. 2009. Optoelectronic switching of addressable self-assembled monolayer molecular junctions. *Chem. Phys. Lett.* 473:189–192.
7. Li, J. C. 2009. Electrical switching and memory phenomena observed in redox dendrimer thin films. *Thin Solid Films* 517:3385–3388.
8. Lenfant, S., D. Guerin, F. T. Van et al. 2006. Electron transport through rectifying self-assembled monolayer diodes on silicon: Fermi-level pinning at the molecule-metal interface. *J. Phys. Chem. B* 110:13947–13958.
9. Collet, J., M. Bonnier, O. Bouloussa et al. 1997. Electrical properties of end-group functionalized self-assembled monolayers. *Microelectron. Eng.* 36:119–122.
10. Marchenko, A., N. Katsonis, D. Fichou et al. 2008. Long-range self-assembly of a polyunsaturated linear organosilane at the *n*-tetradecance/Au(111) interface studied by STM. *J. Am. Chem. Soc.* 124:9998–9999.
11. Ma, L. C., R. Subramanian, H. W. Huang et al. 2006. Electrostatic funneling for precise nanoparticle placement: A route to wafer-scale integration. *Nano Lett.* 7:439–445.
12. Morrill, A. R., D. T. Duong, S. J. Lee et al. 2009. Imagine 3-aminopropyltriethoxysilane self-assembled monolayers on nanostructured titania and tin (IV) oxide nanowires using colloidal silver nanoparticles. *Chem. Phys. Lett.* 473:116–119.
13. Li, J. C., K. Y. Kim, and S. C. Blackstock et al. 2004. Patterned redox arrays of polyarylamines III. Effect of molecular structure and oxidation potential on film morphology and hole-injection in single-layer organic diodes. *Chem. Mater.* 16:4711–4714.
14. Li, J. C., S. C. Blackstock, and G. J. Szulczewski. 2006. Interface between metal and arylamine molecular films as probed with the anode interfacial engineering approach in single-layer organic diodes. *J. Phys. Chem. B* 110:17493–17497.
15. Kushmetrick, J. G., D. B. Holt, J. C. Yong et al. 2002. Metal-molecule contacts and charge transport across monomolecular layers: Measurement and theory. *Phys. Rev. Lett.* 19:0868021–0868024.
16. Lutwyche, M. I., M. Ddspont, U. Drechsler et al. 2000. Highly parallel data system based on scanning probe arrays. *Appl. Phys. Lett.* 77:3299–3301.
17. Gardner Catherine, E., M. A. Chanem, J. W. Wilson et al. 2006. Development of a nanowire-based test bed devices for molecular electronics applications. *Anal. Chem.* 78:951–955.
18. Dadosh, T., Y. Gordin, R. Krahne et al. 2005. Measurement of the conductance of single conjugated molecules. *Nature* 436:677–680.
19. David, P. l., C. H. Patterson, and M. H. Moore. 2005. Magnetic directed assembly of molecular junctions. *Appl. Phys. Lett.* 86:1531051–1531053.
20. Chen, Y., A. A. O. Douglas, X. Li. et al. 2003. Nanoscale molecular-switch devices fabricated by imprint lithography. *Appl. Phys. Lett.* 82:1610–1612.
21. Malave, A., M. Tewes, T. Gronwold et al. 2005. Development of impedance biosensors with nanometer gaps for marker-free analytical measurements. *Microelectron. Eng.* 78:587–592.
22. Zhou, C., M. R. Deshpanda, and M. A. Read. 1997. Nanoscale metal/self-assembled monolayer/metal heterostructures. *Appl. Phys. Lett.* 71:611–613.
23. Majumdar, N., N. Gergel, D. Routenberg et al. 2005. Nanowell device for electrical characterization of metal-molecule-metal junctions. *J. Vac. Sci. Technol. B* 23:1417–1421.
24. Ojima, K., Y. O. Tsuka, and T. J. Kawai. 2005. Printing electrode for top-contact molecular junction. *Appl. Phys. Lett.* 87:2341001–2341003.
25. Szulczewski, G. J., T. D. Selby, K. Y. Kim et al. 2000. Growth and characterization of poly(arylamine) thin films prepared by vapor deposition. *J. Vac. Sci. Technol. A* 18:1875–1880.
26. Selby, T. D., K. Y. Kim, and S. C. Blackstock. 2002. Patterned redox arrays of polyarylamines I. Synthesis and electrochemistry of a P-phenylenediamine and arylamino-appended P-phenylenediamine arrays. *Chem. Mater.* 14:1685–1690.
27. Kim, K. Y., J. D. Hassenzahl, T. D. Selby et al. 2002. Patterned redox arrays of polyarylamines II. Growth of thin films and their electrochemical behavior. *Chem. Mater.* 14:1691–1694.
28. Naito, K. and A. Miura. 1993. Molecular design for nonpolymeric organic dye glasses with thermal stability: Relations between thermodynamic parameters and amorphous properties. *J. Phys. Chem.* 97:6240–6248.

29. Adachi, C., K. Nagai, and N. Tamoto. 1995. Molecular design of hole transport materials for obtaining high durability in organic electroluminescent diodes. *Appl. Phys. Lett.* 20:2679–2681.

30. Shirota, Y., K. Okumoto, and H. Inada. 2000. Thermally stable organic light-emitting diodes using new families of hole-transporting amorphous molecular materials. *Synth. Met.* 111–112:387–391.

31. Thelakkat, M. 2002. Star-shaped, dendrimeric and polymeric triarylamines as photoconductors and hole transport materials for electro-optical applications. *Macromol. Mater. Eng.* 287:442–461.

32. Fave, C., Y. Leroux, G. Trippe et al. 2007. Tunable electrochemical switches based on ultrathin organic films. *J. Am. Chem. Soc.* 129:1890–1891.

33. Haiss, W., R. J. Nichols, S. J. Higgins et al. 2004. Wiring nanoparticles with redox molecules. *Faraday Discuss.* 125:179–194.

34. Kubatkin, S., A. Danllow, M. Hjort et al. 2003. Single-electron transistor of a single organic molecule with access to several redox states. *Nature* 425:698–701.

35. Venkataraman, L., Y. S. Park, A. C. Whalley et al. 2007. Electronics and chemistry: Varying single-molecule junction conductance using chemical substituents. *Nano Lett.* 7:502–506.

36. Gosvami, N., K. H. A. Lau, S. K. Sinha et al. 2005. Effect of end groups on contact resistance of alkane-thiol based metal-molecule-metal junctions using current sensing AFM. *Surf. Sci.* 252:3956–3960.

37. Park, Y. S., A. C. Whalley, M. Kamenetska et al. 2007. Contact chemistry and single-molecule conductance: A comparison of phosphines, methyl sulfides, and amines. *J. Am. Chem. Soc.* 129:15768–15769.

38. Holman, M. W., R. Liu, and D. M. Adams. 2003. Single-molecule spectroscopy of interfacial electron transfer. *J. Am. Chem. Soc.* 125:12649–12654.

39. McGee, B. J., L. J. Sherwood, M. L. Greer et al. 1999. A chiral 2-D donor-acceptor array of a bipyrazine N-oxide and tetracyanoethylene. *Org. Lett.* 2:1181–1184.

40. Renzi, V. D., R. Rousseau, D. Marchetto et al. 2005. Metal working-function changes induced by organic adsorbates: A combined experimental and theoretical study. *Phys. Rev. Lett.* 95:0468041–0468044.

41. Zehner, R. W., B. F. Parsons, R. P. Hsung et al. 1999. Tuning the work function of gold with self-assembled monolayers derived from X-[C_6H_4–C≡C–]$_n$$C_6H_4$–SH($n=0$, 1, 2; X=H, F, CH_3, CF_3, and OCH_3). *Langmuir* 15:1121–1127.

42. Ramanath, G., G. Cui, X. Guo et al. 2003. Self-assembled subnanolayers as interfacial adhesion enhancers and diffusion barriers for integrated circuits. *Appl. Phys. Lett.* 83:383–385.

43. Campbell, J. L., S. Rubin, J. D. Kress et al. 1996. Controlling Schottky energy barriers in organic electronic devices using self-assembled monolayers. *Phys. Rev. B* 54:14321–14324.

44. Crispin, X., V. Geskin, A. Crispin et al. 2002. Characterization of the interface dipole at organic/metal interfaces. *J. Am. Chem. Soc.* 124:8131–8141.

45. Boyen, H. G., P. Ziemann, U. Wiedwald et al. 2006. Local density of states effects at the metal-molecule interfaces in a molecular device. *Nat. Mater.* 5:394–399.

46. Kahn, A., N. Koch, and W. Gao. 2003. Electronic structure and electrical properties of interfaces between metals and π-conjugated molecular films. *J. Polym. Sci. B: Polym. Phys.* 41:2529–2548.

47. Wakayama, Y., K. Ogawa, T. Kubota et al. 2000. Optical switching of single-electron tunneling in SiO_2/molecule/SiO_2 multilayer on Si (100). *Appl. Phys. Lett.* 85:329–331.

48. Liu, C.-Y., H.-L. Pan, M. A. Fox et al. 1997. Reversible charge trapping/detrapping in a photoconductive insulator of liquid crystal zinc porphyrin. *Chem. Mater.* 9:1422–1429.

49. Chotsuwan, C. and S. C. Blackstock. 2008. Single molecule charging by atomic force microscopy. *J. Am. Chem. Soc.* 38:12556–12557.

50. Roth, K. M., N. Dontha, R. B. Dabke et al. 2000. Molecular approach toward information storage based on the redox properties of porphyrins in self-assembled monolayers. *J. Vac. Sci. Technol. B* 18:2359–2364.

51. Haiss, W., C. Wang, I. Grace et al. 2006. Precision control of single-molecule electrical junctions. *Nat. Mater.* 5:995–1002.

52. Li, C., W. Fan, B. Lei et al. 2004. Multilevel memory based on molecular devices. *Appl. Phys. Lett.* 84:1949–1951.

53. Guisinger, N. P., N. L. Yoder, and M. C. Hersam. 2005. Probing charge transport at the single-molecule level on silicon by using cryogenic ultra-high vacuum scanning tunneling microscopy. *PNAS* 102:8838–8843.

54. Jiang, P., G. M. Morales, W. You et al. 2004. Synthesis of diode molecules and their sequential assembly to control electron transport. *Angew. Chem. Int. Ed.* 43:4471–4475.

55. Qu, N., W. Yao, T. Garcia et al. 2003. Nanoscale polarization manipulation and conductance switching in ultrathin films of a ferroelectric copolymer. *Appl. Phys. Lett.* 82:4322–4324.

56. Haque, S. A., J. S. Park, M. Srinivasarao et al. 2004. Molecular-level insulation: An approach to controlling interfacial charge transfer. *Adv. Mater.* 16:1177–1181.

57. Ferri, V., M. Elbing, G. Pace et al. 2008. Light-powered electrical switch based on cargo-lifting azobenzene monolayers. *Angew. Chem. Int. Ed.* 47:3407–3409.
58. Liu, C. Y. and J. B. Allen. 1998. Optoelectric charge trapping/detrapping in thin solid films of organic azo dyes: Application of scanning tunneling microscopic tip contact to photoconductive films for data storage. *Chem. Mater.* 10:840–846.
59. Geppert, D., L. Seyfarith, and R. D. Vivie-Riedle. 2004. Laser control schemes for molecular switches. *Appl. Phys. B* 79:987–992.
60. Sanchez, C., F. Ribot, and B. Lebeau. 1999. Molecular design of hybrid organic-inorganic nanocomposites synthesized via sol-gel chemistry. *J. Mater. Chem.* 9:35–44.
61. Reed, M. A., J. Chen, A. M. Rawlett et al. 2001. Molecular random access memory cell. *Appl. Phys. Lett.* 78:3735–3737.
62. Seminario, J. M., P. A. Derosa, and J. L. Bastos. 2002. Theoretical interpretation of switching in experiments with single molecules. *J. Am. Chem. Soc.* 124:10266–10267.
63. Long, D. P., J. L. Lazorcik, B. A. Mantooth et al. 2006. Effects of hydration on molecular junction transport. *Nat. Mater.* 5:901–908.
64. Ventra, M. D., S. G. Kim, S. T. Pantelides et al. 2000. Temperature effects on the transport properties of molecules. *Phys. Rev. Lett.* 86:288–291.
65. Simmon, J. G. 1963. Generalized formula for electric tunnel effect between similar electrodes separated by a thin insulating film. *J. Appl. Phys.* 34:1973–1803.
66. Kitagawa, K., T. Morita, and S. Kimura. 2005. Molecule rectification of a helical peptide with a redox group in the metal-molecule-metal junction. *J. Phys. Chem. B* 109:13906–13911.
67. Bumm, L. A., J. J. Arnold, T. D. Dunbar et al. 1999. Electron transfer through organic molecules. *J. Phys. Chem. B* 103:8122–8127.
68. Wang, W., T. Lee, M. Kamdar et al. 2003. Electrical characterization of metal-molecule-silicon junctions. *Ann. N.Y. Acad. Sci.* 1006:36–47.
69. Datta, S., W. D. Tian, S. H. Hong et al. 1997. Current–voltage characteristics of self-assembled monolayers by scanning tunneling microscopy. *Phys. Rev. Lett.* 79:2530–2533.
70. Gonzalez, C., V. Mujica, and M. A. Ratner. 2002. Modeling the electrostatic potential spatial profile of molecular junctions: The influence of defects and weak links. *Ann. N.Y. Acad. Sci.* 960:163–176.
71. Lehmann, J., G.-L. Ingold, and P. Hanggi. 2002. Incoherent charge transport through molecular wires: Interplay of Coulomb interaction and wire population. *Chem. Phys.* 281:199–209.
72. Salomon, B. A., D. Cahen, S. Lindsay et al. 2003. Comparison of electronic transport measurements on organic molecules. *Adv. Mater.* 15:1881–1890.
73. Moth-Poulsen, K., L. Patrone, N. Stuhr-Hansen et al. 2005. Probing the effects of conjugation path on the electronic transmission through single molecules using scanning tunneling microscopy. *Nano. Lett.* 5:783–785.
74. Wold, D. J., R. Haag, M. A. Rampi et al. 2002. Distance dependence of electron tunneling through self-assembled monolayers measured by conducting probe atomic force microscopy: Unsaturated versus saturated molecular junctions. *J. Phys. Chem. B* 2813–2816.
75. Cui, X. D., A. Primak, X. Zarate et al. 2002. Changes in the electronic properties of a molecule when it is wired into circuit. *J. Phys. Chem. B* 106:8609–8614.
76. Wold, D. J. and C. D. Frisbie. 2001. Fabrication and characterization of metal-molecule-metal junctions by conducting probe atomic force microscopy. *J. Am. Chem. Soc.* 123:5549–5556.
77. Venkataraman, L., J. E. Klare, C. Nuckolls et al. 2006. Dependence of single-molecule junction conductance on molecular conformation. *Nature* 442:904–907.
78. Ventra, M. D., S. T. Pantelides, and N. D. Lang. 1999. First-principles calculation of transport properties of a molecular device. *Phys. Rev. Lett.* 84:979–982.
79. Majumder, C., H. Mizuseki, and Y. Kawazoe. 2002. Theoretical analysis for a molecular resonant tunneling diode. *Jpn. J. Appl. Phys.* 41:2770–2773.
80. Ramachandran, G. K., T. J. Hopson, A. M. Rawlett et al. 2003. A bond-fluctuation mechanism for stochastic switching in wired molecules. *Science* 300:1413–1416.
81. Morales, G. M., P. Jiang, S. Yuan et al. 2005. Inversion of the rectifying effect in diblock molecular diodes by protonation. *J. Am. Chem. Soc.* 127:10456–10457.
82. Sullivan, T. P., M. L. V. Poll, P. Y. W. Dankers et al. 2004. Forced peptide synthesis in nanoscale confinement under elastomeric stamps. *Angew. Chem. Int. Ed.* 43:4190–4193.
83. Lee, K.-B., S.-J. Park, C. A. Mirkin et al. 2002. Protein nanoarrays generated by dip-pen nanolithography. *Science* 295:1702–1705.
84. Piner, R. D., J. Zhu, and F. Xu. 1999. 'Dip-Pen' nanolithography. *Science* 283:661–663.

85. Jiang, S.-Y., M. Marquez, and G. A. Sotzing. 2004. Rapid direct nanowriting of conductive polymer via electrochemical oxidative nanolithography. *J. Am. Chem. Soc.* 126:9476–9477.

86. Schenning, A. P. H. J. and E. W. Meijer. 2005. Supramolecular electronic: Nanowires from self-assembled π-conjugated systems. *Chem. Commun.* 26:3245–3258.

87. Nyamjav, D. and A. Ivanisevic. 2004. Properties of polyelectrolyte templates generated by dip-pen nanolithography and microcontact printing. *Chem. Mater.* 16:5216–5219.

88. Demers, L. M., D. S. Ginger, and S.-J. Park. 2002. Direct patterning of modified oligonucleotides on metals and insulators by dip-pen nanolithography. *Science* 296:1836–1838.

89. Schneegans, O., A. Moradpour, and F. Houze. 2001. Conducting probe-mediated electrochemical nanopatterning of molecular materials. *J. Am. Chem. Soc.* 123:11486–11487.

90. Snow, E. S., P. M. Campbell, and F. K. Perkins. 1997. Nanofabrication with proximal probes. *Proc. IEEE* 85:601–611.

91. Kapetanakis, E., A. M. Douvas, D. Velessiotis et al. 2008. Molecular storage elements for proton memory devices. *Adv. Mater.* 20:4568–4574.

92. Barman, S., F. Deng, and R. McCreey. 2008. Conducting polymer memory devices based dynamic doping. *J. Am. Chem. Soc.* 130:11073–11081.

93. Okazaki, S., U. Sadahito, and O. Masanori. 1997. Guest-host electro-optic switching in spin-coated polymer ferroelectric liquid crystal film. *Appl. Phys. Lett.* 71:3373–3375.

94. Gofer, Y., H. Sarker, G. Jeffey et al. 1997. An all-polymer charge storage device. *Appl. Phys. Lett.* 71:1582–1584.

95. Inoue, H., H. Sakaguchi, T. Nagamura et al. 1998. Ultrafast optical switching by photoinduced electrochromism in cast films of polymeric 4,4′-bipyridinium salts with di-iodides. *Appl. Phys. Lett.* 6:10–12.

96. Masa-aki, H., M. D. Meser, and R. Arakawa. 1996. Proton-induced switching of electron transfer pathways in dendrimer-type tetranuclear $RuOs_3$ complexes. *Angew. Chem. Int. Ed.* 35:76–78.

97. Ling, Q. D., D. J. Liaw, Y. H. Teo et al. 2007. Polymer memories: Bistable electrical switching and device performance. *Polymer* 48:5182–5201.

Index

T - #0264 - 101024 - C0 - 254/178/14 [16] - CB - 9781420093933 - Gloss Lamination